PETROLEUM SEDIMENTOLOGY

by

Winfried Zimmerle

Visiting Lecturer in Sedimentology and Sedimentary Petrology,
Department of Geology,
Marburg University, Germany

KLUWER ACADEMIC PUBLISHERS

DORDRECHT / BOSTON / LONDON

Library of Congress Cataloging-in-Publication Data

```
Zimmerle, Winfried.
    Petroleum sedimentology / by Winfried Zimmerle.
        p.   cm.
    Includes index.
    ISBN 0-7923-3418-3 (HB : acid-free)
    1. Petroleum--Geology.  2. Sedimentology.   I. Title.
TN870.5.Z56  1995
553.2'8--dc20                                    95-9724
```

ISBN 0-7923-3418-3 (HB)
ISBN 0-7923-3419-1 (PB)

Published by Ferdinand Enke Verlag, Postfach 300366
D-70443 Stuttgart, Germany, in co-publication with
Kluwer Academic Publishers. P.O. Box 17, 3300 AA Dordrecht,
The Netherlands.

Sold and distributed in the U.S.A. and Canada by
Kluwer Academic Publishers
101 Philip Drive, Norwell MA 02061, U.S.A.

Sold and distributed in Austria, Switzerland, Germany,
by Ferdinand Enke Verlag
Postfach 300366, D-70443 Stuttgart, Germany

Sold and distributed in all remaining countries by
Kluwer Academic Publishers
P.O. Box 322, 3300 AH Dordrecht, The Netherlands

Printed on acid-free paper

© 1995 Ferdinand Enke Verlag, P.O. Box 300366,
D-70443 Stuttgart, Germany
Printed in the Netherlands

Enke-ISBN: 3-432-25291-9 (PB)

PETROLEUM SEDIMENTOLOGY

TABLE OF CONTENTS

Chapter 1

INTRODUCTION

This book is essentially directed to students of earth sciences and petroleum engineering. In addition, it is intended as an introduction and a simple text to all earth scientists and engineers interested in the subject.

With this goal in mind the text is kept simple; it avoids purposely scientific discussions and controversial issues.

Sedimentology is "the scientific study of sedimentary rocks and of the processes by which they were formed; the description, classification, origin, and interpretation of sediments" (Glossary of Geology 1987, p. 567). Petroleum sedimentology refers to the above defined study of sedimentary rocks as applied to petroleum exploration and exploitation. It necessarily intercommunicates with the neighboring disciplines (Table 1). Petroleum sedimentology comprises studies from micro- to megadimension (Table 2) from the simple analysis of a mineral powder to entire basin studies. Its ultimate aim is the basin analysis based on as many sedimentological criteria as possible.

The book is divided into three major parts with a total of 16 chapters: (1) methods, (2) rocks important to hydrocarbon exploration, and (3) applied petroleum sedimentology. Photoplates 1–15 illustrate macroscopic and microscopic aspects of source rocks, reservoir sandstones and limestones, and cap rocks. They demonstrate rocks important to hydrocarbon exploration and characteristic examples of certain environments of deposition (e.g. eolian sands, evaporite, etc.). Only selective classic or modern pertinent literature references or comprehensive text books are listed under references.

Successful application of the principles of sedimentology to future search for petroleum will mainly depend on (1) careful observation and reflection, (2) complete reporting of details as well as painstakingly summarizing the essentials, and (3) optimal graphic representation.

Miss E. Lammert, Mrs. U. Waldmann, and Mrs. R. Münch participated in preparing this book. A number of colleagues contributed to its success by encouragement and assistance: Dipl. Ing. H.K. Draxler, Dr. A. Fülöp, Dr. K.-H. Gaida, Prof. Dr. K. Helbig, Dr. R. Koch, Prof. Dr. A. Muller, Mr. R. Schmuhl, Prof. Dr. J. Schröder, Dr. F. Werner, Dr. A. Wetzel, Prof. Dr. Monika Wolf, and Dr. U. Zinkernagel. BEB, DTA, Leitz, Mobil Oil, Schlumberger, and WIAG furnished samples and ready advice.

Table 1.   Petroleum sedimentology and neighbouring disciplines, an 'egocentric' view.

BIOLOGY	GEOLOGY	MATHEMATICS/ STATISTICS
Paleontology	Regional Geology	Computer Science Data Handling
Paleoecology	Stratigraphy	Hydraulics/ Fluid Mechanics
	PETROLEUM SEDIMENTOLOGY	
Oceanography	Sedimentary Petrography	Engineering Geology
Geochemistry	Mineralogy	Geophysics/ Tectonics
CHEMISTRY		PHYSICS

Table 2.   Sedimentological techniques: dimensions, objects, tools and goals.

Dimension	Object	Tool	Goal
Mega (10^6 m)	Basin	Regional description, serial photography, facies mapping, paleocurrents	Basin-wide facies analysis, basin evaluation
Kilo (10^3 m)	Oil field	Facies mapping, paleocurrents, seismics, isopach trend analysis, outcrop/well log descriptions	Distribution and shape of sediment bodies, trends in reservoirs (quality, thickness), regional faciel analysis, trends in cementation, prediction of underground combustion trends
Deca (10^1 m)	Outcrop/ well	Outcrop/well description, photography, large scale sedimentary structures, log interpretation	Local facies analysis, local sedimentary anisotropies
Deci (10^{-1} m)	Hand specimen	Description, photography, binocular microscope, small-scale sedimentary structures	Mineral and rock analysis, depositional environment, age determination, minor sedimentary anisotropies
Milli (10^{-3} m)	Thin section	Petrographic microscope, thin section description, photomicrography, micro-scale sedimentary structures	Primary pore geometry, diagenetic alteration, top and bottom analysis
Micro (10^{-6} m)	Mineral powder	X-ray equipment, electron microscope, electron microprobe, scanning electron micrography	Mineral determination, particle shape analysis, microchemical test, diagenetic alteration of clay minerals

Chapter 2

SEDIMENTOLOGICAL INVESTIGATIONS

In the following sections the major sedimentary rock types and their classification are described as well as the sedimentological methods used. Since examples are taken from the European scenario, a map of the petroleum geology of Europe is shown (Figure 1). The geological atlas of western and central Europe (Ziegler 1990) has become a standard reference for all earth scientists in paleogeography and paleotectonics. The most up-to-date account of North Sea exploration is that by Glennie (1990). Blundell and Gibbs (1990) comment on the tectonic evolution of the North Sea rifts. With respect to North Sea reservoirs, Scotchman and Johnes (1990) reviewed the reservoir-sandstone petrology of the famous Brent field, and Feazel and Farrell (1988) the chalk petrology from the Ekofisk area. Major hydrocarbon plays and traps associated with the Viking graben, North Sea are discussed by Harding (1983, 1984). An example of ship-borne geochemical surface exploration for hydrocarbons emanating from the seafloor of the North Sea is presented by Faber and Stahl (1984).

A N–S cross-section through the oil- and gas-bearing NW German Basin is shown in Figure 2. Pertinent accounts of hydrocarbon exploration in central Europe, especially in the F.R.G. and former G.D.R., are those by Boigk (1981), Bachmann and Grosse (1989) and Eiserbeck et al. (1990). Aspects of Rotliegend reservoirs in Germany are dealt with by Plein (1978), Grotewold et al. (1979), Gast (1988) and Gralla (1988). Details of Jurassic reservoirs of the Gifhorn Trough are contained in Philipp (1961), Philipp et al. (1963), Boigk (1986), and Gaida et al. (1987); a case history of a Cretaceous reservoir (Bentheim Sandstone) in the Emsland is that by Wittenhagen (1980). Gaida et al. (1973) summarized the petropyhsical characteristics of reservoir sandstones in Germany.

Hydrocarbon generation, accumulation, and production in the various producing provinces of Europe are discussed by the following review papers: Hobson and Tiratsoo (1981), Bjørlykke (1989) and Spencer (1991). Pertinent textbooks on sedimentary petrology are Milner (1962), Folk (1972), Pettijohn (1975), Blatt (1982), and Greensmith (1989). Useful texts on sedimentology include Leeder (1982), Brenchley and Williams (1985), Hsü (1989), and Chamley (1989). Data on reservoir sedimentology has been compiled by Tillman and Weber (1987) and Bjørlykke (1989).

Introductions in the field of petroleum geology are given by Hobson and Tiratsoo (1981), Chapman (1983) and Dickey (1986).

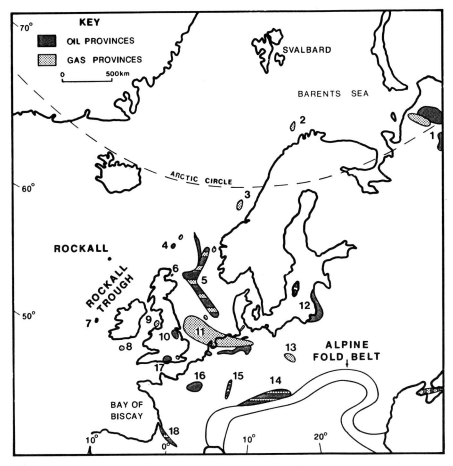

Fig. 1.   Hydrocarbon provinces of NW Europe: 1. Pechora Basin, 2. S Barents Sea, 3. Mid-Norway offshore, 4. W of Shetland, 5. C and N North Sea, 6. Inner Moray Firth, 7. W of Ireland, 8. Celtic Sea, 9. Irish Sea, 10. English Midlands, 11. S North Sea-N German Plain, 12. Baltic Sea-Estonia, 13. Poland, 14. Alpine Foreland, 15. Rhine Graben, 16. Paris Basin, 17. Wessex-Weald Basin, 18. Aquitaine Basin (after Glennie *et al.* 1987).

2.1. Major Types of Sedimentary Rocks and Their Classification

Sedimentary rocks can be divided grossly into two groups: autochthonous and allochthonous sediments. Autochthonous sediments are deposited from material within the environment. They comprise chemical precipitates such as gypsum, rock salt and organic deposits, coal, limestones. The debris of allochthonous sediments was transported from the outside into the environment. They include terrigenous (e.g. clays, sands, conglomerates) and pyroclastic deposits.

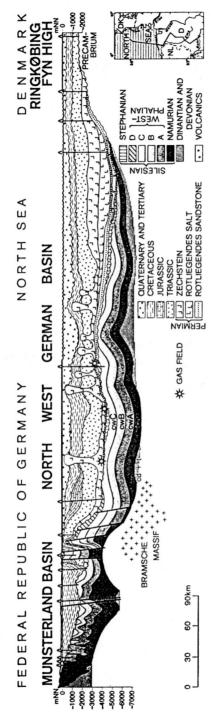

Fig. 2. N–S cross-section through the oil and gas-bearing NW German Basin showing gas-productive horizons (after Stancu-Kristoff and Stehn 1984).

Sedimentary rock types are generally classified by (1) grain size and (2) mineral composition.

Based on grain size the general terms are: conglomerates, sandstones, siltstones, and claystones or mudstones. Names based on the predominating minerals are limestones, dolomites, siliciceous sediments, anhydrites, phosphorites, and clay ironstones. Coals, bituminous rocks, oil shales, or oil sands are classified by their content of organic matter.

Other ways to name sedimentary rocks are artificial word compositions, e.g. characterizing the major components as bio-oo-pel-sparite. Other names refer to geographic localities (ganister, itacolumite, sparagmite). Moreover, some old terms (so-called names of common use) survived attempts of modern classification: e.g. in sandstones (arkose, crystal sandstone, graywacke, gritstone, flagstone, greensand, quartzite), in claystones and shales (argillite, cornstones), in siliceous rocks (gaize, opoka, porcellanite).

Arkoses are light gray, pink or reddish, mostly poorly sorted quartzose sandstones but rich in angular to subangular feldspars, proximal to a granitic source.

Graywackes, an old rock name created by miners in the Harz Mts., are dark greenish gray, firmly indurated, poorly sorted sandstones, mostly rich in rock fragments with varying feldspar content, and with a compact clayey matrix. These rocks reflect rapid erosion, transportation, deposition, and burial (e.g. in orogenic belts).

Classifications are normally of two types: descriptive and genetic. Within a descriptive system the sedimentary rock may be classed without knowing its origin. Recently genetic terms such as turbidites, fluxoturbidites, contourites, or tempestites have been preferred to purely descriptive terms. However, in most cases descriptive terms are used at first. Because most sedimentary rocks are polygenetic, genetic interpretation is more complex. Nevertheless, it is the ultimate objective of any classification.

The classification in this book is essentially descriptive. It is outlined in the following scheme:

Detrital rocks (terrigenous sediments s. str.)	coarse	= Rudaceous rocks (gravels, conglomerates, breccias)
	medium	= Arenaceous rocks (sands, sandstones, siltstones)
	fine	= Argillaceous rocks (clays, muds, shales)
Detrital as well as chemical rocks	carbonate rocks	= limestones and dolomites
Chemical rocks	siliceous rocks	
	evaporites	= (chemical sediments s. str.)
	coals	= (organic sediments s. str.)

Siltstones are not discussed. Neither is discussed the classification of pyroclastic sediments, which occasionally function as reservoir rocks, but are volumetrically insignificant.

Since most sediments and sedimentary rocks consist of few end members (e.g. binary, ternary, or quaternary), the chief rock or mineral constituents may by simply diagrammed, for instance: sandstone-limestone-evaporite. Sediments containing three major consti-

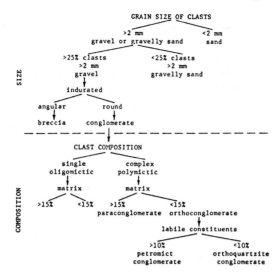

Fig. 3.   Classification of conglomerates by size, texture and composition (courtesy E. G. Williams, after Bourgeois 1978).

tuents can be classified in a triangular diagram, four-component sediment systems in a tetrahedron. Such diagrams demonstrate the rock or mineral composition and/or the textural and structural features most clearly.

2.1.1. CONGLOMERATES

Conglomerates are coarse-grained clastic sedimentary rocks which are composed of rounded to subangular fragments larger than 2 mm in diameter (granules, pebbles, cobbles and boulders). They may be classified according to nature or composition of fragments (dominantly rock fragments), proportion of matrix, degree of size sorting, type of cement, and agent or environment of deposition.

No consistent scheme of nomenclature exists. Three major groups can be distinguished due to composition: (1) volcaniclastic conglomerates (i.e. agglomerates), (2) carbonate conglomerates (= calcirudites), and (3) terrigenous conglomerates (= silicirudites).

Figure 3 lists the descriptive terminology based on the bipartition of texture, composition, and source. Terrigenous conglomerates may be texturally subdivided into grain-supported and mud-supported conglomerates (diamictites or pebbly mudstones). Grain-supported conglomerates are normally deposited with high porosities and permeabilities. Notable examples of diamictites are the Pleistocene boulder clays. Conglomerates can also be classed according to the composition of their pebbles. Polymictic conglomerates are composed of pebbles of more than one type (e.g. pebbles of vein quartz, quartz, and chert), oligomictic (monomictic) conglomerates have pebbles of only a few or one rock type, generally quartzose. Another way of classifying conglomerates is based on the provenance of the pebbles. Thus, extraformational or exotic conglomerates are composed of pebbles from outside the depositional basin. Intraformational conglomerates consist of pebbles originating within the depositional basin.

Breccias are coarse-grained clastic rocks whose particles are angular in contrast to the rounded fragments of conglomerates. They are polygenetic in origin (sedimentary breccia, solution breccia, fault breccia). They occur seldom.

2.1.2. ARENACEOUS ROCKS

Arenaceous rocks include sands and sandstones. Approximately 30 per cent of the sedimentary cover of the earth is made up of terrigenous sand and sandstones. Because of their porosity, sandstones are among the most important hydrocarbon reservoirs. Pertinent textbooks on sands and sandstones are Milner (1962), Folk (1974), Pettijohn, Potter and Siever (1987), and Füchtbauer (1988).

Sandstone is firstly a field term for any clastic sedimentary rock being composed of individual sand grains that are visible to the unaided eye. Sandstone properly defined is a grained clastic rock consisting of rounded or angular fragments of sand size (very fine to very coarse) set in a fine-grained matrix (silt or clay) and cemented by mineral matter (commonly silica, calcite or iron oxide). Sandstone is consolidated sand, intermediate in grain size between conglomerate and shale.

Quartzose terrigenous sands are also termed siliciclastics in order to differentiate them from bioclastic and pyroclastic sediments.

In oil industry it was general practice to describe arenaceous rocks simply as sand or sandstone. At present, on well logs sand/sandstones are described more completely using a format similar to the following: lithology, color, hardness, grain size, grain shape, sorting, mineralogy, fossils (if present), porosity, hydrocarbon shows (i.e. stain, fluorescence or cut).

Details on nomenclature and classification are given by Pettijohn, Potter and Siever (1987), and Füchtbauer (1974, 1988). None of the proposed schemes, however, is used universally. Classifications of sandstones are petrographic, commonly based on microscopic examination. They require approximate, if not accurate, determination of the modal composition. Mineral cements do not enter into modern sandstone classification in contrast to the minerals of the sandstone matrix.

Modern sand/sandstone classifications, both in Europe and the U.S.A., are commonly plotted on end-member triangles, the components being quartz, feldspar, and rock fragments. The matrix content is the modifier in the classification. Another criterion in sandstone classification is the degree of maturity. One might discriminate chemical and physical maturation. Quartz is the most widespread stable mineral; feldspar is an unstable mineral. Thus, the quartz/feldspar ratio is an index of chemical maturity. The grain/matrix ratio, on the other hand, is an index of physical maturation.

The quantitative sandstone classification of Füchtbauer (1974, 1988) is based on the 10–25–50 per cent triangular system (Figure 4) with quartz, feldspar, and rock fragments as end-members. Thus, sandstones with more than 50 per cent feldspar are termed feldspar sandstones and with more than 50 percent rock fragments rock-fragment sandstones. Sandstones with 25–50 per cent feldspar are called feldspar-rich sandstones or sandstones

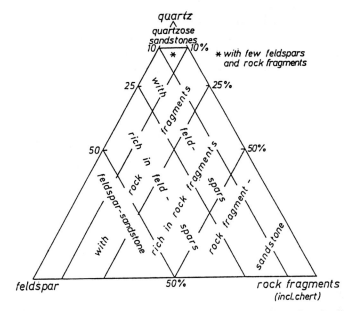

Fig. 4. Sandstone classification after Füchtbauer (1974, 1988) plotted in a triangular diagram.

rich in feldspar; those with 25–50 per cent rock fragments are rock-fragment-rich sandstones or sandstones rich in rock-fragments. Sandstones with 10–25 per cent feldspar or
with 10–25 per cent rock fragments are classed as feldspar-bearing sandstones and rock
fragment-bearing sandstones, respectively. A content of less than 10 per cent feldspar or
rock fragments is not mentioned in the term, or it may be indicated as 'sandstone with few
feldspar or rock fragments'. Sandstones with more than 90 per cent quartz (i.e. less than
10 per cent feldspar and rock fragments) are termed quartz sandstones.

In the upper half of the triangular diagram graywackes from the Paleozoic of Central
Europe are plotted, in the lower half on the left side quartz-free arkoses and on the right
side quartz-free volcanic graywackes both from New Zealand.

Other mineral and rock constituents such as micas, dolomite, volcanic glass, and/or
heavy minerals as well as clay and silt components can be included in this nomenclature
in a similar way.

The sand/sandstone classification of Pettijohn, Potter and Siever (1987) is simple
(Figure 5) and widely used. It relies on framework grains of sand-grain size (quartz,
feldspar, and rock fragments). The classification distinguishes between 'clean' sands or
arenites – sands with less than 15 per cent matrix – and 'dirty' sands or wackes – those
with more than 15 per cent matrix. Among the matrix-poor sands, those with no more
than five per cent of either feldspar or rock particles are called quartz arenites. Those with
25 per cent or more feldspar, which exceeds rock fragments, are arkosic arenites. Those
with 25 or more per cent of rock fragments, but a lesser amount of feldspar, are lithic

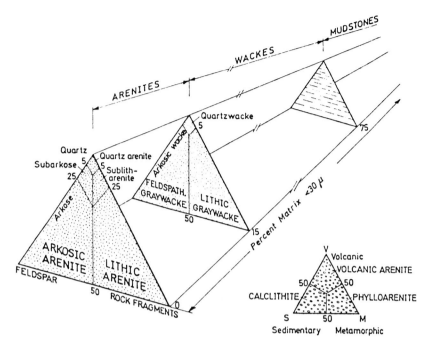

Fig. 5.   Sandstone classification after Potter, Pettijohn and Siever (1987, p. 145).

arenites. Transitional classes are subarkoses and sublitharenites.

The term feldspathic arenite is loosely used. These arenites contain varying amounts of feldspar. Any sandstone with five or more per cent feldspar is 'feldspathic'. Thus, feldspathic sandstones include subarkoses, arkoses, some litharenites, and many graywackes. The lithic arenites (litharenites) are those arenites with 25 or more per cent of rock particles and a minimal matrix content. Most commonly these rock particles are pelitic (shale, siltstone, slate, phyllite, and mica schist). Consequently the litharenites with such phylloid fragments have been termed phyllarenites. Other important types include those in which the dominant rock particles are limestone or dolomite. The term calclithite has been proposed to distinguish this carbonate sand from calcarenite – a carbonate sand produced by biochemical or chemical precipitation. Volcanic arenites are lithic arenites whose rock detritus is derived from desintegration of extrusive rocks.

The transitional sublitharenite has a lesser rock-fragment content. The term subarkose is used for a transitional class of arenites with less feldspar than an arkose and with few or no rock fragments.

Sandstones with a significant matrix content (more than 15 per cent) are called wacke. The wackes are subdivided mainly into two classes: feldspathic graywacke (feldspar prevailing) and lithic graywacke (rock fragments predominating). Quartz wackes constitute a relatively small group within the wacke clan.

This classification, based on mineral composition, is normally not dependent on the environment of deposition. For instance, quartz arenite may have been deposited in a subaerial dune, in a stream, or on a beach.

2.1.3. ARGILLACEOUS ROCKS

The classic standard text in clay mineralogy is Grim (1968). Geological and sedimentological aspects of clay mineralogy are discussed by Millot (1970), Füchtbauer (1988), Chamley (1989), and Heim (1990). Odin (1988) focussed on the significance of green clays (chamosite-glauconite-celadonite) in the marine environment. Weaver (1989) compiled genetically relevant data on clays, muds, and shales. Brindley and Brown (1980), Fripiat (1982) and Moore and Reynolds (1989) describe advanced techniques in clay-mineral analysis. The compaction of argillaceous sediments is treated by Rieke and Chilingarian (1974). The 'Argillaceous Rock Atlas' (O'Brien and Slatt 1990) takes a systematic approach to fabric analysis of shales, irrespective of the provenance of the clay minerals. Kubanek *et al.* (1988) determined the provenance of a shale by integrated analysis of all pertinent parameters of the clay minerals present.

The history of sandstone and carbonate classifications demonstrates that descriptive classifications are generally superior to genetic classifications. Precise description aids interpretation. Genetic assignment, however, depends always on the momentary state of geological knowledge.

The finest-grained clastic rocks, the shales, contain 50 per cent or more of terrigenous, generally argillaceous, clastic components (less than 62 μm in size). Names such as lutite, clay and claystone, mud and mudstone, silt and siltstone, shale, argillite, phyllite, and slate are commonly used. Shale is the generally used name for fine-grained rocks.

Shales are the most abundant of all lithologies, constituting between 45 and 55 per cent of sedimentary-rock sequences. Attempts to classify shales remained unsuccessful, so far. None of these classifications has been widely accepted.

Shales and other fine-grained terrigenous sediments contain major amounts of clay minerals and clay-size carbonate, kerogen, and silica. Potter, Maynard and Prior (1980) use the term shale as major class name and descriptively classify shale based on: state of induration, relative amount of clay-size constituents and silt-sized particles, and bedding and laminae thickness. These features are readily determined in the field and are easily seen in thin section (Table 3). Mineralogical modifiers include calcareous, dolomitic, sideritic, carbonaceous, ferruginous, phosphatic, pyritiferous, quartzose, siliceous, etc. Other modifiers are color, induration, fracture (e.g. conchoidal, hackly, blocky, brittle, splintery, earthy, etc.), specific bedding features, fossil content, bioturbation and trace fossils, and organic constituents.

The subdivision is based on the relative percentages of clay minerals and the clay-size carbonate, kerogen, and silica. This subdivision has both a mineralogical and a textural significance. The size boundary between clay and silt is 4 μm, but most clay minerals are less than 2 μm. Bedding features (beds vs. laminae) are another modifier.

The suffix-stone is used to denote layering greater than 10 mm (beds) and the suffix-shale to denote layering less than 10 mm (laminations). The commonly used term siltstone

Table 3.   Classification of shale (more than 50% grains < 62 μm) (after Potter, Maynard and Pryor 1980).

Percentage clay-size constituents			0-32	33-65	66-100
Field Adjective			Gritty	Loamy	Fat or Slick
NONINDURATED	Beds	Greater than 10 mm	BEDDED SILT	BEDDED MUD	BEDDED CLAYMUD
	Laminae	Less than 10 mm	LAMINATED SILT	LAMINATED MUD	LAMINATED CLAYMUD
INDURATED	Beds	Greater than 10 mm	BEDDED SILTSTONE	MUDSTONE	CLAYSTONE
	Laminae	Less than 10 mm	LAMINATED SILTSTONE	MUDSHALE	CLAYSHALE
METAMORPHOSED	Degree of metamorphism	LOW	QUARTZ ARGILLITE	ARGILLITE	
			QUARTZ SLATE	SLATE	
		HIGH	PHYLLITE AND/OR MICA SCHIST		

is restricted to bedded silt-rich rocks; silt shale is a laminated silt-rich rock.

Metamorphic processes, especially high temperatures, readily alter shales. Argillites are weakly metamorphosed rocks, firmly indurated without fissility or slaty cleavage and with some of the clay minerals and micas reconstituted to sericite, chlorite, epidote or green biotite. Progressive metamorphism of shales generally results in gradual transition from argillite to slate to phyllite to micaschist, associated with an increase in grain size.

The above classification is supposed to be simple and easy to use in the field and laboratory. It also provides a basis for description as examplified in the following:

Mudshale: Dark greenish gray; hard, brittle, platy parting; 10% quartz sand, 30% quartz silt, 60% clay; medium laminae (1.5 mm); small coaly fragments, and small pelecypod impressions; micaceous; sharp basal contact.

As investigations of shales augment, this classification scheme will be further evolved. True shales consist of minute flaky minerals produced by chemical weathering of feldspars and other destructible minerals. Genetically, they have been deposited in various depositional environments: pelagic, shelf, delta, flood plain, lake, eolian, glacial, etc., dominantly under low-energy conditions. Shales are the chief lithology of present ocean basins. Most shales are marine; thick shale sequences were deposited in relatively deep water (e.g. distal turbidites). Other examples of shale deposits are loess, a windblown dust of Pleistocene age, or varve clays, Pleistocene glacial deposits.

Marls consist of clay mixed with calcium carbonate; they are intermediate in composition between shales and limestones.

2.1.4. Carbonate Rocks

Classic texts on carbonate rocks are those by Ham (1962), Bathurst (1975), Flügel (1982), and Carozzi (1989). Flügel (1982) and Carozzi (1989) both summarize data on microfacies analysis. Tucker and Wright (1990) refer to modern work on carbonate sedimentology. The controlling factors in carbonate platform and basin development are discussed by Crevello *et al.* (1989).

Many parameters may be used to define carbonate-rock types, such as chemical composition (e.g. limestone, dolomite), grain size, particle type, nature and amount of porosity, degree of crystallinity, and quantity of mud admixture.

The concepts of present-day carbonate nomenclature are thoroughly discussed by Ham (1962). Two contributions (Folk 1962, Dunham 1962) deserve special mention because they contain a series of terms and groups which are widely accepted and used at present.

Folk divided limestones into three major groups (Figure 6): (1) allochemical limestones composed largely of detrital components (= allochems), (2) orthochemical limestones composed dominantly of micrite (i.e. micrite lime-mud carbonates), and (3) autochthonous reef rocks (i.e. biolithites). The allochemical limestones are subdivided into rocks with dominant sparite cement and rocks with dominant micrite matrix. Folk proposed a bipartite nomenclature for allochemical rocks referring to grain type and cement/matrix. The prefix defines the grain type and the suffix denotes the cement/matrix; e.g. oo-sparite or pel-micrite. If more than one important allochem type is present, two or more prefixes can be used with the most abundant prefix first (e.g. bio-oo-pelsparite). If both micrite and sparite are present, the term can be composed to 'microsparite'.

Dunham's approach to classifying carbonate rocks was different (Table 4). He divided the principal carbonates into four groups according to whether their fabric was grain-supported or mud-supported. Grainstones are grain-supported carbonate sands with no micrite matrix. Packstones are grain-supported carbonate sands with minor amounts of micrite matrix. Wackestones are mud-supported rocks with a significant, but dispersed

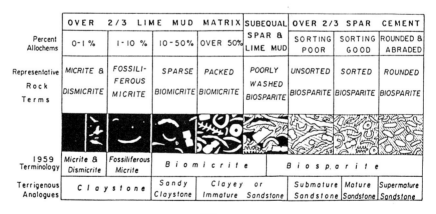

Fig. 6.   Carbonate rock classification after Folk (1962, reprinted with permission).

Table 4.   Classification of carbonate rocks (after Dunham 1962, reprinted with permission).

Depositional Texture Recognizable					Depositional Texture Not Recognizable
Original components not bound together during deposition				Original components were bound together during deposition . . . as shown by skeletal matter, lamination contrary to gravity, or sediment-floored cavities that are roofed over by organic or questionably organic matter and are too large to be interstices.	*Crystalline Carbonate*
Contains mud (particles of clay and fine silt size)			Lacks mud and is grain-supported		
Mud-supported		Grain-supported			
Less than 10 percent grains	More than 10 percent grains				(Subdivide according to classifications designed to bear on physical texture or diagenesis.)
Mudstone	*Wackestone*	*Packstone*	*Grainstone*	*Boundstone*	

amount of grains. Mudstones are carbonate muds. Moreover, Dunham placed the in-situ reef rocks in an own class: the boundstones. Crystalline carbonates whose primary depositional fabric could not be determined were classed separately.

Mud-supported limestones indicate deposition in a low-energy environment. Matrix-free, grain-supported rocks, on the other hand, suggest deposition in a high-energy environment without mud. This concept, successfully applied to siliciclastic rocks, must be used with reservation in carbonate rocks. Micrite is polygenetic. For instance, clean carbonate sand, deposited in a high-energy environment, may subsequently develop into micrite by bioturbation, algal micritization, and by infiltration due to high permeability. The large size of autochthonous carbonate skeletal debris can not be taken eo ipso as an energy index as it is used in terrigenous deposits (e.g. oyster reefs in modern lagoons).

The classifications and nomenclatures of carbonate rocks, proposed by Folk (1962) and Dunham (1962), are practical and, if used in conjunction, include most varieties of limestones. The classifications must be based on parameters determinable in the field as well as under the binocular and petrographic microscope. The nomenclatures are descriptive and emphasize constituents and texture rather than particle size. Some of the textural elements are hints at depositional and diagenetic processes. Grain size, sorting, and matrix content can only be used with caution as indicators of hydrodynamic environment in carbonate rocks. Thus, classification and nomenclature of carbonate rocks remain a much disputed topic.

2.1.5. SILICEOUS ROCKS

Siliceous rocks include purely organic deposits with organic remains embedded in a cryp-tocrystalline silica matrix as well as cherts representing metasomatic replacements of limestones. Intermediate varieties with various proportions of siliceous organic remains can be classed between purely organic and purely metasomatic siliceous rocks. Precambrian iron formations partly consist of chert with not much evidence of biologic influence during deposition. Other chert varieties formed diagenetically from water-laid volcanic ash. The classification of water-laid siliceous rocks requires some further explanation of terms: Unconsolidated organic deposits are called oozes. Sedimentary deposits which have remained unconsolidated are termed earths (e.g. radiolarian earths). The consolidated equivalents of these purely organic accumulations are termed radiolarites, diatomites, and spicularites (sponge-spicule rocks).

Chert is a general term for very fine siliceous sediments of chemical, biochemical or biogenic origin. It is an extremely dense or compact, semivitreous, cryptocrystalline, hard sedimentary rock with a splintery to conchoidal fracture. Most cherts are composed of interlocking quartz crystals less than 20 μm in diameter; they may contain amorphous silica (opal). Only minor impurities of calcite and iron oxide as well as some remains of siliceous and other organisms are included. Chert can be divided into bedded and nodular chert. Bedded cherts are commonly associated with volcanic rocks. Thus, the 'chert problem' has centered on the volcanic versus biogenic origin of the silica. Modern equivalents of ancient bedded cherts, the radiolarian and diatom oozes, cover large areas of the deep-ocean floors. Nodular cherts occur generally in limestones and to a lesser extent

Table 5. The common marine and non-marine evaporite minerals (after Tucker 1981, p. 158, Table 5.1).

Common Marine Evaporite Minerals		Non-Marine Evaporite Minerals	
Halite	NaCl	Halite, gypsum, anhydrite	
Sylvite	KCl	Epsomite	$MgSO_4 \cdot 7H_2O$
Carnallite	$KMgCl_3 \cdot 6H_2O$	Trona	$Na_2CO_3 \cdot NaHCO_3 \cdot 2H_2O$
Kainite	$KMgClSO_4 \cdot 3H_2O$	Mirabilite	$Na_2SO_4 \cdot 10H_2O$
Anhydrite	$CaSO_4$	Thenardite	$NaSO_4$
Gypsum	$CaSO_4 \cdot 2H_2O$	Bloedite	$Na_2SO_4 \cdot MgSO_4 \cdot 4H_2O$
Polyhalite	$K_2MgCa_2(SO_4)_4 \cdot 2H_2O$	Gaylussite	$Na_2CO_3 \cdot CaCO_3 \cdot 5H_2$
Kieserite	$MgSO_4 \cdot H_2O$	Glauberite	$CaSO_4 \cdot Na_2SO_4$

in shales and evaporites. They are mostly diagenetic, having formed by replacement. Certain types of chert have specific names: Flint is a synonym for siliceous nodules in Cretaceous chalks. Jasper refers to a red variety of chert with finely disseminated hematite in Precambrian iron formations. Porcellanite refers to dense siliceous rocks with the general appearance of unglazed porcelain. Gaize is a porous fine-grained micaceous glauconitic sandstone containing much soluble silica. Opoka, a name mainly used in Poland and Russia, is a porous, flinty, and calcareous sedimentary rock consisting of fine-grained, opaline silica (up to 90%), and hardened by the presence of silica of organic origin (radiolaria, sponge spicules, and diatoms).

Silexite is a French term used for chert occurring in calcareous beds. Hornstone is a compact, tough, siliceous rock with subconchoidal fracture. Lydite (Kieselschiefer or Lydian stone), widespread in the Paleozoic of Central Europe, is a flinty stone (silicified shale or slate) consisting of a compact, extremely fine-grained, gray-black variety of jasper. Adinole, originally an argillaceous sediment albitizised by contact metamorphism, is commonly mistaken for chert in the field. Novaculite is a light-colored, dense, hard, even-textured, cryptocrystalline siliceous rock, presumably thermally metamorphosed, occurring in the Paleozoic of the southeastern U.S.A.

Siliceous sediments are also being deposited in lakes.

Since siliceous sediments are not important to petroleum sedimentology at the present time, neither as source rock, nor reservoir, nor cap rock, they are not discussed any further.

2.1.6. EVAPORITIC ROCKS

Evaporites are non-clastic, mainly chemical sedimentary rocks composed primarily of minerals produced from saline solutions due to extensive or total evaporation of the solvent. Principal evaporite minerals are gypsum, anhydrite, and halite. Table 5 lists the chief evaporite minerals. The actual number of all evaporite minerals, however, is much higher.

Salt deposits are normally associated with detrital sediments of desert facies. Partly confined water bodies suffer considerable losses by evaporation, especially in areas of

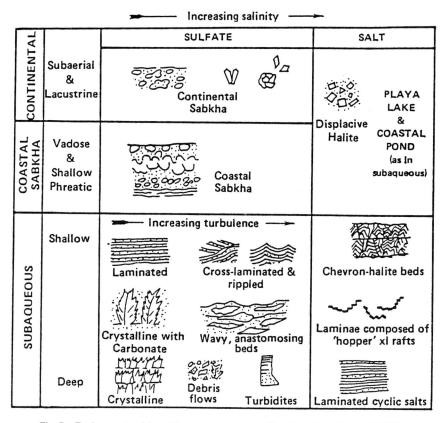

Fig. 7.   Environments of deposition of evaporites (modified from Schreiber *et al.* 1976).

high temperature and semiaridity. Thus, salts in solution become sufficiently concentrated for crystallization to begin. Present sea-water will first precipitate carbonates then, at salt concentrations beyond four times normal, sulfates beyond twelve times normal halite, beyond sixty-four times normal magnesium-potassium salts and, beyond one hundred and twenty times normal bischofite. The bromide geochemistry of salt rocks in respect to their genetic interpretation was discussed by Holser (1966).

Sedimentary features of subaqueous, coastal sabkha and continental evaporites as a function of increasing salinity are shown in Figure 7.

Many evaporite basins show a regular distribution of these evaporitic precipitates in extent and geological time. Calcium sulfates are precipitated usually towards the edge of the basins and form the lower layers within a given evaporite succession. Chlorides are deposited towards the centers of the basins and form the upper layers in given successions. In the potassium-bearing facies the regular distribution may be affected by secondary metasomatic changes caused by residual interstitial brines. The aspects of the exceptional diagenesis in evaporitic rocks are treated in detail by Kulke (1979).

Evaporite beds are important to oil exploration being locally the cap rock of carbonate-rock reservoirs (e.g. Middle East, western Canada) or forming structural traps through salt diapirism (NW Germany).

2.1.7. COALS

A brief review on coal classification is included because coaly particles and coal seams play an essential role as gas source (e.g. Rotliegend gas field of Groningen).

The coal classification presented is taken from the International Handbook of Coal Petrography (1963/1993). Coal is composed of a variety of plant tissues such as cuticles, woody structures, spores, etc. in different states of preservation formed under non-marine or paralic conditions.

One has to distinguish between coal type and coal rank. Coal type refers to the composition of the original plant material, whereas rank is a measure of maturation. Each coal type is composed of various discrete components, termed macerals (the prefix 'mac' implies macerated plant remains, the termination '-erals' indicates an analogon to minerals). Macerals are named with suffix '-inite', to distinguish them from the 'rock' type of coal, terminated 'ain'. The common macerals include: vitrinite, the main constituent of the coal type vitrain, subdivided into two varieties: tellinite with compressed cellular structure and structureless collinite. Suberinite, cutinite, and exinite are macerals formed by coalification of bark, leaf cuticles, and spore jackets respectively. The maceral fusinite is almost pure carbon with cellular structure. Micrinite is the general term applied to aggregates of very fine-grained macerated plant materials, too small to be identified.

A distinction is made between humic coals and sapropelic coals. Humic coals pass through a peat stage with processes of humification. Their major organic component is a lustrous dark brown to black material, derived from the humification of woody tissues, the so-called huminite in low-rank humic coals and vitrinite in bituminous and anthracite coals. Humic coals are commonly stratified. Sapropelic coals, on the other hand, are dull and not stratified. Humic and sapropelic coals are composed of relatively fine-grained organic muds deposited in quiet, oxygen-deficient ponds, lakes, or lagoons. Sapropelic coals contain varying amounts of detrital organic and mineral matter. The organic fraction consists mainly of algal remains. Sapropelic coals are relatively rare. They are subdivided microscopically into boghead and channel coals. Boghead coals are also called torbanites.

After deposition, the plant remains undergo physical, biochemical, and chemical changes of various intensity (diagenesis to catagenesis). Based on the coalification rank, coal series can be established (Table 6). The series begin with unaltered plant material and peat; with increasing rank it continues through brown coal (lignite), bituminous coal and finally to anthracite, a high-rank coal. The carbon content augments progressively with increasing rank; oxygen decreases vice versa. Normally the rank of coals tends to increase with age.

Modern texts on coal petrology include Bustin (1985) and on the sedimentology of coal Rahmani and Flores (1985).

Table 6. Main coalification stages as characterized by chemical and petrographic parameters (after Stach *et al.* 1982).

Rank German	Rank USA	Refl. Rm$_{oil}$	Vol. M. d. a. f. %	Carbon d. a. f. Vitrite	Bed Moisture	Cal. Value Btu/lb (kcal/kg)	Applicability of Different Rank Parameters
Torf	Peat	— 0.2	— 68				
			— 64	— ca. 60	— ca. 75		
Weich-	Lignite	— 0.3	— 60				
			— 56		— ca. 35	7200 (4000)	
Matt-	Sub-Bit. C	— 0.4	— 52				
	Sub-Bit. B		— 48	— ca. 71	— ca. 25	9900 (5500)	
Glanz-	A	— 0.5					
	C	— 0.6	— 44	— ca. 77	— ca. 8–10	12600 (7000)	
Flamm-	B	— 0.7	— 40				
		— 0.8					
Gasflamm-	A		— 36				
		— 1.0	— 32				
Gas-	Medium	— 1.2	— 28	— ca. 87		15500 (8650)	
	Volatile						
Fett-	Bituminous	— 1.4	— 24				
	Low	— 1.6	— 20				
Ess-	Volatile Bituminous	— 1.8	— 16				
		— 2.0					
Mager-	Semi-Anthracite		— 12				
			— 8	— ca. 91		15500 (8650)	
Anthrazit	Anthracite	— 3.0					
		— 4.0	— 4				
Meta-Anthr.	Meta-A.						

(Rank column: "Braunkohle" and "Steinkohle" labeled vertically; "High Vol. Bituminous" labeled vertically in USA column)

Applicability of Different Rank Parameters (vertical labels): hydrogen (d. a. f.), volatile matter (dry, ash-free), reflectance of vitrinite, carbon (dry, ash-free), bed moisture (ash-free), calorific value (moist, ash-free), moist., X-ray, diffr.

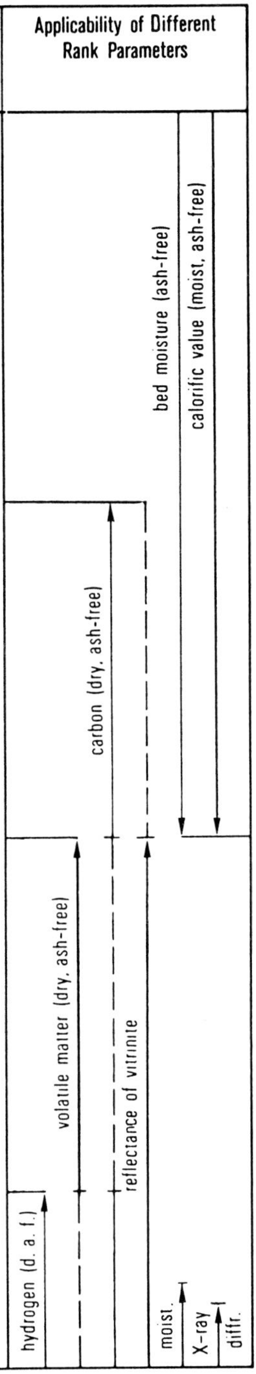

2.1.8. CONCRETIONS

A particular sedimentary rock type are concretions. Their occurrence and composition are used for environmental interpretation and diagenetic evaluation.

Concretions are generally hard, compact mineral aggregates, normally subspherical but also oblate, disc-shaped or irregular with varying outlines. They form by precipitation from aqueous solution around a biogenic nucleus, such as a leaf, shell, bone, or fossil in the pores of sedimentary rocks (e.g. shales, sandstones, limestones, etc.). Their size ranges from small pellets to large spheroidal bodies up to 3 m in diameter. Most concretions were formed penecontemporaneously or during early diagenesis. Synonyms are nodules, geodes, septaria (with shrinking cracks), secretions, and accretions.

Concretions are classed according to their main mineral constituents: carbonate concretions (composed of calcite, dolomite, siderite, rhodochrosite or a mixture of various carbonate minerals), clay ironstone nodules, phosphorite concretions, ferromanganese nodules or chert nodules.

Siderite, rhodochrosite or phosphorite concretions reflect certain pH-Eh conditions during formation. Accretionary phosphorite nodules are commonly indicative of coastal upwelling conditions.

2.2. Methods in Sedimentology

Conventional methods in sedimentology comprise generally megascopic rock description, photography, grain-size analysis, evaluation of sedimentary structures and physical properties of particles, heavy-mineral separation, X-ray diffractometry, microprobe analysis, and cathodoluminescence as well as scanning electron microscopy (Milner 1962, Carver 1971, Lindholm 1987, Füchtbauer 1988). The most important steps studying rock samples in the laboratory are compiled in a flow diagram (Figure 8).

Such conventional methods are applied to investigate sediments as precisely as possible as to their color, structure, and mineral composition in order to reliably analyze the environment of deposition.

Reference books on mineral and rock properties are essential for any work in sedimentology. The handbook of physical properties of rocks (Carmichael 1982/1984) collects pertinent data on mineral composition of rocks, electrical properties of rocks and minerals, their spectroscopic, magnetic and elastic properties. The treatise on applied mineralogy (Jones 1987) refers to sampling of mineralogical materials, fractionation of mineral particles, mineral identification, use of polarizing microscope, and image analysis. A comprehensive compilation of mineralogical methods and equipment with emphasis on classical methods is given by Szymanski (1989).

Most conventional methods are unspecific as to the rock type (e.g. grain-size analysis, thin section petrography, microprobe analysis, etc.); some are specific (e.g. acetate peels in carbonate rocks, peels from loose sediments, differential thermal analysis for clays). In the following the descriptions of the methods proceed from megascopic aspects to microscopic (thin-section and submicroscopic dimensions, X-ray diffraction). Distinction

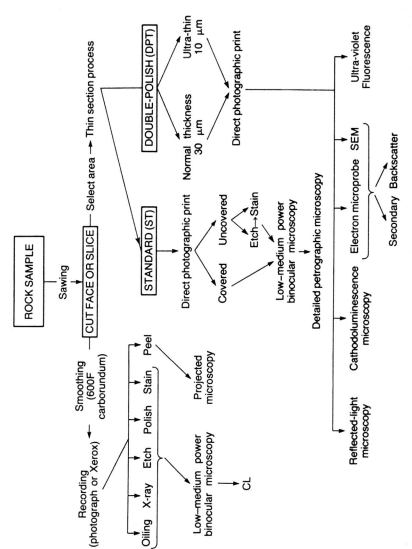

Fig. 8.   Flow diagram showing methods of rock-sample preparation and thin-section studies (after Miller 1988).

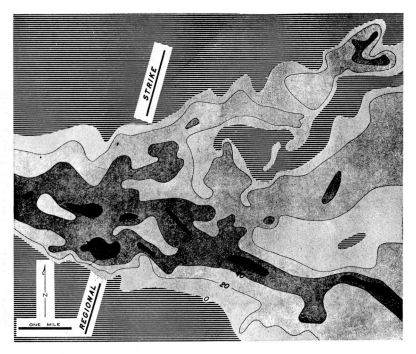

Fig. 9.   Delta-shaped sand body in the Seeligson Field, Texas, U.S.A. One of the classical studies in sedimentology (after Nanz 1954, reprinted with permission).

is to be made between the method proper, the tools applied, and the conclusions drawn.

For further information see pertinent textbooks (Müller 1967, Carver 1971, Tucker 1988). Müller (1967) is rather exhaustive, and Carver (1971) discusses the more common methods.

A classic study (Figure 9) in which various methods have been fully integrated is that of Nanz (1954).

Paleontological methods which considerably assist work in sedimentology are not discussed here. Pertinent textbooks on biostratigraphy, biofacies, and environmental analysis are those by Dunbar and Rodgers (1957) and Geyer (1973, 1977).

2.2.1. THE SUBSIDENCE-TIME PLOT

The subsidence-time plot (Figure 10), also called depth-time plot, was introduced by Philipp (1961) for deciphering the tectonic-structural development of the Eldingen oilfield in NW Germany. Ever since it is being used increasingly in sedimentological studies, in studies of single wells, oil fields or entire sedimentary basins.

The plot depends (1) on the availability of an accurate time scale and (2) on the

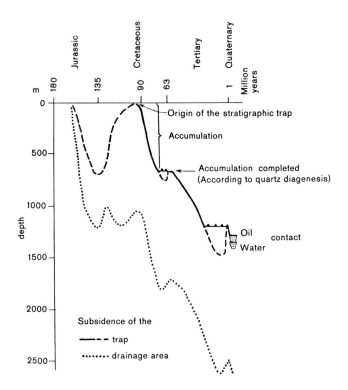

Fig. 10.   Subsidence-time plot, Gifhorn Basin, NW Germany (redrawn from Philipp *et al.* 1963: Fig. 2).

ability to relate local stratigraphy to that scale. Micro-paleontological zonation of marine intervals within the Mesozoic and Cenozoic allows time estimates as precise as 1–2 m.y. Recognition of unconformities is essential for successful use of these plots. The curves reflect the subsidence history and summarize much geological information graphically in a simple way. Parameters pertinent to oil exploration such as rate of sedimentation, paleotemperature, provenance data, key structural events (folding, faulting or fracturing), oil immigration, time sequence of diagenesis, and porosity can be plotted.

Thus, subsidence-time plots are suitable to demonstrate geological processes developing with time such as diagenesis and/or oil migration and emplacement.

2.2.2. MEGASCOPIC ROCK DESCRIPTION

Megascopic rock description and observation of sedimentary sequences are necessary before any other detailed analysis is done.

The field of megascopic rock observation reaches from outcrops, over conventional cores, sidewall cores, cuttings, to single sedimentary particles. The larger the field of observation, the more information can be gained for thorough description. Photographic documentation – either in black and white or in color – must supplement the description of the sedimentological features. Major oil companies standardize their description procedures of well samples by means of check lists, uniform legends, and symbols as well as with other measures. In the 'description of strata' (Schichtenverzeichnis) of oil companies in Germany a great number of sedimentological information is accumulated. Recently it has become possible to transform these written 'descriptions of strata' into lithologs. The aim of all these measures is to make megascopic descriptions of different well site geologists or different companies comparable with each other. Subjective views are thus reduced considerably.

The check list (Table 7) applies for all megascopic sedimentary rock descriptions from entire outcrops to single rock samples. The smaller the sample, however, the less informative and diagnostic sedimentary features can be recognized: e.g. a volcanic rock pebble is a better hint at provenance than a solitary quartz grain. Or: large-scale cross-bedding can never be observed in a cutting.

Color description is best based on some type of color chart (e.g. Der Farbenatlas Ostwald 1921; Rock-color chart 1984).

Emphasis has to be laid upon evaluation and rating of bedding and other sedimentary boundaries (e.g. unconformities). Not each boundary is equally important; they should be differentiated in first-, second-, third-order categories. The sedimentary contact of an unconformity rates much higher than a simple bedding plane. Often breaks in sedimentation are camouflaged and not easy to recognize. Diagnostic criteria might help to better identify such first-order sedimentary boundaries. Second-order sedimentary boundaries might be formed even by diagenetic events (e.g. ancient oil-water contact).

Sedimentary rock samples, recovered while drilling a well, are conventional cores, sidewall cores, and cuttings. A core furnishes a complete and continuous record of a sedimentary sequence; soft strata are normally preserved. Conventional cores are the best means to obtain direct information on all aspects of a sedimentary sequence and its hydrocarbon content. Because of the high cost of coring the sedimentological information contained in a core has to be completely evaluated. Normally, conventional cores are cleaned after recovery and slabbed by a diamond saw into segments that are fixed on a board. The slabbed surfaces may be sprayed or covered by thermoplastic to make the surface smoother and details better visible. It is very important to label the core properly and to mark the bottom of a core by an arrow. Core loss or differences between log and core depth are determined by careful descriptions and comparison with the well logs or by core-gamma-ray measurement in the laboratory, and its comparison with the conventional gamma-ray well log. A core furnishes a complete and continuous record of a sedimentary sequence; soft strata are normally preserved. Disadvantages of coring are high cost, occasional loss of more friable or more soluble rocks, and transport and storage of the cores. Cores are described individually from the top down. If the entire core is homogeneous, it may be described as one unit. However, if it is heterogeneous, each lithological unit requires separate description. In general, conventional core description of a well report or in a composite well log comprises the following properties in the

Table 7.   Check list of megascopic rock description (after Krumbein and Pettijohn 1938, p. 8/9).

External form of the rock unit
  Dimensions, persistence, regularity

Color
  Wet or dry, on basis of accepted color scheme

Bedding
  Sharp or transitional
  Plane, undulatory, or ripple-marked
  Thickness
    Constant or variable
    Rhythmic or random
  Attitude and direction of bedding surfaces
    Horizontal, inclined or curved
    Parallel, intersecting, or tangential to other beds
    Relation of particle properties to attitude and direction
  Markings of bedding surfaces
    Mudcracks, rain prints, footprints, etc.
  Disturbances of bedding
    Folding or crumpling
    Intraformational conglomerates

Concretions
  Kinds, size
  Condition and distribution
  Orientation with respect to bedding
  Form, size, composition
  Internal structure
  Boundary against country rock
    Sharp or transitional
  Relation to bedding
  Distribution
    Random or regular

Organic constituents
  Kinds, size
  Condition
    Whole or broken
  Distribution
  Orientation with respect to bedding

order: Rock NAME – texture – COLOR – luster – HARDNESS – cohesion – fracture feel – BEDDING – sedimentary structures – concretions and minerals – special features – FOSSILS – OIL and GAS – DIP.

Core workshops are most effective for working out subsurface sequence stratigraphy and for depositional modelling of hydrocarbon reservoirs (Siemers *et al.* 1981, Weimer *et al.* 1985, Feazel and Farrell 1988, Mitchell *et al.* 1988).

Conventional cores are also taken for measurement of porosity, permeability, and fluid content, which are petrophysical parameters for the evaluation as reservoir rock. Permeability can be determined only in cores. All these petrophysical parameters are essentially controlled by sedimentological parameters.

Sidewall cores with a diameter of about 22 mm are taken from the wall of the bore hole for stratigraphic control (Beckmann 1976, p. 130) when conventional coring is not possible. The sidewall cores are well localized as far as depth is concerned, an advantage when compared with cuttings; one disadvantage is their small size. Also the texture of the sidewall-cored sediments is frequently distorted or completely destroyed by the firing process; grains are thoroughly fractured. Thus, conventional grain-size analysis by sieving is useless; microscopic grain-size analysis, however, is feasible. Sidewall cores are commonly contaminated by drilling mud. Thus, petrophysical measurements are often not possible. Sidewall cores are of less value than conventional cores, but serve well for micropaleontological purposes.

In most cases cuttings are the only rock samples to be obtained from wells. They are usually taken every two meters in a wildcat well or every five meters in production wells. Cuttings practically give a 'menu' of a sedimentary sequence and contain not only material from the formation just drilled through but – to varying extents – also cavings from higher horizons. Thus, cuttings are less suited for biostratigraphic dating than conventional cores. However, with careful picking and simultaneous depth control by means of well logs the analysis of cuttings can be optimized in regard to depth.

Cuttings samples have to be covered with water when examined under a low-power binocular microscope. The well-site geologist has to determine the various lithologies of the cuttings. Complete descriptions save time and money, while mediocre descriptions normally require re-descriptions. Cuttings samples can be described in two principal ways: (1) By the interpretative method the well-site geologist picks out and describes those cuttings he thinks to be representative of the formation penetrated. This type of description brings out formational changes; however, it tends to mask minor variations of the lithology. (2) In a percentage description all cuttings in the sample are described, except for obvious cavings and artifacts. In regions of marked lateral changes this second method of description is more satisfactory.

Description of porosity, permeability, and oil stain of cuttings samples is an important analytical procedure of well-site geology. Especially in carbonate rock sequences intergranular porosity can be well distinguished from vuggy or fracture porosity.

Consolidated cuttings can be examined by thin sections or by acetate peels (Beckmann 1976, p. 120). In fact, periodic thin-section control is recommended to check proper sample identification and description. The advantage of episodic thin-section examination of cutting samples are listed in Table 8. Koch (1991) demonstrated convincingly that even facies analysis can be carried out by an integral study of cuttings.

Effort should be made to evaluate cuttings to an optimum. Beware of contaminations and artifacts such as gel flakes, other plugging material, iron particles, etc. The quality of

Table 8.   Advantages of thin-section analysis of cuttings.

- Only small quantity of material required.
- Selection of fragments to be analyzed.
- Individual analysis of each fragment.
- Averaging of rock types over a certain interval.
- Detailed information on rock composition, structure, texture, and diagenetic alteration is obtained.

cuttings samples depends on drilling circumstances and lithology. The mechanical stress on cuttings may vary; poor cuttings frequently are the result of modern rotary tools, even to such a degree that the microfauna is extinguished. Examples of bit metamorphism, i.e. the thorough alteration of cuttings by the drilling bit, have been reported (Taylor 1983).

Standardized sample-description sheets for cuttings with appropriate abbreviations are recommended. The following colors are conventionally used in making cuttings samples logs more readable: Light blue = calcareous limestone, dark blue = dolomite, red = red shale, gray = gray shale, yellow = sandstone, green = salt, and purple = anhydrite.

No rigid rules should be given for guidance in describing cuttings samples. Each geological province presents its own problems.

2.2.3. SAMPLING AND SAMPLE PREPARATION

Since it is impossible to analyze an entire sedimentary sequence it is necessary to work with samples. A sample is considered to be representative of a geological formation at the point of sampling, rarely of the entire formation. Type, size, and number of samples depend on the geological problem to be solved (e.g. lithology, diagenesis, fabric analysis, petrophysics, environment of deposition).

Among outcrop samples one can choose between grab samples, serial samples, channel samples, and compound samples. The decision for compound sampling instead of discrete sampling depends on scale and type of the investigation. For taking oriented samples dip and strike should be properly and correctly marked in place before the specimen is broken off.

The size of a sample to be collected depends on the coarseness of the sediment and on the analytical procedures envisaged. Sample containers such as cloth bags, paper bags, plastic bags, glass jars, or waterproof cartoons should be suited for transportation and storage. Wet samples are to be stored only in cloth bags or plastic bags. Complete labeling and precise numbering at the sampling locality are indispensable. Number all samples serially during a given sampling expedition, regardless of their composition or locality.

Sample preparation and laboratory investigations are described in detail by Müller (1967). The procedures comprise drying of samples; removal of water-soluble salts through repeated washing; desintegration of consolidated sedimentary rocks for sieve

Table 9.   Sedimentary parameters deducable from electric well logs.

1.	Types of bedding boundaries

	First order	(differences in lithology)	} Primary boundaries
	Second order	(differences in grain size and mineral composition)	
	Third order	(differences in diagenetic consolidation)	} Secondary boundary

2. External properties of beds
   – Thickness
   – Surface of beds
   – Lateral extent

3. Internal properties of beds
   – Massive bedding (no internal structure)
   – Lamination
   – Graded bedding
   – Cross-bedding
   – Conglomerates
   – Imbrication of pebbles
   – Reef structures
   – Homogeneous dolomite
   – Bioturbation
   – Clay fabric
   – Schistocity

4. Sequences – rhythms – cycles

5. Environment of deposition

or heavy mineral analysis by dissolution of the carbonate components or by chemical or ultrasonic dispersion of argillaceous and silty sediments (separation of microfossils or clay minerals); desintegration of sandstones with different cements (clay, carbonate, silica, iron compounds); sample splitting; and impregnation of friable, brittle or wet sediments with synthetic resins.

2.2.4. STRUCTURAL AND TEXTURAL ANALYSIS

The term textural analysis is generally applied to smaller features, i.e. size, shape, and geometric arrangement of constituents of a sedimentary rock. Textural analysis is done in outcrops or in cores and subsequently, when required, in thin sections or by X-ray diffractometry. Numerous sedimentary parameters can also be determined by properly evaluating electric well logs (Table 9). Texture should always be reviewed in three dimensions. Smoothened surfaces of consolidated cores or hand specimens facilitate textural analysis.

Large-scale textures of unconsolidated sediments are analyzed in outcrops; small-

Table 10.   Sedimentary structures: process and structure (after Pettijohn, Potter and Siever 1987, p. 98).

CURRENT	DEFORMATIONAL
Depositional	desiccation (mud-crack casts)
beach cusps	eruption (sand volcanoes and spring pits)
graded bedding	founder and load structures
parallel lamination (parting lineation)	impact (spray, hail, and spring pits)
sand waves (ripple mark and cross-bedding)	injection (neptunian dikes and sills)
wave and swash marks	slump (folds, faults, and breccia)
CURRENT	BIOGENIC
Erosional	Animal
channels	crawling trails
obstacle scours	feeding trails
rill marks	grazing trails
scour marks	residence structures
	resting trails
Tool marks	Plant
bounce, brush, prod, and skip marks	impressions
roll marks	rootlets
slide and groove marks	
striations and grooves	CHEMICAL
	cementation (sand crystals)
	crystallization (salt and ice)
	diffusion (color banding)
	pressure solution (stylolites)
	replacement (nodules)

scale textural features (e.g. original grain arrangement), however, can be preserved by impregnation techniques. Various methods of impregnating unconsolidated sediments were described by Bouma (1969): Sedimentary peels, lacquer peels of clay material, peels of wet unconsolidated sands, lacquer-resin peels, epoxy peels, and acetate peels.

The term structure is generally used for larger features of sedimentary rocks. Sedimentary structures are of both inorganic and organic origin (Table 10). Inorganic sedimentary structures are bedding, cross-bedding, and bedding deformations; organic sedimentary structures are mainly various types of bioturbation. The majority of sedimentary structures is not dependent on the rock type: e.g. bedding is ubiquitous; cross-bedding occurs in sandstones and carbonate rocks, slumping is recorded from siltstones, sandstones, and limestones.

Size and areal extent of sedimentary textures and structures vary as a function of grain size. Boulder conglomerates display large-scale features, whereas claystones and micrites reveal a smaller-scale inventory which is easily overlooked. Textural analysis of very fine-grained rocks of varying composition is a promising field for further research on depositional environments. The compendium of igneous and metamorphic rock textures (Bard 1980) is useful when comparing sedimentary textures, which are seen under the microscope, with igneous and metamorphic textures.

2.2.4.1. Description of Sedimentary Structures

Sedimentary structures are features observed within sediment layers or along the sediment/fluid interface in response to different processes active prior to lithification. Their sizes range from millimeter to kilometer dimensions.

Inorganic structures are classified as primary or secondary sedimentary structures. Primary sedimentary structures are produced during deposition, e.g. bedding, cross-bedding, graded bedding, current marks, and sole marks. Secondary sedimentary structures are post-depositional, produced by processes such as compaction and diagenesis. Standard texts on sedimentary structures are those by Allen (1982) and Pettijohn, Potter and Siever (1987).

Primary sedimentary structures range from large-size sedimentary bodies (e.g. shoestring sands) to small-scale structures occuring in exposures or in cores. The bed is the smallest rock-stratigraphic unit or the sedimentation unit (Otto 1938) that presumably formed under essentially constant conditions of deposition. Bedding is a ubiquitious feature of sedimentary rocks. The term bedding is synonymous with stratification (Latin: stratus, a layer); lamination is the finest stratification or bedding, typically shown by shales and fine-grained sandstones.

Fundamental bedding features are depicted in Table 11. A layer of 1 cm thickness or more is arbitrarily defined as a bed, a layer of less than 1 cm thickness as a lamina. Bedding varieties comprise tabular, lenticular, trough, wedge-like, rippled, irregular, and contorted bedding. The time required for the deposition of a lamina or a bed ranges from seconds for coarse clastics to hundreds of years for pelagic sediments.

Cross-bedding (cross-lamination, false bedding) consists of an internal arrangement of subsets of strata transverse or oblique to the main bedding planes. It is found only in granular sediments such as siltstones, sandstones, conglomerates, calcarenites, etc. McKee and Weir (1953) classified cross-bedding based on its geometry and on the type of the lower boundary. Allen (1963) revised the above classification. Cross-bedding is a significant sedimentary structure for paleocurrent analysis and basin analysis (Pettijohn, Potter and Siever 1987); the dip angles indicate the direction of paleocurrent flow within a basin.

Graded bedding is the upward decrease of particle size within a bed. Graded beds vary in thickness from a few mm to 1 m or even more. They occur in sequences of several units, frequently with erosive contacts at the base of each unit. Thick graded beds may be correlated over many kilometers.

Graded beds are commonly found in turbidites from early Precambrian to the present. The Bouma sequence (Figure 11) describes a complete turbidite bed characterized by a vertical succession of five units: (1) graded or massive unit, (2) lower unit of parallel lamination, (3) unit with current ripple lamination, (4) upper unit of parallel lamination, and (5) pelitic unit. Mostly one or more of the units may be absent, either at the top

Table 11.   Fundamental bedding features in alluvial channels
(after Kennedy 1978).

BED CONFIGURATION	BARS
Bed geometry	Sand waves
Forms of bed roughness	Banks
Bed form	Sand banks
Bed regime	Deltas
Bed phase	RIPPLES
Bed irregularities	Dunes
Sand waves	Sand waves
Bed material forms	Ripple marks
Bed shape	Current ripples
FLAT BED	DUNES
Smooth bed	Ripples
Plane bed	Sand waves
BED FORM	Sand bars
Bed irregularity	TRANSITION
Bed wave	Sand waves
Bed feature	Washed-out dunes
Dune	ANTIDUNES
Ripple	Standing waves
Sand bar	Antiripples
Gravel bar	Sand waves (also applied
Sand wave	to water waves)
	CHUTES AND POOLS
	Violent antidunes

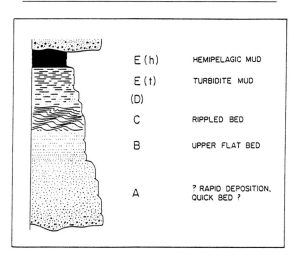

Fig. 11.   Complete Bouma sequence divided in five sedimentological units: A, B, C, D and E (after Walker
1984, p. 173).

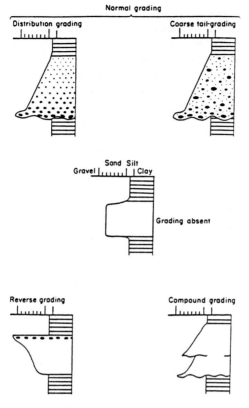

Fig. 12.   Various types of graded bedding in turbiditic sequences (after Selley 1988, p. 127).

or at the bottom. Various types of graded bedding from turbidite sequences are shown in Figure 12. Graded bedding may also form through processes such as the activity of burrowing organisms, the effect of storms (e.g. tempestites), or by daily or seasonal variations in sedimentation (e.g. varved sediments).

Inverse grading is typical of high-concentration gravity-flow deposits (grain-flow and debris-flow deposits). It occurs in turbidites, but also in beach and stream gravels, fluvial deposits as well as in pumice-fall, lahars, and other pyroclastic-flow deposits.

Consequently, caution is recommended using graded bedding as a top-and-bottom criterion.

Bedding planes may exhibit a wide variety of sedimentary structures (Pettijohn, Potter and Siever 1987). On the upper surface of a bed mud cracks, raindrop imprints, pseudomorphs after halite as well as swash, rill, and ripple marks may occur. These features are clues to the transport medium and to the environment of deposition. Sole marks are found on the underside of beds and represent molds of current marks, scour marks, or tool marks. The long axes of linear current marks are usually parallel to the main current direction. Thus, numerous paleocurrent studies are based on measurements of the preferred orientation of current marks.

Other primary sedimentary structures are caused by sediment loading and penecontemporaneous deformation or by the movement of water trapped in the unconsolidated sediment (water-escape structures). The deformational structures comprise load casts, ball-and-pillow structures, convolute bedding, slump structures, as well as clastic dikes and related structures (Pettijohn, Potter and Siever 1987, pp. 275·ff.).

Secondary sedimentary structures are internal and external sedimentary structures formed after deposition by processes related either to physical stress or to chemical changes, commonly through the interaction between connate water and the sediments. Folds, faults, cone-in-cone structures, and stylolites result from physical stress. Caution is recommended not to confuse some folded structures like drag folds with convolute bedding. Similar complex fold structures are observed in evaporitic rocks (enterolithic structures, Gekrösegips). They are caused by the transformation of anhydrite into gypsum. Remobilization of mineral matter leads to the formation of geodes, septaria, concretions, loess dolls, chert nodules, and color banding ('Liesegang rings'). Selective dissolution of mineral substance causes boxwork structures (e.g. Rauhwacke or Zellendolomit), dissolution breccia (subrosion), vugs, and corrosion surfaces.

Biogenic sedimentary structures comprise bioturbation by borings, tracks and trails, casts and prints, as well as fecal pellets and coprolites. The importance of these features for environmental analysis will be discussed in the following section.

2.2.4.2. X-Ray Photography

X-ray photographs can be obtained from consolidated sedimentary rocks, unconsolidated dry or wet sediments, and impregnated samples. A plane-parallel slice of sediment, as thin as possible (0.2–2 cm), is commonly used. Thick slices furnish too many superimposed features. Non-parallel samples should be embedded in fine sand in order to compensate thickness variations. The X-ray photograph is a shadow image caused by differential X-ray absorption of various parts of the sediment (Hamblin in Carver 1971).

The X-ray apparatus used should meet the specification of low kilovoltage (30–150 kV range). A higher kV range gives more penetration, but also produces more scattering. The tube current needs not to exeed 4 or 5 mA. The distance between X-ray tube and specimen should be about 1 m; the radiation cone should cover the whole object. Lead letters can be used for identification. This method is fast and easy, and – compared with other time-consuming techniques – inexpensive. The specimen is not destroyed, damaged, or altered in any way. Also X-ray microphotography is feasible.

X-ray photography reveals sedimentary structures of so-called 'homogeneous' sediments that are not seen by visual inspection and provides details of internal structures. Thus, it greatly enhances studies of microfacies, paleocurrents, and environment of deposition, and furnishes information on the distribution of minerals, fabric, stratification, and bioturbation. X-ray photography is especially useful in studying cores (Figure 13). Geologically young, soft sediments, especially of the pelitic grain-size range, give generally better pictures than compacted, geologically older sedimentary rocks.

Computerized tomography (CT), another non-destructive X-ray imaging technique,

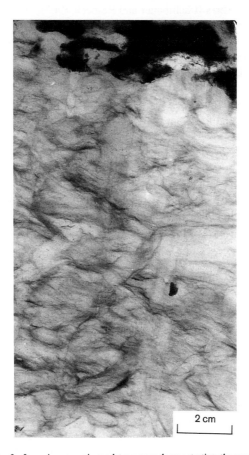

2 cm

Fig. 13. X-ray photograph of a Jurassic reservoir-sandstone core demonstrating thorough bioturbation, Preetz 5 well, N Germany.

was applied by Swennen *et al.* (1991) to better characterize reservoir rocks and reservoir behavior. CT registers mineral variations and different types of porosity and helps to observe multiphase flow in porous rocks such as sandstones.

2.2.4.3. Sedimentary Fabrics

Sedimentary fabrics refer to the spatial arrangements of sedimentary particles and to the orientation of rock components. Details on this subject are found in Müller (1967), Carver (1971), Potter, Maynard and Pryor (1980, pp. 66 ff.), and Pettijohn, Potter and Siever (1987).

Packing is the spacing pattern of mineral grains in a sediment. It correlates to porosity. Primary porosity and its changes during diagenesis and burial have been thoroughly studied in sand/sandstones (Füchtbauer and Reineck 1963); less information is available on lime mud/limestones and mud/shales. An accurate prediction of porosity, however, is – except for the simple model of packing of spheres – impossible, because of the interaction of numerous textural variables. Packing is determined occasionally in gravel outcrops and – more often – in thin sections cut perpendicular to the bedding.

Particle orientation refers to the two- or three-dimensional orientation of linear or planar sedimentary particles. Imbrication is the 'shingling' effect of flat pebbles and elongate sand grains produced by a current. Many individual particles are inclined in the current direction. Parting lineation, a typical bedding-surface feature of thin-bedded sandstones, displays internal particle orientation on a large scale. Even heavy mineral streaks, when depicted by X-ray photographs parallel to the bedding, may indicate the paleocurrent direction.

Particle orientation is determined directly or indirectly. Direct determination relies either on the dimensional orientation of elongated particles (pebbles, sand grains or fossils) or on the optical crystallographic orientation of single mineral particles such as detrital quartz.

Azimuth and imbrication are plotted in a structural diagram. Measurements can even be carried out on photographs of gravel accumulations. Microscopic measurements in sands and sandstones are carried out by means of oriented thin sections parallel and perpendicular to the bedding.

Statistics are indispensable for handling both two-dimensional and three-dimensional data. Quartz orientation in the bedding plane of even silty claystones might be determined by means of low-power scanning electronmicrographs.

Measurements of the optical-crystallographic orientation of minerals in sandstones under the petrographic microscope, either on single grains or within the entire field of view, are feasible in quartz arenites only. Indirect measurements of internal particle orientation are bulk measurements of physical parameters on cores. They correlate physical properties of a mineral such as the dielectric or the acoustic anisotropy with the dimensional orientation of such minerals. A number of factors such as polymineralic composition of detritus and cements, microfracturing, unrecognized cross-lamination, etc., however, seems to interfere. Currents also affect the orientation of small magnetic particles ($0.1–30~\mu$m) at the time of deposition.

The interrelation between the sedimentary fabric anisotropy and the anisotropy of physical properties is a promising field for further research.

Type and degree of preferred orientation and of imbrication vary considerably. The relationship to the paleocurrent direction is complex. However, preferred orientation of sand grains tends to be parallel to subparallel to paleocurrents determined by other features like cross-bedding, ripple marks, or parting lineation.

Preferred orientation of sedimentary particles and their imbrication can be used as a hint at the direction of paleocurrents, at the setting of paleoslopes, as well as at the environment of deposition. Moreover, the direction of preferred grain orientation commonly parallels the direction of maximum permeability in reservoir sandstones. Sedimentary fabric anisotropies, primary as well as secondary ones, the latter produced by differential cementation, will influence enhanced oil recovery.

More systematic observation on sedimentary fabrics of recent sediments from various environments might reveal environment-specific patterns of particle orientation. Automatic image analysis will enhance and facilitate future research in sedimentary fabrics.

2.2.4.4. *Physical Properties of Sedimentary Particles*

Physical properties of sedimentary particles like size, shape, and partly surface textures of particles are not mineral-specific. Other physical properties such as color, hardness, crystal structure, cleavage, fracturing, and optical properties are mineral-specific. These properties are attained during the transport of the sedimentary particles and/or through diagenetic alteration. Modifications of these properties by diagenetic processes are the greater, the smaller the grain size is. They comprise decrease of grain size by dissolution (e.g. silt), increase of grain size by secondary overgrowth, or transformation of pitted grain surfaces into plane crystal faces if a detrital grain such as quartz grew diagenetically.

2.2.4.4.1. Size. Size, usually defined as the particle or grain diameter, is one of the main properties of a sedimentary particle. Standard texts on grain-size determination, grain-size parameters, and grain-size evaluation are Köster (1964) and Müller (1967).

Based on particle-size sediments are broadly classed into gravels (conglomerates), sands (sandstones), silt (siltstones), and clay (claystones). Particle size controls porosity, permeability, and other physical rock properties. Mineral composition normally changes with particle size; rock fragments predominate in conglomerates, clay minerals in shales. Particle size and the parameters defining particle-size distribution (mean, sorting, skewness, kurtosis) characterize a sediment and give valuable hints at the environment of deposition.

Various grade scales have been proposed which arbitrarily divide sediments into specimens of size classes (Figure 14). The classification commonly used in the F.R.G. corresponds to the DIN Norm (DIN 4022). The Wentworth grade scale is generally used by Anglosaxon geologists. Moreover, the Phi (ϕ) scale retains the Wentworth grade names, but converts the grain boundaries into Phi values (e.g. the logarithmic transform: $\phi = \log_2 d$ (d = diameter)). The figure furthermore demonstrates the range of the applicability of different techniques of particle-size analysis.

Particle-grain size distribution of a sediment is determined by particle-size analysis. Because particle size may range from less than one micron (e.g. clays) to several meters (e.g. boulders) in diameter various methods must be applied for particle-size analysis: direct measurement, sieving and settling methods, or thin-section analysis. If particle size is measured directly, the number of grains within a given fraction is determined (grain percentage).

A first approximation of the particle size can be obtained either from visual inspection or from the particular behavior of the particle between ones teeth (silt = crunchy, clay = plastic).

Boulder, cobbles, and gravels are measured manually with a tape or ruler. Grain size in the sand range can be roughly determined by means of a grain-size magnifier with a built-in size scale.

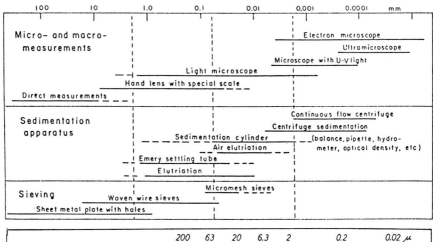

Fig. 14.   Grain-size classifications and methods of grain-size analysis (modified after Müller 1967).

Proper sampling is important for grain-size studies. Fifty to one hundred grams of sand are sufficient for either sieve or sedimentation tube analysis.

Particle-size distribution in sands, friable sandstones, and pebble conglomerates in the particle-size range from 0.063 to 125 mm is commonly determined by sieve analysis which is fast and efficient.

Settling methods (e.g. Atterberg tube, pipette method, sedimentation balance, hydro-meter and centrifuge method) are used for grain-size determination below 0.063 mm; settling analysis approximates grain transport by suspension better than sieve analysis.

Settling methods using the settling velocity of particles are based on Stokes' law which defines the settling velocity of a sphere as:

$$w = \frac{(P_1 - P)g}{18 \text{ mPa.s}} \, d^2$$

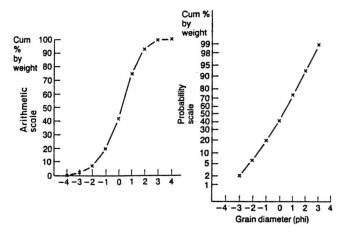

Fig. 15. Cumulative frequency-distribution curves plotted on arithmetic and probability scales (after McManus 1988).

where w = settling velocity, $(P_1 - P)$ = density difference between particle and fluid, g = acceleration due to gravity, mPa.s = viscosity, and d = particle diameter.

Particle-size analysis by means of thin section may give a more realistic grain-size distribution than sieve analysis, especially in tight-cemented sandstones or in shattered sandstone sidewall cores. Methods for analysis are described in detail by Müller (1967, pp. 58 ff.).

Siltstones and claystones are normally not amenable to particle-size analysis by means of thin sections. Nevertheless, in thin sections of claystones the larger particle size of the precursory sediment may be revealed. Modern computer methods allow transformation from grain-size analyses obtained from thin sections into a grain-size distribution obtained by sieving (Schäfer 1982).

Particle-size distributions are usually shown either by a cumulative frequency curve (Figure 15) and/or by numerically specifying their central tendency (mean, mode, or median) and their form (sorting, skewness and kurtosis). A cumulative frequency curve is described by the percentile. Previously only two percentiles (ϕ 25 and ϕ 75), called the first and third quartiles, were used to determine sorting (So). Later as many as five or more quartiles were taken to define additional parameters such as bimodality index, skewness and kurtosis. Table 12 summarizes the most common graphic and moment measures.

Ratios of sand, silt, and clay are commonly plotted on triangular diagrams.

The C–M diagram (Figure 16), introduced by Passega (1957, 1964) plots C, the maximum particle size (99th percentile) against M, the median size (50 percentile). Used for both modern sands and ancient sandstones, they show areas for pelagic suspensions, uniform suspensions, graded suspensions, bed load suspensions, and turbidites. Graphic plots on arithmetic-cumulative probability paper (Doeglas 1946) are designed to show only a limited number of particle-size distributions, called F, F–S, M, M–S, M–C, and C–S–C types. The F, M, and B curves are for sands, the C and S curves for clays. These

Table 12. Graphic and moment measures (after Pettijohn, Potter and Siever 1987, p. 74).

Name	Graphic formula[a]	Moment formula[b]	Remarks
Mean	$Me_\phi = \dfrac{(\phi_{16} + \phi_{50} + \phi_{84})}{3}$	First moment $$\bar{x} = \sum_{i=1}^{n} f_i m_{i\phi}/100$$	All three measures of central tendency reflect average kinetic energy of depositing medium plus size distribution of available sediment.
Median	$Md_\phi = \phi_{50}$		
Mode	M_ϕ = Midpoint of most abundant class interval		
Bimodality index	$Mi_\phi = 1 + \dfrac{(\phi_f - \phi_c)}{2\phi}$		Measure of bimodality, if present: ϕ_f is midpoint of finest mode; ϕ_c is midpoint of coarsest.
Sorting	Inclusive graphic standard deviation $$s_l = \dfrac{\phi_{84} - \phi_{16}}{4} + \dfrac{\phi_{95} - \phi_5}{6.6}$$	Second moment $$s_\phi = \left[\sum_{i=1}^{n} f_i(M_{i\phi} - \bar{x}_\phi)^2/100\right]^{1/2}$$	Measures dispersion, which is dependent upon velocity variations plus bimodality.
Skewness	Inclusive graphic skewness[c] $$SK_l = \dfrac{\phi_{84} + \phi_{16} - 2\phi_{50}}{2(\phi_{84} - \phi_{16})} + \dfrac{\phi_{95} + \phi_5 - 2\phi_{50}}{2(\phi_{95} - \phi_5)}$$	Third moment $$3_\phi = \sum_{i=1}^{n} \dfrac{f_i(M_{i\phi} - \bar{x}_\phi)^3}{100s_\phi^3}$$	Measures asymmetry, the direction of 'tails', which are widely believed to have environmental significance. Skewness varies from +1.0 (positive) to 0.0 (symmetrical) to -1.0 (negative).
Kurtosis	$K_G = \dfrac{(\phi_{95} - \phi_5)}{2.44(\phi_{75} - \phi_{25})}$	Fourth moment $$4_\phi = \sum_{i=1}^{n} \dfrac{f_i(M_{i\phi} - \bar{x}_\phi)^4}{100s_\phi^4}$$	Measures peakedness. Graphic measure is ratio of sorting of centered 90 percent to centered 50 percent. $K_G = 1.0$ for normal curve, $K_G > 1.0$ for peaked curve, and $K_G < 1.0$ for flattened curve.

[a] All after Folk and Ward (1957) except bimodality index after Sahu (1964).

[b] f_i = fraction of total weight in each class interval; $m_{i\phi}$ = midpoint of each class interval in phi units.

[c] Also $SK_l = \dfrac{\phi_{84} - \phi_{50}}{\phi_{84} - \phi_{16}} - \dfrac{\phi_{50} - \phi_5}{\phi_{95} - \phi_5}$ according to Warren (1974).

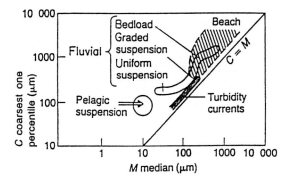

Fig. 16. Main C–M diagram after Passega (1957, reprinted with permission).

curves reveal size differentiation and mixing more clearly than other methods. Sindowski (1957), using logarithmic probability plots, recognized 10 types of cumulative curves by their shapes.

Grain sizes are used to characterize the depositional environment of modern and ancient sediments. The granulometry of modern rivers, beaches, and dunes has been extensively investigated. Statistical coefficients (e.g. skewness) allow occasionally to differentiate between beach and dune sands. However, factors such as mixed sediments from different environments and/or inherited grain-size distributions complicate the evaluation of grain-size analyses.

Evaluation of grain-size analyses in ancient sedimentary rocks is even more complicated for diagenetic and/or methodical reasons: diagenesis may cause mechanical infiltration of clay minerals after deposition, neoformation of clay minerals and/or decay of unstable components into clay minerals. Granulometric analysis of a fossil sediment by sieving may disaggregate clay clasts, pellets, and/or pel-aggregates into their constituent clay particles. Some shales may originally have been deposited as sandstones completely composed of fine to coarse-grained clay clasts or pellets; only careful examination of thin sections could reveal this complete inversion of grain sizes.

Grain-size analysis is, notwithstanding the lack of standardized methods and nomenclature, of great significance for environmental analysis.

2.2.4.4.2. Shape. The shape of pebbles is best defined (Zingg 1935) by the ratio between length, breadth, and thickness, i.e. with terms like special or equant, oblate (disc or tabular), blade, and prolate (roller) shape. It is controlled by the parent-rock type and by the subsequent transport history. Pebbles from slates and schists are normally tabular or bladed, those from isotropic rocks (e.g. quartzite) equant to subspherical. Attempts have been made to relate pebble shape to the depositional environment (e.g. fluvial versus littoral). The shape of sand-size particles is best defined by their elongation (a:b); the sand-size particles are not suited for determining long, medium and short axes.

Sphericity is the degree to which the shape of a sedimentary particle approaches that of a sphere. A perfect sphere has a sphericity of 1.0; all other particles have sphericity values less than 1.0.

Roundness (angularity) is the ratio of the average radius of curvature of the several corners or edges of a particle to the radius of curvature of the maximum inscribed sphere. A perfectly rounded particle has a roundness value of 1.0; less-rounded particles have values less than 1.0. Roundness is independent of shape. It must not be confused with sphericity. The roundness of sand grains in thin sections is best determined by means of the matching chart of Russel–Taylor–Pettijohn (Schneiderhöhn 1954). For visual comparison the degree of roundness can be rated (Pettijohn 1975, pp. 58–59) as angular, subangular, subrounded, rounded, and well rounded. Experiments showed that eolian transport was more efficient as rounding mechanism than water transport over an equivalent distance. Chemical solution seems to be unsignificant as rounding agent.

2.2.4.4.3. Surface texture. Surface textures of sedimentary particles were being repeatedly examined in order to relate them to depositional processes. Cailleux (1952) studied sand grains, 0.3–2.0 mm in size, under the binocular and classified them as: unworn, polished (émoussé-luisant) or frosted (rond mat). Polished grains with clear translucent surfaces are considered to originate from subaqueous and frosted grains from eolian transport.

Megascopic striations on pebbles are usually considered as evidence of glacial transport. Pebbles from a desert environment display occasionally a shiny surface, the so-called 'desert varnish'.

With the introduction of scanning electron microscopy the study of surface textures has greatly enhanced thanks to the great depth of focus of the microscope. Together with observations by the transmission electron microscope the following surface features, i.e. abrasion patterns, have been observed on detrital quartz grains (Figure 17) and attributed to the following major depositional environments:

–   Subaqueous: medium to high relief patterns with V-shaped indentations, oriented (i.e. chemical etching) and randomly oriented (chemical or mechanical origin); straight or slightly curved grooves and scarps.
–   Aeolian: low relief patterns with flat-pitted surfaces, meandering ridges, and oriented fractures.
–   Glacial: very high relief patterns with conchoidal breakage varying in size and parallel striations of different size.
–   Pedologic: medium to high relief patterns with pitted surfaces, rich in silica globules and pellicles.

Two factors, however, make environmental interpretation of surface textures difficult: (1) surface textures might have been inherited from the source rock prior to transport and (2) original grain surfaces might have been weathered or diageneticaly altered by dissolution or secondary overgrowth.

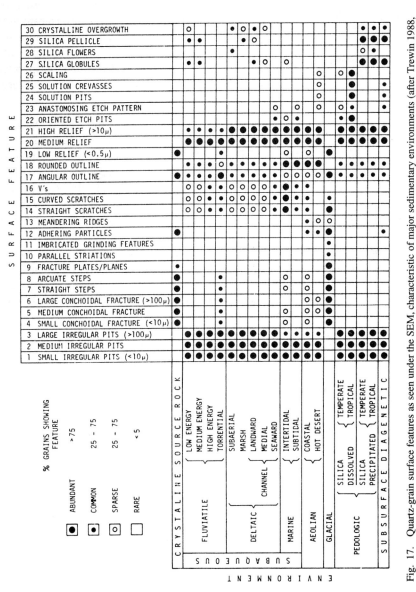

Fig. 17. Quartz-grain surface features as seen under the SEM, characteristic of major sedimentary environments (after Trewin 1988, redrawn from Higgs 1979).

Table 13.   Staining methods for minerals in sediments: in hand specimen, thin sectic

Mineral/ Substance Groups	Mineral/Substance	Procedure/Reagent	Stain
	Dispersed clay minerals as such	Preferential absorption of a Zyglo ultra- violet sensitive dye on clay minerals	bright blue and yellow fluorescence
CLAY MINERALS	Montmorillonite Illite Kaolinite	Powdered and acid-extracted clay stained with malachite green	yellow green blue
	Montmorillonite Illite Kaolinite	Powdered and acid-extracted clay stained with safarine T	violet reddish brown raspberry red
	Aragonite	Feigl's solution (solid Ag_2SO_4 dissolved in $MnSO_4 \cdot 7H_2O$)	black
	Mg-calcite	Alizarine red S in alkaline solution	purple
	Calcite	Lemberg solution (ferric chloride) Alizarin red S	brown → black deep red
CARBONATE MINERALS	Dolomite	Amaranth stain	deep red
	Fe-calcite/ Fe-dolomite	K-ferricyanide or alizarine red S – K-ferrycyanide solution	mauve–purple– light blue
	Siderite, rhodochrosite, smithsonite, cerussite, witherite	Combined application of alizarin red S, Feigl's solution, Na-thiocyanate, benzidene, Magneson solution, NaOH and HCl	dark brown/black–blu colorless–dark red brown–orange red
	Marble	Marble staining mechanism	
FELDSPAR MINERALS	K-Feldspar	HF etching and Na-cobaltinitrate stain	yellow
	CaNa-Feldspars	HF etching and K-rhodizonate stain	pink
BARIUM SILICATES	Ba-Feldspars Ba-Micas	Rhodizonate reagent	pale pink
	Gypsum	Alizarine red S, methanol and NaOH	deep purple
	Gypsum and anhydrite	Nitric acid solution and mercuric nitrate	yellow
EVAPORITE MINERALS	Ca-sulfates	Optical staining of the various $CaSO_4$ hydrates under phase contrast by means of an immersion method	
	K-minerals	Mg-dipikrylamine solution	bright orange-red
PHOSPHATE MINERALS	Apatite	Ammonium molybdate and benzidine	yellow
	Fossil bone hydroxyapatite	Haematoxylin/eosin and other stains	yellow-brown black
ORGANIC MATTER	Polysaccharide mucus	Periodic acid-Schiff (PAS) reaction stains mucus linings of fecal pellets and of some trace fossils	dark reddish- purple
	Leaf cuticulae	H_2O_2 and KOH (1–10%), Sudan III black/safranine stain	dark colored

and individual grains (modified after Friedman 1978 and extended).

in outcrop and hand specimen	in thin section	in solitary grain	References	Remarks
●	●	●	Miller 1988	
			Vahl 1956	
		●	Grim 1968	
			Allman and Lawrence 1972	
		●		
●	●	●	Warne 1962	
●	●	●	Müller 1967, pp. 162-164	
			Feigl and Anger 1972	
●	●	●	Friedman 1978	
●	●	●	Miller 1988	
●	●	●		
●	●	●	Warne 1962	
●			Hunt 1991	Lasa type marble
●	●	●	Bailey and Stevens 1960 Van der Plas 1966	Well polished samples
●	●	●	Müller 1967, pp. 164–167 Houghton 1980 Miller 1988, pp. 98–99	essential for good stain
	●		Fisk 1985	Applicable only to more soluble Ba minerals
●	●	●	Friedman 1978, p. 765	
●	●	●	Hounslow 1979	Extremely
			Miller 1988, p. 99	poisonous
		●	Lanitz 1965	Industrial quality-control method
●	●	●	Kühn 1955 Ludwig and Rosenbaum 1966	Industrial quality-control method; health hazard
●	●	●	Feigl and Anger 1972	
●	●	●	Garland 1989	
●	?		Risk and Szczuczko 1977 Miller 1988, p. 101	Beware of contamination (e.g. fingerprints!)
●			Sander *et al.* 1993	

2.2.4.5. Staining Techniques

Staining is among the most useful techniques in the analysis of sedimentary rocks, also in fabric analysis.

Staining techniques are used in sedimentology for discriminating between minerals of the same group (chlorides, carbonates, sulfates, and feldspars) or for reliability determining the identity of a mineral such as phosphate or of trace fossils (Risk and Szczuczko 1977). Most techniques can be applied to walls in outcrops, to hand specimens, on polished surfaces, thin sections, and peels. Hamblin (1962) applied alizarin red S to stain argillaceous portions of sandstones in order to reveal hidden sedimentary structures. Staining techniques are reviewed in detail by Cayeux (1916, pp. 95–170), Müller (1967, p. 162), Friedman (1978, p. 764), Houghton (1980), and Miller (1988). Suggestions for applying less known staining techniques to sedimentological problems see Feigl and Anger (1972).

Common staining techniques for minerals in sediments are summarized in Table 13. Each set of samples will behave differently during mineral staining. Some staining procedures, especially those for discriminating different carbonate minerals, are in common use.

Chloride and sulfate staining are applied for underground mapping in salt domes and salt caverns or for examining cores from evaporitic sequences. Carbonate staining is used largely for carbonate petrography and sedimentology, i.e. discrimination of metastable aragonite, Mg calcite, calcite or dolomite, as well as for identifying sandstone cements. Feldspar staining assists in proving presence and determining frequency of detrital feldspars in sedimentary rocks. Staining for plagioclase feldspar has not become a routine procedure in petrography (Houghton 1980). Thin sections stained for both types of feldspar, however, are an invaluable aid in petrographic research. A modified phosphate staining is recommended to discriminate the phosphorus content in cryptocrystalline sediments, rocks, and concretions that is difficult to determine by means of the polarizing microscope alone.

2.2.5. MINERAL ANALYSIS

This section treats the analysis of microscopic and submicroscopic minerals. Analytical methods referring to whole-rock analysis by thin section are discussed first, subsequently the separation into single mineral constituents, in both light and heavy minerals, by various separation methods.

2.2.5.1. Microscopic Preparation Methods of Whole Rocks

Microscopic preparation methods of whole-rock specimens comprise impregnation of loose sediments and friable sedimentary rocks, making of thin sections, and visual recording of the mineral content.

The advantage of this whole-rock method is that structural as well as compositional details of a given sedimentary rock are preserved for complete evaluation.

Impregnation of loose sediments and friable sedimentary rocks such as claystones, shales, tuffs, sands, and sandstones is a prerequisite for thin-section preparation. Impregnation of porous reservoir rocks such as sandstones and limestones with an ordinary dye, fluorescence dye, or Wood's metal will facilitate microscopic determination of porosity.

Preparation of thin sections is a fully automatized procedure. Ultra-thin sections ($\pm$5–10 μm thick) facilitate microscopic examination of very fine-grained sediments (micrites, shales, and cherts). Recommended is the preparation of highly polished thin sections without cover glass which serve for all analytical procedures alike: thin-section, microprobe, scanning electron microscope, and cathodoluminescence analysis.

Quantitative recording of the mineral content is based on the geometric rock analysis, i.e. modal analysis, by means of point counting (Müller 1967, pp. 142 ff.; Carver 1971). By this statistical method the frequency of a mineral, rock fragment or fossil in a sample is determined by counting the number of times it occurs at specified intervals throughout the sample. A computer-controlled scanning electron microscope in combination with an energy-dispersive X-ray spectrometer (Minnis 1984) even allows automatic point-counting for determining the relative content and distribution of minerals in rock samples. For sole estimation of mineral frequencies comparison charts for visual percentage estimation serve best. Image analysis using differences in color, birefringence, translucency, and/or special sample preparation can simplify quantitative mineral determination.

2.2.5.2. Conventional Thin-Section Microscopy

The determination of sedimentary textures such as grain shape, roundness, and grain sizes as well as staining techniques discussed in the preceeding sections, are also conventional methods applied in thin-section microscopy.

The most essential aspect of thin-section microscopy is the determination of rock composition by conventional optical procedures. Best summaries on optical mineralogy and thin-section petrography are those by Tröger and Braitsch (1967), Kerr (1977), and Philpotts (1989). The tables by Tröger (1979) are a convenient supplement.

Thin-section petrography of argillaceous rocks is a promising approach in sedimentology (Zimmerle 1991). The chief precondition for thin-section petrography, especially of argillaceous rocks, is that absolutely flawless thin sections can be prepared. Conventional thin-sectioning is summarized by Murphy (1986) and Miller (1988).

Thin sections, however, are not always the best means to identify a mineral properly: heavy minerals are examined best in grain mounts, clay minerals by X-ray methods, phosphates by chemical means.

The most widespread and frequent detrital constituents of sandstones are rock fragments and minerals.

Rock fragments (lithics) are defined as sand grains that consist of at least three individuals of the same mineral or at least two crystals of different minerals, inclusions excluded. In sandstones grain size, provenance, maturity, and age determine the amount of detrital rock fragments.

The proportion of rock fragments augments with increasing grain size. Texturally mature sands normally have a lower content of rock fragments than immature sands.

Major types of rock fragments in sediments are (1) the argillaceous group with shales, slates, pyllites, and schists, (2) volcanic rocks including volcanic glass, and (3) the silica group with polycrystalline quartz grains and cherts. Of local importance only are carbonate, metamorphic or plutonic rock fragments. It is difficult to decide which detrital particle to report as a 'rock fragment' and how to name it, especially in the case of polycrystalline quartz grains. Detrital rock fragments are best hints at the provenance of a sediment and useful for stratigraphic correlation.

Quartz is the most common mineral in sediments, foremost in sandstones, because of its high chemical stability and resistance to abrasion. Varietal features such as inclusions, undulose extinction, color, twinning, trace elements, shape, roundness, surface texture, etching behavior, radiation color, cathodoluminescence, and thermoluminescence (Figure 18) can give clues to the provenance of quartz. Le Ribault (1977) used thin-section and scanning-electron microscopy to determine provenance of sedimentary quartz.

Feldspars, widespread and rather common in sandstones, especially in arkoses, are less resistant to mechanical abrasion than quartz and are chemically less stable. Feldspars comprise mostly acid and intermediate members of the plagioclase group as well as potassium feldspars. In thin sections they are identified by their optical properties, twinning (if present), staining, cathodoluminescence, and element composition. During diagenesis they can be altered by kaolinization or sericitization.

Detrital feldspars may transform diagenetically into albite. Thus, provenance based on the composition of feldspars must be interpreted with caution.

Micas, chlorites, and clay minerals, closely related in chemical composition (i.e. hydrous aluminosilicates) and crystal structure (phyllosilicate), are the chief constituents in pelitic rocks; in sandstones they are normally minor constituents. As essential constituents of the clay matrix and as argillaceous rock fragments they include all major groups of clays: Kaolin group (kaolinite, dickite, nacrite), micas (muscovite, illite, glauconite), smectite group, chlorite group, and mixed-layer group (corrensite, etc.). Only a rough grouping of clay minerals based on particle size, texture, optical properties, and color is feasible in thin sections. Precise identification of clay minerals by conventional optical methods is difficult; phase microscopy, however, makes identification more feasible.

Quantitative determination of the relative abundance of the various clay minerals must rely on separation, size fractionation, and X-ray diffraction methods.

The proportion of detrital versus authigenic clay and mica minerals can be determined best by combining size fractionation, thin-section analysis, and scanning electron microscopy.

Carbonate minerals (mainly calcite, dolomite, and siderite) are rock-forming in carbonate rocks and concretions; they are admixtures in shales, and detrital constituents as well as cementing material in sandstones. As mostly biogenic constituents of calcareous sands or sandstones they comprise skeletal fragments, oolites, and fecal pellets. Cayeux (1916), Majewske (1969), and Horowitz and Potter (1971) provided a useful guide to the recognition of skeletal debris in thin sections. Abraded carbonate grains derived from source regions outside the basin are a significant constituent of Molasse sandstones. Diagenetic calcite, dolomite, and/or siderite are abundant in sandstones as pore fillings and replacement cement. In thin sections calcite is normally anhedral consisting of a crystal

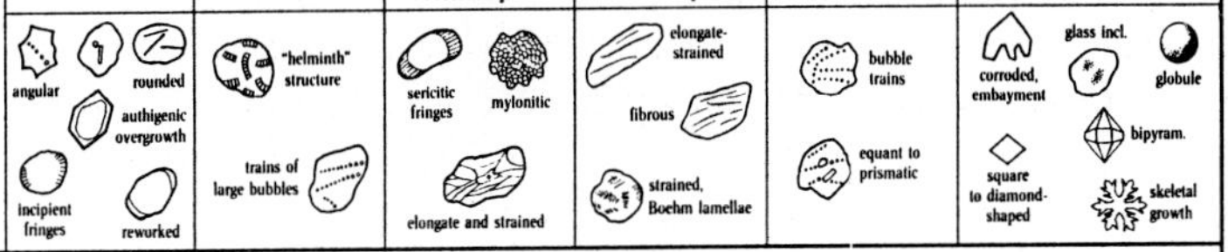

	Sedimentary	Hydrothermal	Low-grade Metamorphic	High-grade Metamorphic	Plutonic	Volcanic
Characteristic Detrital Grain Types (Diagrammatic)	angular, rounded, authigenic overgrowth, incipient fringes, reworked	"helminth" structure, trains of large bubbles	sericitic fringes, mylonitic, elongate and strained	elongate-strained, fibrous, strained, Boehm lamellae	bubble trains, equant to prismatic	glass incl., corroded, embayment, globule, square to diamond-shaped, bipyram., skeletal growth
Properties						
Primary Shape	angular to rounded, rarely idiotopic	anhedral to euhedral	anhedral, elongate	anhedral, elongate, equant	mostly anhedral, exceptionally euhedral	euhedral, bipyramidal, "fenêtres volcaniques"; magmatic corrosion embayments; square or diamond-shaped keratophyric quartz
Secondary Shape (due to volcanic shattering)	├───────── ├──────── ├──────── ├──────── ├─────────→ Intratelluric Quartz Shattered to splinters					
Diagenetic Grain Surface Features	A variety of possible etch patterns; little researched; not yet a reliable provenance indicator					
Twinning	Secondary Dauphiné twinning commonly generated by stress and thus indicative of deep burial and/or deformation (SIMANOVICH 1978). Twinning in detrital quartz potentially helpful in provenance studies.					
Deformation Features	Undulatory extinction of quartz and other deformation features (mosaic structure, Boehm lamellae, strain rosettes) are diagnostic of the grade of deformation or increasing burial depth and not of a specific rock type (such as granite, metamorphic rock, etc.). Compactional fractures — Tectonic fracturing					
Mineral Inclusions	equant to rounded, fibrous (illite)	"helminth" structure and ore minerals common		fibers and needles common (sillimanite, kyanite, rutile, tourmaline, garnet)	equant to prismatic inclusions dominant, fibrous inclusions rare	inclusions of volcanic glass common

	Sedimentary	Hydrothermal	Low-grade Metamorphic	High-grade Metamorphic	Plutonic	Volcanic
Fluid Inclusions	various generations in detrital quartz, in fractures, fillings and overgrowths	large bubble trains			bubble trains common, abundant in miarolitic quartz	locally large bubble trains
Isotopic Composition	Oxygen isotope composition ($\delta^{18}O$ in ‰): igneous quartz: 9.4–10.3, migmatitic quartz: 9.9–12.7, metamorphic quartz: 8.4–22.3, hydrothermal / pneumatolitic quartz 7.4–19.9. Generally, $\delta^{18}O$ is higher at low temperatures than at high temperatures.					
Trace Elements	The following trace elements occur in quartz: Al, B, Ba, Ca, Co, Cu, Fe, Ga, Ge, Li, Mg, Mn, REE, Ti, Zr. They may fingerprint quartz from different source rocks. Ti, Mg, Fe may vary in plutonic, volcanic and metamorphic rocks on a local scale. Ge content of granite and gneiss quartz 0.8–1.0 ppm, in hydrothermal and pegmatitic quartz are characterized by a very low trace-element content. The Al content of quartz increases with temperature of formation.					
Electron Paramagnetic Resonance (EPR)	The following EPR centers are known in crystalline quartz: Al^{3+}, Fe^{3+}, Ge^{3+}, Ti^{3+}, Li, Na, H, OH, Al-O., -SiO$_3$, Al-O-P, -O-M-O-, -Fe-O-, NH_3. These centers may turn out to be helpful in provenance studies. They may be characteristic of ore deposits and certain quartz-bearing igneous rocks.					
Cathodoluminescence (CL)	none (low diagenesis) to greenish (high diagenesis), evaporitic quartz brown violet	zoning common; greenish, bluish, reddish, yellow or ochrous colors		initially brown to brownish violet, later bluish hue	high-temperature, quartz bright violet blue	phenocrysts of high-temperature quartz bright blue; quartz in groundmass commonly reddish
Irradiation Color	Degree of darkening under the same saturation dosage of irradiation:	pegmatitic 20 %		metamorphic ≤ 5 %	granitic 40 %	rhyolitic 70 %

Fig. 18. Varietal features of detrital quartz indicative of provenance (mainly after Zimmerle 1994).

mosaic of individual crystals ranging from a few microns to several centimeters. In contrast dolomite ordinarily assumes a rhombohedral form and siderite a dog-tooth shape. Acid etching and/or carbonate staining are routine procedures to thin sections containing carbonate minerals.

Sulfate minerals (gypsum, anhydrite, and barite) are rock-forming in evaporitic rocks, common admixtures in carbonate rocks, and cementing material in sandstones. Perfect

Table 14.   Thin-section criteria which can be diagnostic to determine the environment of deposition.

Rock Type
Macrofossil Content
Large-Scale Sedimentary Structures
Microfossil Content
Segregation of Detrital Minerals
Small-Scale Sedimentary Structures
Grain-Shape and Grain-Size Distribution
Grain-Surface Textures
Redox Potential (Eh) and pH
Mineral Cements

cleavage, xeno- to hypidiotopic crystal contours, and specific staining are typical features of sulfates in thin sections.

Other minerals, occurring in minor amounts and only locally, are sulfides (marcasite and pyrite), biogenic and concretionary phosphates (collophane), grains of green iron silicates (chamosite and glauconite), iron and titanium oxides, and zeolites (associated with volcanic debris). Most of them are synsedimentary or early diagenetic. Presence, crystal habit, and/or spatial arrangement of these minerals may serve as useful clue to the environment of deposition and diagenesis. For instance, heavy minerals often occur in placers, or abundant phosphate may indicate upwelling conditions, or early diagenetic pyrite is normally present in form of framboids.

Glauconite, a dull-green earthy or granular mineral, is the most common sedimentary (diagenetic) iron silicate found in marine sediments from Cambrian to present. As an indicator of marine environment and very slow sedimentation it deserves special mention; it is also commonly used for radiometric age determination of sediments. A modern case history of glauconite formation is given by Jeans *et al.* (1982).

Thin sections can also be successfully applied to contribute to the analysis of the environment of deposition. Thin-section criteria which can be diagnostic of a certain environment of deposition are listed in Table 14.

An abridged data sheet as used by petroleum company research laboratories is shown in Figure 19.

The scope and usefulness of reflected-light microscopy for studies of siliceous, calcareous and phosphatic sedimentary rocks was recently demonstrated by Braun (1994).

2.2.5.3. *Mineral Separation and Preparation Methods of Mineral Fractions*

Mineralogical and textural analyses of individual mineral grains or of mineral groups require separated disaggregation and specific separation techniques by physical means. Disaggregation of consolidated sedimentary rocks into their components is done by crushing and subsequent grain-size separation either by simple elutriation or by sieve analysis.

Location/well   ....

Elevation/depth   ....

Grain supported   ☐

Matrix supported   ☐

Carbonate minerals	%
Calcite	....
Fe-calcite	....
Dolomite	....
Fe-dolomite	....

Terrigenous minerals

Quartz	....
K feldspar	....
Plagioclase	....
Mica	....
Others (biotite, amph. etc.)	....

Other minerals

Anhydrite	....
Gypsum	....
Halite	....
Chert	....
Others	....
	....
	....

Cement   ....
....
....
....
....

Comments

Diagram/photomicrograph

Rock name ....

Sedimentary structures

Skeletal components

Matrix

Porosity

Fig. 19. Abridged data sheet as used by some petroleum company research laboratories (after Harwood 1988).

Methods of disaggregation are outlined by Müller (1967, pp. 35–41). Separation methods (Müller 1967, pp. 115 ff; Parfenoff *et al.* 1970; Boenigk 1983) comprise manual selection, heavy-liquid separation, magnetic separation, flotation, dielectric and electrostatic separation, and chemical isolation. The separation methods to be used depend on grain size, physical and chemical nature of the specific mineral or mineral group, and minimum amount of material available.

Manual selection of grains of a specific mineral species for further mineral analysis is facilitated by a pair of tweezers, a needle or some suction device, a perforated sorting tray, and by a large magnifying glass or binocular microscope.

Heavy-liquid separation is the individual separation of a single mineral or the mass separation of minerals of the same species or of a group of various mineral species in a liquid with a given specific gravity. For standard heavy mineral separation bromoform

(2.89) is commonly used. Other heavy liquids in wide use are tetrabromethane (2.97), methylene iodide (3.32), and Clerici solution (4.2). Heavy liquids are expensive and very toxic. Details on mean density values of heavy minerals and heavy liquids are found in Müller (1967, p. 116), Carver (1971), and Boenigk (1983).

Magnetic separation of individual minerals or of sedimentary rock fragments is based on differences in mass susceptibility of paramagnetic minerals or mineral mixtures. Concentration of ferromagnetic minerals (magnetite, pyrrhotite) can be achieved by a hand magnet. Complete separation of paramagnetic minerals is achieved best by the 'Frantz Isodynamic Magnet Separator'.

Flotation, ordinarily applied in ore dressing, uses differences in wetting ability of minerals. The sample is suspended in water to which chemicals and/or foaming agents have been added. Poorly wetted particles rise with the foam and can be removed with the foam.

Dielectric and electrostatic separation methods can be applied to separate quartz-feldspar mixtures or to isolate glauconite.

Chemical isolation of minerals or mineral mixtures can be achieved by dissolving one or more constituents in water, acid, or in a melt depending on the mineral association, e.g. gypsum by water from a halite-gypsum assemblage, insoluble residues by acid from carbonate rocks, or quartz by hydrofluoric acid from a quartz-feldspar mixture. This is an inexpensive and rapid procedure.

2.2.5.3.1. Heavy minerals. Heavy minerals are sedimentary minerals with a specific gravity higher than a standard (usually bromoform = 2.85). They are normally a minor constituent of a sediment (in sands less than 1%, in shales and limestones much lesser). Pertinent textbooks on heavy mineral analysis are those by Milner (1962), Parfenoff *et al.* (1970), Carver (1971), Boenigk (1983), and Mange and Maurer (1991).

Heavy minerals include: magnetite, ilmenite, leucoxene, zircon, tourmaline, rutile, biotite, apatite, hornblende, augite, garnet, kyanite, and sphene. They are characterized by their crystal habit, color, and varietal features.

Special attention is being focussed on zircon and tourmaline, two of the most common stable heavy minerals. Varietal features of zircon such as color, crystal habit, type of mineral inclusions, growth irregularities, length/width ratio, degree of radioactivity, fluorescence, and fluorescence spectra have been successfully used to trace source rocks and to differentiate sediments. For instance, purple or rose-pink zircon ordinarily originates from Precambrian granites and gneisses as an ultimate source. And growth impedance seems to be diagnostic of volcanic zircons. Krynine (1946) used color varieties of tourmaline to track provenance; yet no simple relationship exists between color and composition of tourmaline, and the type of igneous or metamorphic rocks it is derived from.

As each heavy mineral variety shows another size spectrum in crystalline source rocks (Feniak 1944), a wide size range, ordinarily 0.2–0.065 mm, covering all heavy-mineral size spectra, has to be examined in heavy-mineral analysis.

Heavy minerals have been formed primordially as accessory minerals in crystalline rocks: most in acidic rocks, some in basic rocks. Only a few heavy minerals are diagnostic of a specific provenance (Table 15). Zircon, tourmaline, and rutile are more widespread

Table 15.   Detrital mineral suites (both heavy and light) of sands and sandstones characteristic of specific groups of source rocks (modified from Heinrich 1956).

1. Pegmatites and hydrothermal veins		Chromite	Talc
Microcline	Rutile	Ilmenite	Chlorite
Albite	Anatase	Anthophyllite	Pyrope
Muscovite	Brookit	Enstatite	Magnesite
Biotite	Columbite-tantalite	*6. Contact-metamorphic rocks*	
Tourmaline (especially	Cassiterite	Andalusite	Dravite
indicolite and other	Garnet	Hypersthene	Tremolite
varieties except dravite)	Apatite	Diopside	Wollastonite
Dumortierite	Monazite	Corundum	Vesuvianite
Spodumene	Hematite (specular)	Cordierite	Axinite
Topaz	Sphalerite	Scheelite	
Barite	Wolframite	*7. Low- to medium-grade regional metamorphic*	
Fluorite	Gold	*rocks*	
2. Granites, granodiorites, syenites, rhyolites,		Muscovite	brown with
quartz latites, trachytes		Biotite	graphite inclusions)
Orthoclase	Zircon	Chlorite	Albite
Microcline	Monazite	Epidote-clinozoisite	Actinolite
Oligoclase	Xenotime	Spessartine	Talc
Hornblende	Rutile	Chloritoid	Piedmontite
Biotite	Sphene	Tourmaline (black, blue)	
Muscovite	Magnetite	*8. High-grade regional metamorphic rocks*	
Apatite	Ilmenite	Kyanite	Hematite (specular)
Tourmaline	Pyrite	Silimanite	Hornblende
Allanite		Almandite	Tourmaline
3. Tonalites, diorites, monzonites, dacites,		Staurolite	Oligoclase-andesine
andesites, latites		Biotite	Magnetite
Zircon	Hornblende	Diopside	Zircon
Oligoclase-andesine	Biotite	Epidote-clinozoisite	Glaucophane
Magnetite	Sphene	Rutile	
Ilmenite	Apatite	*9. From sediments (authigenic)*	
4. Basalts, diabases, gabbros		Pyrite	Celestite
Augite	Ilmenite	Barite	Collophane
Hypersthene	Apatite	Glauconite	Leucoxene
Hornblende	Pleonaste	Hematite (earthy)	Limonite
Epidote-clinozoisite	Olivine	*10. Volcanic ash falls*	
Magnetite	Leucoxene	Apatite	Hornblende
5. Periodites, serpentinites		Augite	Zircon
Magnetite	Leucoxene	Biotite	
Picotite	Serpentine		

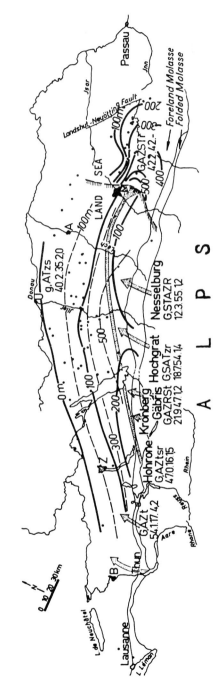

Fig. 20.   Regional heavy-mineral case history: Upper Tertiary – Lower Aquitanian Molasse Basin. Dots = wells; arrows = direction of transport; heavy minerals: A(patite), G(arnet), R(utile), S(taurolite), T(ourmaline), Z(ircon). Order and majuscles/ minuscles indicate frequency. Lower line: First number = feldspar, second number = chert, third number = carbonate, fourth number = calcite/dolomite) (redrawn from Füchtbauer 1964, Fig. 14d).

than other heavy minerals. Hubert (1960) introduced the zircon-tourmaline-rutile maturity index (ZTR index).

Heavy-mineral studies are carried out to reconstruct the paleogeography of a sedimentary basin (Figure 20) or to correlate the lithostratigraphy of non-fossiliferous sedimentary sequences. Also the gradual unroofing of a source area from a sedimentary to a metamorphic/igneous terrain can be traced by the heavy minerals. More accurate analysis of minerals and their varieties, sophisticated statistical evaluation, and new ways of plotting and mapping will enhance the applicability of heavy-mineral studies in the future.

2.2.5.3.2. Insoluble residues. Insoluble residues are the remaining materials of a sample dissolved in a solvent, normally an acid, e.g. hydrochloric, acetic, and formic. Dissolved are most commonly carbonate rocks. Siliceous rocks can be dissolved with hydrofluoric acid, some evaporitic rocks simply with water. The principal components of insoluble residues from limestones and dolomites are chert, siliceous oolites, quartz, pyrite, siliceous fossil remains, and aggregates of clay, silt, and/or sand. Gypsum, anhydrite, glauconite, mica, and iron oxides are accessories. Seventy percent of insoluble microfossils such as arenaceous foraminifera, conodonts, graptolites, chitinozoans, and spores are recovered by some dissolution method.

The insoluble residue techniques, the choice of solvents, general procedures, plotting of residue data, and application are described in detail by Ireland (1971).

Insoluble residues can be used for correlation, subdivision, identification, and description of thick carbonate sections, especially in non-fossiliferous well sections.

2.2.5.3.3. Grain mounts. For further mineralogical determinations isolated or separated minerals or rock particles have to be mounted temporarily or permanently in mounts of either one single grain or numerous grains.

Oil and resin mounts are normally used for optical identification of minerals or two-dimensional shape analysis of sedimentary particles. Immersion oils are used for the determination of the refractive index with a temporary mount. For permanent mounts of heavy minerals, low-risk resins of high refractive index such as Araldit B ($n = 1.595$), Aroclor 4465 ($n = 1.66$), Mountex ($n = 1.67$), Meltmount ($n = 1.704$), or Hyrax ($n = 1.71$) are available. Three-dimensional analysis requires more complicated procedures.

Details on the glass to be used, the mounting media (oils and resins), and preparation procedures are found in Parfenoff *et al.* (1970), Carver (1971, p. 499), and Boenigk (1983).

2.2.5.4. Other Analytical Methods

Other analytical methods – mostly in the microscopic or submicroscopic range and employed either separately or supplementing thin-section work – are fluorescence, cathodoluminescence, thermoluminescence, confocal laser scanning microscopy, X-ray diffraction analysis, X-ray fluorescence spectroscopy, infrared absorption spectroscopy, differential thermal analysis (Moenke and Moenke-Blankenburg 1965), scanning electron microscopy, image analysis, and absolute age dating.

Fluorescence (FL) is a type of luminescence in which the emission of light stops

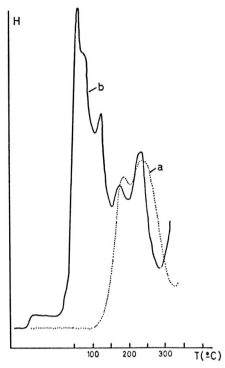

Fig. 21. Thermoluminescence (TL) glow curves of a quartz sandstone (Cherichira Sandstone, Tunisia). Natural (a) and induced (b) macro TL. The sharp 60°C peak provides the standards of induced TL (after Charlet 1971, Fig. 3).

when the external incitation ceases produced by electromagnetic radiation of ultraviolet and blueviolet light of a wavelength from 300 to 600 nm. Several minerals such as carbonate minerals, zircon, and apatite display characteristic fluorescence under ultraviolet light. Immature organic matter shows fluorescence too, however, mature organic matter shows none. Non-fluorescent mounting media are indispensable for reliable fluorescence analysis.

Cathodoluminescence (CL) is the emission of characteristic visible luminescence by a mineral substance that is under bombardment of electrons. Sedimentary minerals such as quartz, carbonates, kaolinite, and apatite display specific CL, especially in thin sections (Photoplates 1–2). CL facilitates quick identification of certain minerals and their distribution in thin sections. It also helps to decipher quartz provenance, diagenetic silicification processes, and carbonate diagenesis (Zinkernagel 1978, Marshall 1988).

Thermoluminescence (TL) is the property possessed by numerous mineral substances of emitting light when heated; it results from release of energy stored as electron displacements in the crystal lattice. TL is used in whole-rock analysis, e.g. of sandstones, to obtain specific TL glow curves (Figure 21) that are diagnostic 'fingerprints' of a certain provenance (Charlet 1971).

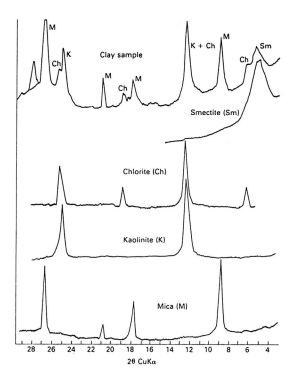

Fig. 22. X-ray diffraction powder diagrams of a clay sample and of component clay minerals (smectite, chlorite, kaolinite and mica) (after Gibbs 1967).

Confocal laser scanning microscopy (CLSM) is a microscope technique (Liedmann and Quader 1991) which extends the possibilities provided by the conventional fluorescence and bright-field microscope. Due to an exceptionally short depth of field continuous optical sections can be obtained in thin section.

X-ray diffraction analysis is an indispensible tool (Kaelble 1967, Klug and Alexander 1974, Brindley and Brown 1980, Bish and Post 1989) in analytical sedimentology, especially for clay-mineral analysis of shales and of sandstone cements. It mainly comprises X-ray diffraction analysis and X-ray fluorescence spectroscopy, besides X-ray photography (Section 2.2.4.2). X-ray diffraction (XRD) is the diffraction of a beam of X-rays (very short wavelength radiation in the 0.1–100 Å interval) by the three-dimensional periodic array of atoms in a crystal. It is either registered by film (Debye–Scherrer) or by diagram (counter-tube method). Either single mineral substances or whole-rocks are analyzed by XRD. The samples are disaggregated, then sedimented on slides, or packed in holders. Also thin-section blanks can be used. Advanced automated powder diffractometer systems with improved d-spacing resolution (Figure 22) made XRD analysis an inexpensive routine tool in petroleum sedimentology. Sample preparation and other procedures are described in Carver (1971), Klug and Alexander (1974), and Tucker (1988). XRD analysis is applied to mineral identification proper (powder method), to the determination of the degree of crystallinity (e.g. of illite) and orientation of clay minerals, to size determination of crystallites, and to semiquantitative mineral analysis of sedimenary rocks. The crystallographic fundamentals necessary for optimum interpretation of XRD data are well summarized in Kleber (1974).

X-ray fluorescence spectroscopy (XRF) is a type of X-ray emission spectroscopy in which the diagnostic X-ray spectrum of a mineral substance is produced by using X-rays of short wavelength to induce the mineral substance in order to emit X-rays of a longer wavelength. It is used for elemental analysis of mineral substances.

Infrared absorption spectroscopy (IR) is the observation of an absorption spectrum in the infrared frequency region (0.7 μm to about 1 mm). A transparent disc of the finely ground mineral powder to be analyzed mixed with spectrally pure KBr is brought into the path of an infrared beam and compared spectrometrically with a tablet of pure KBr. Sedimentary minerals such as silicates, carbonates, and sulfates show characteristic absorption bands. For instance, different members of the plagioclase group or potassium feldspars can be distinguished by IR. Moenke (1962, 1966) catalogued IR spectra of 505 minerals.

Differential thermal analysis (DTA) (Moenke and Moenke-Blankenburg 1965, Langier-Kuzniarowa 1973) is carried out by constant heating or cooling of a mineral sample or a mineral mixture that undergoes chemical and physical changes. Simultaneously a reference material (e.g. metakaolinite) that undergoes no changes is heated or cooled in identical fashion. The temperature difference between the sample and the reference material is measured. Exothermic and endothermic reactions caused by water release, recrystallization, changes in structure, oxidation, and reduction are recorded. DTA has been increasingly used in combination with X-ray analysis to study clay minerals, clays, and shales. It is a means to determine the degree of crystallinity rather than to identify a specific mineral substance. Illites may show identical X-ray patterns, but very different DTA curves. Also carbonates can be easily distinguished by means of DTA.

A powerful tool in oil exploration and exploitation is the scanning electron microscope (SEM). A finely focused beam of electron is moved across the specimen to be examined and the emitted and reflected electron intensity is measured and depicted on a screen, sequentially building up an image. The ultimate magnification reaches $\times 100\,000$. Transparent as well as opaque objects can be examined and a great depth of field is obtained. In conjunction with an energy-dispersive X-ray analyzer, compositional data can be obtained with a spatial resolution of 0.5–1 μm. The successful application of thin-section analysis by the SEM combined with integrated energy-dispersive XRF analysis was shown by Nöltner (1988). Examples of scanning electron micrographs from typical sandstone and carbonate reservoirs are depicted in Photoplates 6–14. Up-to-date accounts on principles and geological applications of the SEM were published by Whalley (1979), Goldstein *et al.* (1981), Smart and Tovey (1981, 1982), Grabowska-Olszewska *et al.* (1984), and Welton (1984).

Image analysis of sedimentary particles and rocks (Carver 1971, Oosthuyzen 1980,

Schäfer 1982, Serra 1982) will find increasing application. Still difficult to achieve are only the various steps of preparation, if the sedimentary rocks examined are cemented. Quantitative image analysis can also provide a rapid method for direct determination of porosity, specific surface of pores, pore size, and pore shape from rock sections (Ruzyla 1986).

The absolute age of a fossil organism, rock, mineral cement, or geologic feature or event, usually expressed in years, is indispensable for sedimentological investigations. Physical and chemical absolute age determination methods and their application are discussed by Geyh and Schleicher (1990).

2.2.6. CHEMICAL ANALYSIS

Inorganic as well as organic chemical analyses are applied to determine the composition of brines, gaseous, liquid and solid hydrocarbons, and sedimentary rocks.

Analysis of brines, i.e. pore fluids in deep sedimentary basins such as oil-field waters, is important to define the present chemical environment of sedimentary rocks, especially in sandstone or limestone reservoirs, and to allow prediction of pore system behavior after the drilling bit disturbed the existent chemical equilibrium. Berner (1980) presents a new geochemical classification of sedimentary environments.

Gaseous, liquid, and solid hydrocarbons are essential components of source rocks and the pore filling of reservoir rocks. Thus, they are subject to subtle analysis. With increasing application of organic geochemistry in petroleum exploration organic geochemical procedures reached a sophisticated state (Hunt 1979, Tissot and Welte 1984, Waples 1985).

Radioactive and stable isotope analysis by means of an isotope mass spectrometer is a common tool in the inorganic and organic chemistry. It includes radiometric age dating as well as the determination of origin, mechanisms, and conditions of geologic processes. Detailed accounts on isotope analysis and its geological application are found in Wetzel (1983), Aswathanarayana (1985), and Bowen (1988).

Analytical methods for geochemical exploration are summarized by Rose *et al.* (1979) and Van Loon and Barefoot (1989). Seim and Tischendorf (1990) compile data and summarize it in numerous tables and graphs, which are most useful for geochemical studies.

2.2.6.1. Inorganic-Chemical Analysis

Whole-rock and trace-element analysis of bulk samples have been sparingly used to characterize single sedimentary rock specimens, suites of sedimentary rocks from the same area, or from various sedimentary provinces. It has been applied to distinguish chemically different sandstone types (Pettijohn 1963) or shales from different environments (Table 16).

Pertinent textbooks on inorganic-chemical analysis and various aspects of geochemistry are those by Degens (1968), Wedepohl (1969–1978), Yariv and Cross (1979), and Möller (1986).

Table 16. Chemical analyses of glauconitic, phosphatic, and calcarenaceous sandstones (after Pettijohn 1963, Table 9).

	A	B	C	D	E	F
SiO_2	57.40	50.74	75.95	45.43	48.85	51.32
Al_2O_3	6.89	1.93	2.91	0.03	11.82	2.92
Fe_2O_3	11.98	17.36	10.29	2.92	1.83	0.72
FeO	3.04	3.34	–	–	1.22	
MgO	2.41	3.76	1.37	0.61	0.45	0.58
CaO	1.78	2.86	0.10	26.21	12.85	24.70
Na_2O	1.11	1.53	0.35	0.34	0.47	–
K_2O	4.85	6.68	2.99	0.16	0.64	–
H_2O^+	5.36	9.08	5.40	2.78	2.75	–
H_2O^-	4.46	–	–	–	–	–
TiO_2	0.29	–	0.20	0.11	trace	0.30
P_2O_5	0.22	1.79	–	16.05	10.70	–
CO_2	–	0.88	–	3.12	3.40	20.00[4]
MnO	0.03	–	–	0.02	–	–
SO_3	0.45	–	–	0.86	–	trace
F	–	–	–	1.87	2.86	–
Total	100.29[1]	99.95	99.56	101.25[2]	97.84	100.54
Less 0				–0.79	–1.20	
				100.46[3]	96.64	

(1) Includes BaO, 0.02; (2) includes C, 0.45, FeS_2, 0.29; (3) given as 99.01 in original; (4) by calculation.

A. 'Greensand' (Middle Eocene), Pahi Peninsula, New Zealand (Ferrar, 1934, p. 47).

B. 'Greensand marl' (Upper Cretaceous), New Jersey, U.S.A., R. K. Bailey, analyst (Mansfield, 1920, p. 553).

C. Greensand, opal-centered (Thanetien), Angre, Belgium (Cayeux, 1929, p. 130).

D. Phosphate sandstone, 'Upper phosphorite stratum' (Cenomanian), Kursk, Schchigri, U.S.S.R. (Bushinsky, 1935, p. 90). About 38 percent quartz, 45 percent phosphorite, 5 percent glauconite.

E. Phosphatic sandstone, Saint Pôt, Boulannais, France (Cayeux, 1929, p. 191).

F. Calcarenaceous sandstone, Loyalhanna Formation (Mississippian), Pennsylvania, U.S.A. (Hickok and Moyer, 1940, p. 464).

Analytical methods comprise wet chemical analysis, atomic absorption spectrometry (AAS), and X-ray fluorescence spectroscopy (XRF) or a combination of all three methods. Trace elements are normally determined by XRF or by electron microprobe analysis. Certain trace elements or elemental ratios, especially in pelitic rocks, are well suited for chemo-stratigraphic correlation and hints at the environment of deposition, e.g. Sr and Na contents in limestones, B or Ti in claystones. Certain elements also help to trace

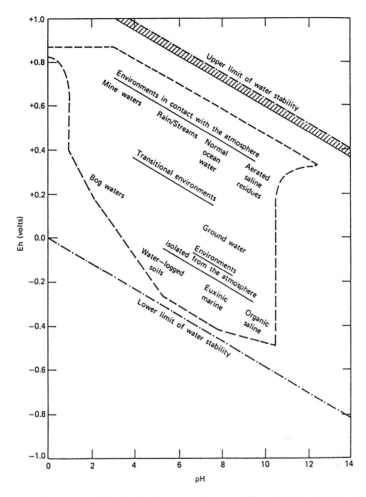

Fig. 23. Position of some natural chemical environments in terms of Eh and pH (modified after Garrels 1960, p. 201, from Carver 1971).

diagenetic changes.

The electron microprobe is an efficient tool to analyze elemental composition and areal elemental distribution within a single mineral grain in a polished thin section, or in a polished rock surface. A finely focussed beam of electrons (as small as 1 μm) excites X-ray emission from selected portions of a sample. From the X-ray spectrum emitted the composition of the sample at the point of excitation is determined with a sensitivity around 50 ppm.

Also Eh and pH measurements have been used (Carver 1971, p. 597) to characterize chemical environments within the hydrosphere (Figure 23). However, great care is necessary in order to obtain reproducible and quantitatively meaningful measurements of surface waters, ground waters, or sediments in the field. The abrasion pH, defined as the pH of a slurry produced by water, is a method for field identification of minerals. It approximates the chemical environment in which a mineral has been formed. Stevens and Carron (1948) determined the abrasion pH of 280 minerals.

Microchemical spot tests (Feigl and Anger 1972), made on minute mineral grains or polished surfaces under a microscope, deserve more attention as an auxiliary to routine thin-section analysis for a better identification of sedimentary minerals.

In the field of inorganic-chemical analysis stable isotope analysis is usually applied to the following elements: $^{18}O/^{16}O$, $^{34}S/^{32}S$, $^{13}C/^{12}C$.

The isotopes of an element have slightly different physical and chemical properties, owing to their mass differences determined by means of mass spectrometer. Stable isotope analysis serves to determine paleotemperatures, as well as relative time, trends, and origin of diagenetic mineral formation.

2.2.6.2. Organic-Chemical Analysis

Organic chemical analysis refers to the chemical analysis of solid, liquid or gaseous components dispersed or concentrated in sedimentary rocks as well as dissolved in oil field brines.

Analytical procedures include organic carbon determination, elemental analysis, as well as pyrolysis of organic matter, chromatography, and stable isotope analysis besides optical methods such as the determination of vitrinite reflectance and fluorescence microscopy (Tissot and Welte 1984, Hollerbach 1985, Brooks and Welte 1984, 1987).

Because carbon is the most abundant element in organic matter organic carbon analysis provides a sensitive test for the abundance of organic matter in sedimentary rocks, especially in potential source rocks. Nitrogen and phosphorus are less abundant. A common method is to combust the sediment sample in a high-frequency induction furnace, e.g. LECO, in an atmosphere of dry, CO_2-free oxygen at temperatures exceeding 1 500°C.

Since organic matter also contains other elements such as nitrogen, sulfur, oxygen besides carbon and trace metals elemental analysis became a rather standardized analytical method.

Pyrolysis is the chemical decomposition of organic matter by heat. Rock-Eval or other modern apparatuses automatically registers the decomposition plotting a decomposition curve. It reports – analogous to the differential thermal analysis – exothermic reactions as the organic sample is heated up to 650°C. As compared to simple elemental analysis, pyrolysis is more meaningful because it reflects the geological history of the organic substance analyzed.

Chromatography, a term for processes separating organic components, is widely applied as analytical technique in hydrocarbon exploration. A gaseous or liquid hydrocarbon sample is moved through a column using adsorption, partition, ion exchange, or other properties in such a way that the different components become separated (gas and liquid chromatography). Chromatograms (Figure 24) are used to 'fingerprint' gaseous or liquid hydrocarbons or hydrocarbons extracted from sedimentary rocks. Chromatography is also

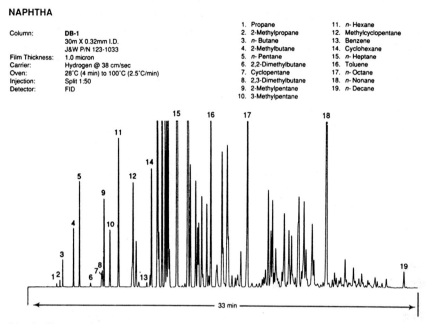

NAPHTHA

Column:	DB-1
	30m X 0.32mm I.D.
	J&W P/N 123-1033
Film Thickness:	1.0 micron
Carrier:	Hydrogen @ 38 cm/sec
Oven:	28°C (4 min) to 100°C (2.5°C/min)
Injection:	Split 1:50
Detector:	FID

1. Propane		11. *n*- Hexane	
2. 2-Methylpropane		12. Methylcyclopentane	
3. *n*- Butane		13. Benzene	
4. 2-Methylbutane		14. Cyclohexane	
5. *n*- Pentane		15. *n*- Heptane	
6. 2,2-Dimethylbutane		16. Toluene	
7. Cyclopentane		17. *n*- Octane	
8. 2,3-Dimethylbutane		18. *n*- Nonane	
9. 2-Methylpentane		19. *n*- Decane	
10. 3-Methylpentane			

Fig. 24. Capillary gas chromatogram of low-molecular hydrocarbons as fingerprint for hydrocarbon composition. The main organic compounds present are numbered (Fisons Catalogue).

employed to identify alteration processes such as biodegradation, i.e. the decomposition of hydrocarbons by microorganisms.

Stable isotope analysis of carbon ($^{13}C/^{12}C$), deuterium, nitrogen, another important exploration tool, is applied to characterize gaseous or liquid hydrocarbons and to determine their origin and maturity (Fuex 1977, Hoefs 1987). Carbon-isotope analysis of methane allows to distinguish bacterial and thermal methane.

Biological markers (biomarkers) are diagnostic components such as sterane, triterpane, aromatics and sulfur compounds as well as porphyrins and their distribution patterns in hydrocarbons and sediments. Porphyrins are likely derived from chlorophylls in algal or photosynthetic bacteria. Biomarkers may serve as maturity parameters for both oils and sediments as well as for oil/oil or oil/source rock correlation.

2.2.7. GRAPHIC REPRESENTATION, EVALUATION, AND REPORTING

The graphic representation of all data is done best in graphic logs, analogous to sample logs, depicting the sequence of sedimentological characteristics of the rocks found in outcrops or taken from a well. The information, e.g. megascopic description, sedimentary textures, mineral composition, and/or chemical composition, is referred to the depth of sample location. It can be plotted on a strip log form as an interpretative log, percentage log, etc.

Parfenoff *et al.* (1970) and Carver (1971) refer in greater detail to graphic representations such as used in heavy-mineral or insoluble residue analysis.

Semiquantitative or quantitative analytical data is evaluated best by mathematical-statistical methods, proper computer processing, and by automatic plotting. Berners *et al.* (1983) analyzed the grain-size distribution of 3 000 samples from four different Mesozoic stratigraphic intervals by means of large-scale Q-mode factor analysis in order to determine the depositional environments.

A good example of an exhaustive sandstone description including field relations, sample identification, description of hand specimen and thin section, interpretation, and economic importance is given by Pettijohn, Potter and Siever (1987, p. 525).

A good report clearly separates description from interpretation.

An important, but difficult undertaking is to compose a report, interim or final, on a single analysis or on a regional study. It has to contain statements on the essential findings, interpretation, and conclusion in a logical order. The summary must be concise, but stressing the high lights. Only concise reporting with an informative summary will help to transmit own observations and thoughts to fellow scientists and/or management.

2.3. Biogenic Sedimentary Structures as Environmental Indicators

Plant rootlets, vertebrate tracks, invertebrate trails, soft sediment burrows, hard rock borings, and fecal pellets are referred to as trace fossils or biogenic sedimentary structures (Lebensspuren) in contrast to inorganic sedimentary structures. Their study is termed ichnology, respectively palichnology. Pertinent texts include those by Seilacher (1964, 1967), Häntzschel (1975), Frey (1975), Rodriguez and Gutschick (1970), Crimes and Harper (1970, 1977), Basan (1978), Chamberlain (1975, 1978), Wetzel (1981), Werner and Wetzel (1982), the SEPM Spec. Publ. No. 35 (1985), Ekdale *et al.* (1984), Bromley (1990), Frey *et al.* (1990) and Wetzel (1991). The disturbance of sediment by organisms as expressed by the whole of biogenic structures is called bioturbation.

The study of trace fossils allows correlation between characteristic trace-fossil assemblages (ichnofacies) and the depositional environment, important for sediments where no body fossils are preserved. Figure 25 compiles various environmental distribution charts of selected trace fossils covering all environments from continental to abyssal plain.

The digestion of a sediment by sediment-feeding organisms is the first and important phase of diagenesis of numerous marine sediments because of the reworking under biogenic conditions (Berner 1980). Fecal pellets form; pyrite and/or phosphate are precipitated. The soil-mechanical behavior of a sediment is changed by intense bioturbation. Intense bioturbation frequently leads to complete obliteration of original sedimentary structures in marine sediments. Other sediments may retain some relict lamination between discrete 'lebensspuren' systems. Well-laminated or cross-bedded sediments either accumulate too rapidly (e.g. varve-forming conditions, turbidites), or are deposited in oxygen-free environments hostile to burrowing animals (euxinic facies of black shales or some hypersaline lagoon facies). Intensive burrowing leads to the progressive disruption of bedding until a uniformly mottled clay-sand mixture is left. The term 'mottled struc-

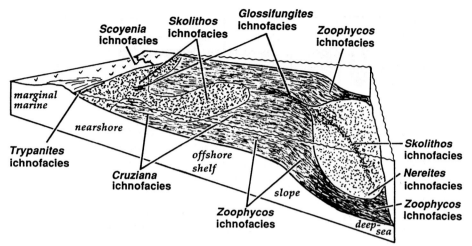

Fig. 25.   Schematic distribution of ichnofacies. Marine ichnocoenoses are related to dynamic environmental factors rather than to water depth or distance from shore (courtesy A. Wetzel – drawing inspired by Frey *et al.* 1990).

ture' (Moore and Scruton 1957) is applied to thoroughly bioturbated sediments diagnostic of the shelf environment which normally is not characterized by distinct 'lebensspuren' i.e. 'deformational biogenic structures'. This structure is frequently found in intertidal and subtidal elements. Thorough bioturbation in a well sequence can be detected by the continuous dipmeter log (CDM) because of the complete lack of dips.

Vertical burrows in interlaminated sands and shales may increase the vertical permeability of such beds. This may be important for aquifers or hydrocarbon reservoir formations.

2.3.1. PHYTOGENIC TRACE FOSSILS

Rootlet structures are phytogenic structures. They are restricted to terrestrial and very shallow water deposits, notably in underclays beneath coal seams (Wurzelböden). Black, carbonaceous tube linings of round to oval cross section and root-like length section indicate generally a phytogenic rather than a zoogenic origin (e.g. in Middle Europe: Late Triassic Schilfsandstein, Early Cretaceous Obernkirchen Sandstein). Varicolored (reddish-violet) mottled sediment structures normally point to phytogenic bioturbation; they are diagnostic of paleosols.

2.3.2. ZOOGENIC TRACE FOSSILS

Trace fossils produced by animals are from millimeter to meter in size. They occur in marine and paralic sediments since late Precambrian; their distribution ranges from continental to abyssal marine deposits. In bioturbated sediments distinct burrow styles and types are recognized.

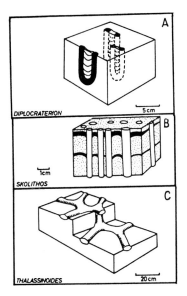 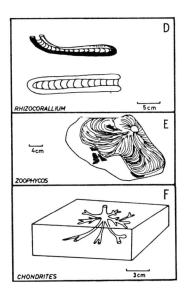

Fig. 26.   Simplified view of some common trace fossils: A. *Diplocraterion*, B. *Skolithos*, C. *Thalassinoides*, D. *Rhizocorallium*, E. *Zoophycos*, F. *Chondrites* (rearranged from Sellwood 1978).

Trace fossils have been grouped according to their ethology (feeding burrows, dwelling burrows, grazing traces, crawling traces, and resting traces) or according to their topology. Feeding burrows and dwelling burrows, however, are often difficult to distinguish. An individual morphological type of a trace fossil is called ichnogenus. These trace fossil 'genera' may represent the burrowing styles of often unrelated animals. Similar forms of burrows simply reflect comparable modes of life of their constructors. Note that similar ichnogenera may be produced by a variety of organisms and one organisms can frequently produce different types of burrows. Consequently, the various types of ichnogenera cannot be grouped in a phylogenetic way.

A selection of important and widespread ichnogenera in marine sediments are (Figure 26):

Diplocraterion are U-shaped burrows (Figure 26A) oriented perpendicular to bedding, with sediment reworked between the arms of the U. It is believed to be a dwelling/feeding burrow of a part-suspension feeding organisms in an environment with unstable sedimentation rates (sandy shore). It is common where oscillatory ripples occur. The vertical translation of the U-shaft which may be sometimes controlled by the negative or positive displacement of the sediment surface produces a 'spreite', i.e. a series of traces of the dwelling tubes left by the animal. *Diplocraterion* is identified in core as a spreite-filled burrow, occasionally only U- or J-shaped.

Skolithos – similar forms are *Monocraterion*, *Tigillites*, and *Sabellarifex* – are vertical, tubular burrows with structure-less, sandy fillings and thin, clayey burrow lining (Figure 26B). It is believed to be produced by suspension-feeding polychaetes under neritic conditions. The burrow may occur in great numbers in sandy sediments (e.g. Cambrian *Skolithos* Sandstone).

Thalassinoides, produced by decapod crustaceans, are mainly horizontal burrow systems extending from an often very deep vertical shaft and following bedding planes (Figure 26C). The burrows are 2–20 cm in diameter and typically show Y-shaped branching patterns forming polygons. Filling may consist of fecal pellets and may show concave/convex laminae. In sands, characteristic of littoral to infralittoral, moderate to relatively high-energy conditions. It also occurs in neritic shales. Vertical elements are common in hardgrounds. In shales burrows can show the greatest vertical extension in a sediment (1–2 m).

Rhizocorallium, also 'spreiten' burrows, are U-shaped burrows oriented parallel or oblique to bedding planes (Figure 26D). The sediment between the arms is reworked. The walls of the tube are often covered with reticulate ridges; septa may also bear ridges. Systems may be up to 30 cm long. *Rhizocorallium* are probably arthropod feeding/dwelling burrows in sandy sediments of the sublittoral. In split cores, *Rhizocorallium* appears as a horizontal spreite-filled ribbon, with the tube present at one end. Rarely, both tubes are visible.

Zoophycos – similar forms are *Spirophycon* – is a complex burrow system with swirling form (helical spiral) (Figure 26E) produced by the progression of an obliquely inclined burrow in a circular path with one entrance remaining fixed. The burrows also may attain large vertical and horizontal extension. It is frequently found at sand/shale interfaces as well as in marl beds of hemipelagic character. In recent sediments it is generally associated with quiet water conditions (abyssal plain, continental slope). In ancient sediments it occurs also in shallower water depth (shelf environments). In core sections, *Zoophycos* appears as horizontal to subhorizontal ribbons with an average diameter of 8 mm, characterized by alternating light and dark spreiten. The trace fossil is constructed by an unknown group.

Chondrites – often also called *fucoids* – are believed to be dwelling/feeding burrows produced by unknown animals; they are plant-like branching burrow systems with a vertical central shaft (Figure 26F). The tunnels are primarily circular in cross section, of constant diameter, and with smooth burrow walls. Branching is lateral, never equal. The diameter of the burrow tubes ranges between 1 and 6 mm. They often occur within thin single beds of shales which probably offer favorable living conditions to the generating organism. It is commonly found in neritic shales. *Chondrites* may represent a behavioral response to a low oxygen content in a sediment. In split cores, *Chondrites* is recognized as clusters of sharp walled, very small, oval burrows.

Cruziana is characterized by a bilobate trail, generally referred to the action of trilobites. *Cruziana*, however, has been found in post-Paleozoic strata, even in formations of fluviatile origin. The environmental significance of this particular ichnogenus has to be interpreted with caution.

Planolites, unlined, rarely branched, straight to tortuous, smooth to irregularly walled burrows of some mm to about 20 mm diameter. In core, *Planolites* appears as ovals of contrasting sediment, usually lighter in color than the host sediment. It is frequently found in marine sediments of different environments.

2.3.3. TYPE OF ORGANISMS

Burrowing animals may belong to nearly each of the macrozoan groups:

Coelenterata	-
Polychaeta	++
Sipunculids	++
Priapulidae, Enteropneusta	+
Echinodermata	++
Bivalves	++
Arthropoda	+
Gastropods	(+)
Cephalopoda	-
Vertebrata	+

The major part of burrowing animals are not known as body fossils, although burrow structures are common in Phanerozoic sediments. The reason is that the majority of burrows are generated by non-skeletal invertebrates. As an exception, molluscs are frequently found within their burrows. Infaunal bivalves and gastropods may frequently be found in their burrowing positions. Burrowing bivalves may ascend/descend in response to sediment accretion/erosion, and their burrows may then give some indication of high sedimentation rates, e.g. in tidal sediments. *Polychaetes*, as non-skeletal organisms, are never preserved in bioturbated sediments, where they probably abounded. Burrowing arthropods particularly crustaceans have thin carapaces, and only the more strongly calcified chaelipeds are commonly preserved. Arthropods dig with their appendages, and their burrows frequently bear characteristic scratch marks (e.g. *Rhizocorallium*). However, examples of fossil burrows containing their crustacean constructors are extremely rare.

Borings are trace fossils consisting of an etching, groove, or hollow, made by plants or animals in hard substrates. Small-scale borings are produced by algae, fungi, and lichens; bryozoans, phoronids, sipunculids, and polychaete annelids. Piercing of shells by turbellarian, brachiopods, cephalopods, and gastropods as well as rock, wood, and shell borings of gastropods, chitons, bivalves, isopods, amphipods, various cirrepeds, echinoids, and sponges are large-scale borings. Fossil borings are indicative of hard substrates.

2.3.4. DEFORMATIONAL STRUCTURES

Besides of the 'lebensspuren' of characteristic forms corresponding to programmed behavior activities of their producers, there exist numerous biogenous structures of no definite topological characteristics. They are referred to as 'deformational structures' because they replace the primary sedimentary structures they have destroyed. They are recognized as patches, eddies, cloud-like structures or horizontal streaks of sand or clay of a pseudo-stratification. This type of structures can be produced by molluscs or polychaetes which dig horizontally through the sediment, inducing sediment liquefaction and sorting.

2.3.5. ICHNOFACIES AND THEIR DISTRIBUTION

Trace fossils are suited as environmental indicators because they reflect a high-order adaption to ecological factors such as mechanical properties of the substrate, oxygen content, distribution and quality of nutrients, etc. This means that a distribution of trace fossils according to paleowater depths cannot directly be inferred. Nevertheless, the integrated influence of these ecological factors will also show correlations to the bathymetry. Therefore, trace-fossil assemblages were tried to be grouped according to (relative) water depth as shown in Figure 25.

The continental ichnofacies consists largely of vertebrate tracks (footprints of birds and terrestrial animals) and burrows (e.g. the spectacular *Xenohelix*, a rodent burrow). Dinosaur tracks are particularly well-studied, although their preservation potential is low. They are normally found on dried-up lake beds, river bottoms, and tidal flats. Paleosols are habitats of continental ichnofacies, too (Chamberlain 1975).

A well-defined ichnofacies occurs in the tidal zone, the *Skolithos* assemblage that is dominated by this ichnogenus. In this environment the sediment substrate is subjected to scouring current action which normally erodes and reworks the sediment.

Thus, some invertebrates of the intertidal zone, e.g. worms, bivalves or crabs, etc., live in crawling, dwelling, and feeding burrows. These burrows are ending at the sediment/water interface, but go down deep enough to provide shelter for the small animals during erosive phases. Burrows may be simple vertical tubes (*Skolithos*), vertical U-tubes with spreiten (*Diplocraterion yoyo*) or without spreiten (*Arenicolites*), or complex networks like those made by crabs (*Ophiomorpha*).

In subtidal and shallow marine environments, a trace fossil association predominates which was defined as '*Cruziana* ichnofacies' by Seilacher (1962) after the characteristic *Cruziana* type. In this zone of less destructive marine action invertebrates crawl over the sea floor to feed in shallow grooves. They also make shallow burrows oriented obliquely or subhorizontally. Other trace fossils of the *Cruziana* ichnofacies include subhorizontal burrows (*Rhizocorallium* and *Harlania*), but also many of the previously mentioned ones.

In the continental slope environment the '*Zoophycos* ichnofacies' (Seilacher 1962) dominates. In this environment weak erosion takes place. Burrows such as *Planolites*, *Chondrites*, *Teichichnus*, *Zoophycos* and sometimes *Scolicia* may constitute this association at varying amounts.

Surface trails are not observed, because they are mostly destroyed by deeper burrowing organisms during the accretion of the sediment column.

In abyssal plains, the realm of turbidite deposition, the '*Nereites* ichnofacies' is characteristic. In this environment, burrows of organisms living on or near the sediment surface have the chance of being preserved by episodic turbidite events. Meandriform traces are typical, they include *Nereites*, *Helminthoida*, and *Cosmorhaphe*. Polygonal reticulate trails (*Paleodictyon*) are also diagnostic. All these trace fossils constitute the Flysch facies.

In the abyssal, but non-turbidite deep-sea environment an association similar in character to the '*Zoophycos* ichnofacies' occurs in continuously deposited and bioturbated deep-sea sediments. *Zoophycos*, *Teichichnus*, and *Chondrites* are typical structures of this assemblage.

The concept of environment-restricted ichnofacies assemblages is useful as auxiliary tool in facies analysis. Although trace fossils are small enough to be studied in cores, even in cuttings, their thorough analysis is hampered due to the difficulties which the three-dimensional reconstruction of burrowing structures cause at such a small scale. Different lithologies (shales, sandstones, limestones) may require different preparation techniques for the cores to bring out trace fossils in an optimum way. Slabbing of the cores is indispensable; impregnation of friable cores advantageous. Simple treatment of the cut-core surface by selected liquids or greases may make trace fossils to stand out better for description. X-ray photography of cores from recent and ancient sediments (Hamblin 1971) facilitates observation of trace fossils (see Figure 13). Staining of trace fossils, i.e. staining of the mucus linings and infillings of burrows (Risk and Szczuzko 1977), may enhance observation and thus interpretation.

2.3.6. EXAMPLES OF SEDIMENTOLOGICAL INTERPRETATION

Biogenic structures are being used to appraise relative rates of sedimentation. Seilacher (1962) showed that quickly deposited graded sandstones from Tertiary Flysch in Spain suppress the pre-existing endofauna in relation to their thickness. Ascending and descending structures in *Diplocraterion* were attributed to a response by the burrower to deposition or erosion at the sediment/water interface. The intensity of bioturbation is frequently negatively correlated to the rate of sedimentation. In fine-grained sediments, bioturbation may trigger thixotropic reactions within the top few centimeters of the sea floor. The complex history of lithification can be unraveled from examining burrows and borings present.

Exhumed concretions may provide hard surfaces bored by cirrepedes, molluscs, algae, and worms. In hardgrounds shells and other hard material are penetrated by the borers. Borings may yield evidence of the time required for the formation of concretions. On the other hand, concretions allow to study the original fabric of biogenic structures because they have preserved their original pre-compactional state.

Trace-fossil associations can also display characteristic vertical zonation, the so-called tiering, within the substrate. When burrows of a particular tier become truncated by turbidites or other catastrophic events, they provide a quantitative means for estimating the incompleteness of a stratigraphic sequence (Wetzel and Aigner 1986).

Kemper (1968) has given an environmental interpretation for the productive Bentheim Sandstone of Early Cretaceous (Valanginian) age in NW Germany as a shallow marine deposit by means of ichnofossils including more than 20 ichnogenera. Trace fossils from the productive Upper Cretaceous chalk in northern Europe and their environmental significance have been studied by numerous authors (e.g. Nygaard *et al.* 1983, Ekdale and Bromley 1984). Several trace-fossil associations are found within the chalk. Porous reservoir intervals, associated with proximal allochthonous beds and debris flow deposits, are characterized by particular ichnofacies.

2.4. Sedimentological Interpretation of Well Logs

A log is a continuous record of observations and/or measurements of physical parameters made on rocks and/or fluids of geologic section in a borehole, usually graphic and plotted to scale on a narrow paper strip or on film as a function of depth. It is recorded during or after drilling. Some logs are relatively simple, identifying the rocks drilled or the time consumed in drilling each meter/foot of hole. Other logs are more complex. Well logs are essentials in oil exploration. By means of the continuous geological information furnished through logs, stratigraphic and lithologic sequences as well as lateral facies changes can be fully registered.

Modern accounts on the sedimentological aspects of well logging are published by Pirson (1983), Desbrandes (1985), Serra (1985), and Ellis (1987). Numerous geological applications of wireline logs are contained in Doveton (1986), Hurst, Lovell and Morton (1990), and Hurst, Griffiths and Worthington (1992).

2.4.1. LOG TYPES

Early logs include the 'driller's log', sample log, mounted log, drilling-time log, and caliper log. They furnish valuable information on the lithology and on the borehole. Modern logging systems used today record electrically or electronically resistivity, the spontaneous potential, natural and induced radioactivity, acoustic properties, and chemical composition of the formation penetrated. Information of specially sedimentological significance can be derived from these 'wireline logs'. Wireline logs need cable for connecting sensor downhole to recording system on surface.

Service companies (Birdwell, British Plaster Board, Dresser Atlas, Edcon, McCullough, Robertson Research International, Schlumberger, and Halliburton Logging) specialized in wireline logging techniques record these logs. The total costs of running logs in a well mostly amount to less than 10% of total costs for drilling the well (usually exploration 10–13%, development 5–6%). Considering the multitude of information obtained from well log measurements, this is not expensive.

2.4.1.1. Early Logs

The 'driller's log' was the first log; it consisted of lithological identification and depth figures stored in the mind of the driller. Subsequently they were written down in a 'log book' kept by the driller. The three most common sedimentary rocks were distinguished: shale = 'soapstone', sandstone = 'sand', and limestone = 'lime'.

Sample logs are plotted strip logs. They show the identification made by a geologist who examined the cuttings in more detail with the binocular microscope. Less common rock types, such as dolomite and anhydrite, are included. Facies changes can be better studied. A modification of the sample log is to glue cuttings on log strips (mounted logs). Breaks in lithology and color variations are more noticeable.

Drilling-time logs consist of a curve plotted on a time-depth basis. Abrupt breaks mark contacts between rocks of unequal rate of penetration (ROP) or drillability. A major factor influencing the drilling speed is also a function of equipment and operational factors (design and sharpness of bit, weight on bit (WOB), rotation per minute (RPM), characteristics and viscosity of the drilling mud, and skill of the driller). In spite of these variables the drilling-time log is normally a reliable index of the nature of a sequence penetrated. Time is logged automatically on the rig.

The borehole geometry can be determined by measurements from electrically powered tools run on cables. Two conventional techniques of measuring the borehole geometry are (1) continuous determination of borehole orientation (deviation with respect to the vertical and the direction of the deviation with respect to magnetic or geographic north) and (2) the caliper log determining the borehole diameter from bottom to top. Because of changes in brittleness, cohesion, and solubility between different layers, the borehole normally does not display smooth vertical walls. Some sedimentary sequences tend to 'cave', i.e. to easily erode. Boreholes through limestones and consolidated sandstones are slightly larger in diameter than the diameter of the bit. In soft, unconsolidated shales, however, the diameter may be considerably larger. Caliper logging is used to better define a stratigraphic succession or to register effects of fracturing and acidizing. It also hints at amount and source of cavings.

2.4.1.2. Modern Logging Systems

Modern electric logging systems furnish details on lithology, mineral content and chemical composition, support subsurface stratigraphic correlation, determine strike and dip, and assist to identify tectonic structures. They also furnish data on porosity, permeability, and fluid content of the rocks penetrated. The first electrical log has been recorded on September 25, 1927 in the Diefenbach 2589 well near Pechelbronn (France). Since this day these methods have become indispensable; they are now routine procedures.

Electrical logs measure physical rock and formation fluid properties (i.e. spontaneous potential, resistivity, propagation of acoustic waves, radioactivity, etc.). They display the following advantages: (1) electric logging is rapid, (2) continuous recording from bottom to top of a borehole in contrast to the discontinuous release of conventional samples by cores, sidewall cores and/or cuttings, (3) precise location of the measurements as to their depth in the borehole, (4) the electrical measurements are reproducible, and (5) the costs are insignificant as compared with the total cost of drilling.

A combination of different electric logs furnishes additional information and presents a means of control. Integrated log evaluation, culminating in the faciolog at present, largely approximates determination of lithology, mineral composition, and petrophysical parameters. It even hints at the distribution and degree of permeability.

Early electrical logs basically recorded two types of physical parameters: (1) the spontaneous potential (SP), which is under favorable conditions a good shale-sand indicator, and (2) one or more resistivities as continuous curves. The resistivity log patterns permit well-to-well correlation, even in massive shales. Resistivity also points at probable hydrocarbon-bearing zones.

While logging, a sample of the drilling-mud should be taken and measured in order to determine the true mud resistivity. Each borehole-logging program depends on the

Table 17. Classification of electrical log measurements (after Serra 1984).

1. *Natural or spontaneous phenomena*			
Basic equipment – Single detector (passive system)			
Spontaneous potential	SP	Temperature	T
Natural gamma-ray activity		Hole Diameter	CAL
Total	GR	Deviation	DEV
Spectrometry	NGS* (spectralog)		

2. *Physical properties measured by inducing a response from the formation*	
Basic equipment – source (or emitter) + detector (s).	
Resistivity:	
(a) Long-spacing devices:	
non-focused:	N(ormal), L(ateral), ES
focused:	LL*, SFL*
(b) Micro-devices:	
non-focused:	ML*
focused:	MLL*, PL*, MSFL*
(c) Ultra long-spacing devices:	ULSEL
Conductivity	IL
Dielectric constant (electromagnetic propagation)	EPT*
Hydrogen index (using neutron bombardment)	N, NE, NT*, SN(P)*
	CN(L)*
Neutron capture cross-section, or thermal neutron	
decay time (neutron lifetime)	TDT*, NLL, GST*
Photoelectric absorption cross-section	LDT*
Electron density	FDC*, LDT*, D, CD
Relaxation time of proton spin (nuclear magnetic resonance)	NML
Elemental composition (induced gamma-ray spectroscopy)	IGT*, GST*, HRS*
Acoustic velocity	SV, SL, LSS*, BHC*, WST*
Formation dip	DM, CDM*, HDT*, SHDT*
Mechanical properties (amplitude of acoustic waves)	A, VDL*

* Mark of Schlumberger

different types of mud (e.g. oil-based mud, fresh-water mud, salt-water mud). Moreover, the resistivity of the mud filtrate is important for reservoir evaluation.

The following basic types of electric logs (Table 17), the technical details of which are described by Serra (1984), are applied at present:

Resistivity logs respond to the formation resistivity. Conventional resistivity logs (Electrical Survey – ES) are obtained by using two, three or four borehole electrodes. Certain electrodes emit a constant current and others record voltages. In this class belong Normal, Lateral devices, and the Microlog.

Focusing electrode systems include Laterolog, Microlaterolog, Proximity log, and Micro-Spherically Focused log. In these systems the current is forced to flow directly into the formation. By changing the electrode configuration and frequency, the depth of penetration can be modified. Multicurve recording gives indications of invasion depth and provides better values to determine true resistivity. Also, bed resolution is enhanced. These logs are successfully applied in high-resistivity rocks like carbonates and hydrocarbon-bearing reservoirs or zones with fresh formation water.

Induction logs are obtained by sending high-frequency alternating currents through a transmitter coil and measuring the currents induced in the formation in a receiver coil. This signal is proportional to the conductivity of the formation. Its basic scale is millimho/m (C) and the equivalent resistivity is 1 000/(C) Ω m. Bed boundaries are located at the inflection points of the conductivity curve. This logging system has been developed to log in holes drilled with oil-based mud or with air.

Some logs are made up of combinations of these three types. The induction electric log includes an induction conductivity curve, an induction resistivity curve (reciprocal conductivity), and a short normal recording on two sensitivities. The amplified short normal is often used to enhance visual correlation patterns in shales. Induction tools are normally used in sand/shale sequences.

Note that the often used names or very numerous abbreviations like DLL, MSFL, Proximity, SFL, MSFL, CNL, SNP, FDC, HDT, etc. are Service Marks or Registered Trademarks of the service companies.

The focused resistivity (SFL) has about the same penetration depth as the short normal, but registers sharper details. In lower-porosity formations, a wider range in resistivity is needed, and there is the likelihood for deeper invasion. When the hole has been drilled with freshwater muds, the Dual Induction-Laterolog 8 is frequently used. Three curves of varying depth of penetration are recorded.

For low-porosity formations drilled with brine or salt-saturated muds, the Dual Laterolog and the Micro SFL are combined. In such a combination, gamma-ray detection is commonly used instead of SP to identify shales. Separation of the resistivity curves indicates the flushing of hydrocarbons. The three measurements permit invasion diameter and true resistivity determinations.

The Spontaneous Potential log (SP) registers naturally occurring potentials within a borehole, referred to as spontaneous potentials, with a single downhole electrode and an electrode on the surface. The potential difference between the two electrodes is recorded as a function of depth.

The SP curve differentiates between impermeable, electrically conductive beds (e.g. shales) and permeable beds. In shale-free, thick, porous formations, the SP deflection from the shale base line is related to the contrast of the salinity of the formation water to that of the fluid in the borehole.

The SP is useful for detecting and registering the thickness of permeable beds, providing correlation, computing formation-water salinity, and indicating shaliness. The SP is routinely recorded together with resistivity logs. It cannot be measured when oil-base mud is used.

Porosity logs respond to variations in porosity, fluid content, and lithology. One or a combination of some porosity-sensitive devices may be needed, depending on the problem. Complex lithologies with changing shaliness require the combination of at least three log types. It permits efficient evaluation of porosity and lithology.

The measurement of the speed of sound through the formation was used as the first effective porosity log. As the first acoustic device the sonic log was introduced in 1958. The transit time of an ultrasonic compressional wave of frequencies between 20 to 40 kHz over two feet along the borehole wall is recorded.

This porosity tool is more effective in consolidated and compacted sedimentary sequences. In consolidated rocks, effects from hydrocarbon saturation are usually negligible. The matrix travel time is well known for the main sedimentary minerals.

Porosity derived from acoustic measurements represents effective porosity as vuggy or intragranular porosity does not influence the sound transmission.

The formation-density tool consists of a sidewall pad device which has a source of gamma ray and two receivers. The gamma rays emitted from the source collide with matter in the formation. Compton scattering is the interaction sensitive to the electron density which is directly related to the bulk density of the formation.

Hydrocarbons present may interfere, especially gas in high-porosity formations. In consolidated formations the hydrocarbon interference is small to negligible.

FDC is a very good porosity tool. It measures total porosity and, therefore, needs correction for shaliness and gas effect. Used in formation evaluation it provides significant data on lithology, shale content, fluid type, and porosity.

Especially, the information on lithology has been enhanced by the introduction of the Litho-Density Tool. With this tool the photoelectric adsorption cross section is recorded. This 'PEF' measurement is a valuable lithology indicator. With this log the differentiation of porous carbonates between limestone and dolomite can be made.

Two types of neutron log tools are in use; one with a single source and single receiver and the more common type with one source and two receivers. The detectors measure the neutron flux at a particular energy level. Porosity sensitivity is based mainly on the response to the hydrogen content.

When fast high-energy neutrons are emitted from a source they are slowed by collisions with the atoms in the formation. Collision between a neutron and a hydrogen nucleus (which has an equal mass) results in rapid reduction in neutron energy to the level of complete loss and capture. At that stage gamma rays are emitted and can be recorded. High hydrogen concentrations can be due to porous rocks, filled with oil or water, or to shale. Gas in the pore space, or absence of porosity in shale-free rocks, result in low hydrogen content and, thus, a low neutron porosity index (ϕN). Calibration for neutron porosity tools is done in a standard limestone sequence. Current logs are recorded and calibrated on linear porosity scales, assumed that the formation is water-bearing limestone (or sandstone in other regions).

Natural gamma-ray logs respond to the gamma rays emitted by the radioactive decay of potassium ^{40}K, the uranium, and the thorium series from naturally occurring minerals or rocks. Clays and shales, containing normally radioactive elements abundantly, are characterized by high gamma-ray counting rates. Moreover, concentrations of potassium feldspar, muscovite and biotite, or radioactive heavy minerals may also produce a high gamma-ray response.

Intercalations of potassium salts can easily be detected in evaporitic sequences that are

Table 18.   K–U–Th spectral ratios of clays and micas (after Merkel 1979, reprinted with permission).

	K (%)	U (ppm)	Th (ppm)
Bauxite	–	3–30	10–130
Glauconite	5.08–5.30	–	–
Bentonite	< 0.5	1–20	6–50
Montmorillonite	0.16	2–5	14–24
Kaolinite	0.42	1.5–3	6–19
Illite	4.5	1.5	–
Mica			
Biotite	6.7–8.3	–	< 0.01
Muscovite	7.9–9.8	–	< 0.01

normally barely radioactive.

The gamma-ray log is used for lithostratigraphic correlation and for distinguishing clays and shales from clean sandstones, carbonates, and evaporites.

Gamma-ray logs are generally recorded in combination with other logs, normally with porosity logs. With caution, the gamma-ray curve can be substituted for an SP as a shale detector if the SP is not available (empty hole, oil-based mud, salt-water muds, cased hole).

The log is not affected by the presence of hydrocarbons in reservoir rocks.

Natural gamma-ray spectroscopy logging is based on gamma-ray counting rates for several energy bands. Energy levels for potassium, the uranium series, and the thorium series are well known. By computer processing, the concentrations of these constituents are determined. This additional information allows a better evaluation of the clay content and its composition. K/U, K/Th, or U/Th ratios commonly give unique signatures for particular clay types (Table 18) and other minerals such as mica or feldspar. Also pelitic rock types can be discriminated by means of K/U ratios as to their composition and environment. Thus, bentonites normally show a high thorium content.

Anomalously high uranium readings can be indicative of fractured zones, vein mineralization, or the presence of organic matter.

Dipmeter Logging. The currently used high-resolution dipmeter is a four-pad wall-contact tool whose finely detailed micro-resistivity curves are correlated in order to measure depth offsets relative to each other. These are focused to achieve penetration through the mud cake and still give a sharp vertical resolution in the order of 1 or 2 cm. Simultaneously, the hole dipmeters from two independent calipers as well as tool position, pad orientation, borehole inclination, and orientation towards magnetic north are recorded. The essential data is recorded continuously on film in the form of curves, and simultaneously digitized and recorded on magnetic tape.

The output from sophisticated computer programs consists of an arrow plot which shows azimuth and dip angle, drift and direction of the borehole and, in some cases, a resistivity correlation curve derived from the dipmeter resistivity data. The correlation curve is used for depth control because it correlates well with a laterolog curve. If a good vertical resolution is required for dip computation a different technique, called GEODIP, is used.

As the dipmeter is a resistivity measurement tool it has only very limited application in oil-based muds.

The redoxomorphic well log (Pirson 1983) measures continuously the reduction-oxidation (redox or Eh) potential of sedimentary formations in a borehole filled with water-base mud (fresh or salt water). It is a modification of the SP curve. The potential measurement between two electrodes and the same level (an inert one of platinum and a reference one of lead) indicates the proportions of oxidized and reduced minerals and ions contained in the formation water of geological formations. This potential is supposed to depend on the depositional environment of the sediments. It also hints at the winnowing action by waves or at high-energy environments.

Other logging tools, such as a full wave sonic recording, borehole televiewer, the borehole gravity meter, and the neutron decay tools (Serra 1984) are used to tackle specific problems.

Tool combinations – Gamma ray and/or SP and caliper logs are commonly recorded on the first 2 1/2 'wide track' on the left hand side of the API logging paper. This log combination is basically used to determine permeable zones, shaliness, and borehole rugosity.

Tool combinations that become increasingly standard include dual induction log with laterolog or micro-spherically focused log, dual laterolog with gamma-ray, and induction electric log, and a density-neutron combination. Table 19 lists these tool combinations and their geological significance.

2.4.2. SEDIMENTOLOGICAL APPLICATIONS OF WELL LOGGING

The present section focuses on the sedimentological aspects of well logging. Well logs are applied in sedimentology for correlation, mineral and rock identification, accessory mineral identification, detection of oil shales, evaluation dipmeter, and interpretation of log curve shapes. Certain logs such as the SP, the gamma-ray, the spectral gamma ray, the litho-density, and – to some extent – the shallow resistivity are used more frequently in geologic and lithologic analysis than others. Typical log responses to specific lithologies are shown in Figure 27. Responses, however, are not necessarily unique to particular lithologies. Because neutron, density, and sonic logs are essentially porosity logs, the responses in sandstone, limestone, and dolomite depend largely on their porosity.

Logs represent the only continuous, detailed record over the complete length of a borehole. They are depth-controlled and can be recorded in open or cased holes. The estimation of in-situ petrographical parameters is possible.

Geological and sedimentological aspects of well log analysis are discussed in more detail by Pirson (1983), Serra and Abbott (1982), Serra (1985, 1986), and Desbrandes (1985).

Table 19.   Logging-tool combination in geological interpretation (after Serra and Abbott 1980).

Log type	Composition	Texture	Sedimentary Structure	Fluid
Resistivity	* *	* * *	* *	* * *
SP	*	* *	* * *	* * *
EPT (Propagation time)	* *	*	*	* * *
EPT (Attenuation)	* *	*	*	* * *
GR	* *	*	*	*
NGS	* * *	*	*	*
CNL	* *	* *	*	* *
FDC · LDT (ρb)	* * *	* *	*	* * *
LDT (Pe)	* * *	*	*	*
TDT (Σ)	* *	* *	*	* * *
BHC (Δt)	* *	* * *	*	* *
BHC (Attenuation)	*	* *	* *	* * *
GST	* * *	* *	*	* * *
HDT or SHDT	*	* *	* * *	*
CAL	*	* *	*	*
HRT	* *	* * *	* * *	* *

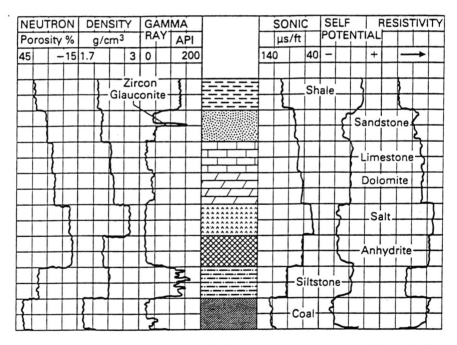

Fig. 27.   Typical wireline log responses to specific lithologies (courtesy Al-Murami and Stow, after Reading 1986, p. 4).

2.4.2.1. Well-to-Well Correlation

Patterns of electric log curves in shale, sand-shale, sandstone or limestone sequences can be quite similar from well to well. Thus, strata with similar physical properties can be correlated according to their log response (log correlation, in contrast to litho-correlation or chrono-correlation).

Correlation is the demonstration of the equivalence of two or more geologic features in different areas (i.e. neighboring wells). Log correlation is more reliable than that by cuttings, because electric logs are continuous and depth-controlled. Short-spacing resistivity logs are most frequently used for correlation. Depending on the lateral scale of a correlation (oil field study versus regional study) the optimum tool combination may be different.

Well-to-well correlations in oil fields are commonly aimed at determining the subsurface position and thickness of marker beds, their strike and dip, fault locations, and facies changes. Both SP and resistivity logs permit reliable bed identification and placement of faults. If closely spaced well control exists, refined analysis and correlation of SP curves combined with a dipmeter are useful to delineate the geometry of sand units. In oil fields with carbonate-rock correlation is best achieved by the combination of gamma-ray and resistivity logs or of gamma-ray log and a porosity log.

For regional correlation in sedimentary basins, SP and short-spacing resistivity logs are used. For facies studies SP, gamma-ray and lithodensity logs are well suited being sensitive to the sand/clay ratio. The gamma-ray log stressing depositional conditions of shale is normally valuable for long-distance correlation. The regionally widespread volcanic ashes with high gamma-ray readings are good time markers.

Well-to-well correlation, however, can be troublesome in environments such as in deltas or in turbidites.

In evaporite series the best correlation tools are gamma-ray, sonic, and litho-density logs because their response shows remarkable differentiation. In coal-bearing strata the sonic, density, resistivity, and natural gamma-ray spectrometer provide excellent correlation possibilities.

2.4.2.2. Mineral and Rock Identification

The principal porosity-sensitive logs (sonic, density, and neutron) are influenced by the mineral composition and the mineral content of the formation. Major changes in mineral composition (e.g. quartz to clay, calcite to dolomite, calcite to anhydrite) are simultaneously changes from one rock type into another (e.g. sandstone to shale, limestone to dolomite, limestone to anhydrite). Because each mineral species with specific physical properties shows a specific response to neutron bombardement or acoustic waves. Combining the sonic, density, and neutron logs permits the computation of the mineral composition of a rock and of its porosity. The significance of a specific mineral composition is demonstrated by the matrix effect of the most common sedimentary minerals (e.g. calcite, quartz, dolomite, halite, anhydrite, etc.).

Cross-plots of common log combinations (density-neutron log, sonic-neutron log,

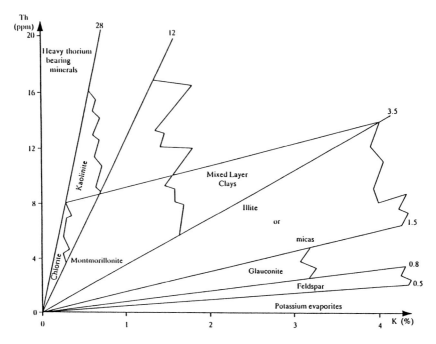

Fig. 28.   Plot of potassium content against thorium content taken from the Natural Gamma Ray Spectrometry Tool (NGT) for mineral identification (after Alexander 1982, p. 85).

density-sonic log) help to best identify mineral composition and, thus, the sedimentary rock type. The log data is cross-plotted on charts for specific logging tools. Effective cross-plot charts are the density-neutron cross-plot, the Th-K cross-plot (Figure 28) from the natural gamma-ray spectrometer, and the Uma (photoelectric cross section normalized for porosity) versus grain density cross-plot from the litho-density log. Density-sonic cross-plots are applied for identifying evaporites and/or carbonates plus anhydrite and evaluating shaliness in sand-shale intervals.

In high-porosity formations or in presence of gas with undefined lithology the cross-plot might be misleading. Therefore, the use of porosity-unaffected cross-plots like the sonic-density (M–N) plots or matrix identification (MID) plots are recommended. Secondary porosity is also detected by comparing porosities computed from the sonic tool with those deduced from the neutron-density combination (Figure 29). Thus, the data from the three principal porosity devices can be evaluated for mineral identification, provided the formations are shale-free. Confirmation of the mineral composition of a formation can be obtained by direct mineral analysis of samples recovered by conventional coring or sidewall coring. Such samples have the advantage over cuttings of being depth-controlled and largely uncontaminated.

The clay content affects the various logs in a different way. The density log is the least

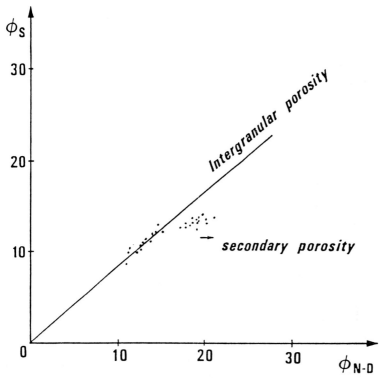

Fig. 29.   Recognition of secondary porosity by comparing porosities deduced from the sonic and the neutron-density log (modified from Serra 1985, p. 131).

affected. In the sonic log the travel time is normally higher in shales than in reservoir rocks. Moreover, the sonic response depends on whether a shale is bedded or a clay is disseminated. The neutron logs respond to the clay content because of the high hydrogen index of clays (intercrystalline water).

2.4.2.3. Accessory Mineral Identification

Admixture of accessory minerals like pyrite, other ore minerals, heavy minerals, feldspar, mica, and glauconite, even in minor amounts, can affect well logs significantly, and, thus, distort porosity and saturation analysis. These minerals are more frequent in sandstones than in other rocks. Each mineral species affects well logging by specific physical or chemical parameters. Ore minerals and transparent heavy minerals are characterized by an anomalously high grain density (> 3). As little as five volume percent of these minerals will affect the porosity calculation markedly. High amounts of potassium feldspar, muscovite, and biotite as well as illite and radioactive heavy minerals (e.g. zircon, monazite) rise the normal gamma-ray counting rate. Sandstones high in potassium feldspar, detrital micas, and/or radioactive heavy minerals, e.g. numerous Jurassic sandstones from the North Sea, appear to be shales according to gamma-ray logs. In such cases gamma-ray spectroscopy helps in discriminating potassium, uranium, and thorium; high potassium indicates mica and/or potassium feldspar; thorium marks shales. The litho-density log combined with the natural gamma-ray spectrometer is successfully applied in determining mineral type in complex lithology reservoirs.

2.4.2.4. Detection of Oil Shales

Oil shales are commonly characterized by a high amount of organic matter, normally bitumen. Uranium is commonly found in clays of reducing environments, particularly in the presence of carbonaceous material. Consequently, dark colored to black, organic-rich shales are highly radioactive and show high gamma-ray log counting rates as well as spectral gamma log responses with high potassium, thorium, and uranium readings. Such shales are ordinarily good source rocks for hydrocarbons.

The density log may support the evaluation of oil shales. Also the redoxomorphic well log can be applied for mapping the extent of potential hydrocarbon source beds and of their degree of devolatization as a measure of the magnitude of hydrocarbon volumes generated.

2.4.2.5. Dipmeter Evaluation

Strong variation in features on conventional resistivity logs can be an indication of structural disturbance or changing sedimentary environment. This is best evaluated by the use of a dipmeter survey. The determination of dip and strike on a continuous basis provides insight to a geological sequence drilled. The arrow plots from the dipmeter computation indicate sedimentary features such as stratification of any kind or tectonic events (unconformities, faults, overthrusts, veins, etc.). Interpretation of those arrow plots, i.e. the dip vectors, from the computer programs CLUSTER and GEODIP requires sound knowledge in both sedimentology and structural geology. In the GEODIP program dips are identified and rated by direct pattern recognition of the four dipmeter resistivity curves with an excellent vertical definition of about 5 centimeters. Figure 30 depicts a series of idealized dipmeter patterns and their most likely interpretation. This interpretation should conform as much as possible with other geological data, particularly with the interpretation of other logs.

In interpreting data on the dip-vector logs and the various dip patterns the following features should be observed: quality of dips as rated by GEODIP, changes in azimuth, increasing/decreasing dip angles, incoherent dips, lack of dips, breaks or continuity, degree of planarity, nature of contacts (abrupt – flat or warped – or gradual).

In general, the vectors are first grouped according to four basic patterns which are marked by corresponding colors on the log:

–   Red patterns are groups of vectors showing augmenting dip angle with increasing depth; associated with deposition upon unconformities, within channels, and overlying barrier bars or reefs; also point to lateral variations in thickness of lithologic

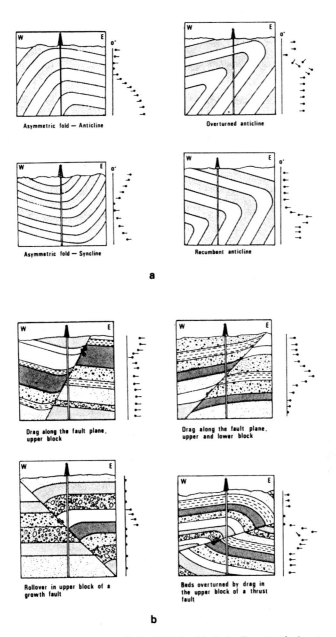

Fig. 30. Idealized dipmeter patterns (arrow plots) of folded and faulted sediment stacks (courtesy Schlumberger, after Serra 1986).

units (e.g. drape, creep, slump). In addition, red patterns very clearly indicate faults and the associated drag.

— Blue patterns are groups of vectors showing decreasing dip angle with increasing depth; associated with current bedding that indicates direction of sediment transport; also found below unconformities and associated with drag below faults. Very pronounced blue patterns are found in eolian environments at the base of dunes.

— Green patterns are groups of vectors displaying consistent dip angle and azimuth representing structural dips.

— Yellow patterns indicate zones with random dips or random events. Such results point to high-energy sedimentation (conglomerates and fanglomerates), thoroughly bioturbated claystones or tectonically strongly disturbed sections. Featureless structures like reef bodies could also produce random dips.

Vectors grouped together must exhibit a consistent dip azimuth (Figure 31).

Azimuth-frequency diagrams, i.e. polar diagrams divided in $10°$ segments, show preferred dip directions for a given interval. Azimuths from red and blue patterns are noted separately. An abrupt change in azimuth, normally associated with change in dip angle, marks a first-order event, mostly caused by tectonics (fold, fault, unconformity) or by differential compaction. No change in azimuth but in dip angle, abrupt or gradual, suggests a stratigraphic anomaly such as disconformity, channelling, sand bars, cross-bedding, etc.

A group of successive dips maintaining the same azimuth over a short interval indicates cross-bedding. If corrected for the structural dip, the average azimuth marks the current direction like cross-bedding measured in outcrops.

The dipmeter provides data on internal structures of sedimentary sequences, on the direction of transport in sandstones, and eventually on the direction in which sand bodies thicken.

The following environments of deposition, well preserved in regressive sequences, can better be traced by continuous dipmeter evaluation: eolian deposits (dip max. $30°$); fining-up fluvial valley fills or marine and turbidite channels (increasing dip angles with depth, azimuth frequency diagrams with two concentration centers $90°$ apart); coarsening-up barrier bars and offshore bars (decreasing dip angles with depth); distributary front deposits (low-angle foresets combined with high-angle foresets and increasing dip angles with depth), tidal channels (red and blue patterns with $180°$ opposite azimuths); turbidite sequences (abrupt increase in dip with depth over small intervals, progressive dip azimuth rotation approaching slump faults, random dip patterns indicating contorted beds).

Paleowind directions that led to the deposition of the eolian Rotliegende sandstones in NW Europe (Figure 105) have been determined by means of GEODIP and CLUSTER. In turbidity-channel sandstones dip patterns allow the delineation of erosional unconformities, channel sequences, slump faulting, contorted or massive bedding, and sedimentary drape.

Secondary dolomites commonly display random dip patterns of poor quality or lack of computed dips at all. Integrated dipmeter evaluation, combined with geological data and sedimentological considerations, may even one enable to extrapolate sand trends on isopach maps.

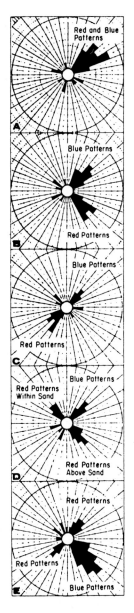

A: Bar-like sand: Red and blue patterns within sand interval point same direction. This diagram indicates sand trends N35W-S35E, thins N55E.

B: Channel or trough fill sand: Red patterns within interval point towards axis. Blue patterns point in direction of sediment transport. This diagram indicates sand trends N45E-S45W, thickens S45E, and was transported N45E.

C: Bar-like sand: Red patterns above sand, blue patterns within sand interval 180° apart. This sand trends N45W-S45E, thins S45W, and was transported N45E.

D: Channel-fill sand: Red patterns above sand 180° from red patterns within sand. Both show this sand thickens N45W, trends N45E-S45W. Blue patterns indicate transport was N45E.

E: Trough-fill sequence: Some red patterns 180° from other red patterns within fill sequence, with blue patterns at 90° to both. This well near axis of N45W-S45E trending trough. Sediments were transported S45E.

Fig. 31. Dipmeter patterns of various sand deposits (courtesy Schlumberger, after Pirson 1983).

A more sophisticated evaluation of the continuous dipmeter by determining degree and orientation of the resistivity anisotropy within given bedding planes is made feasible by a FORTRAN II computer program (Pirson 1983, pp. 130 ff., Appendix 1). The interrelation between grain orientation and dipmeter patterns, however, is not as clear-cut as mentioned

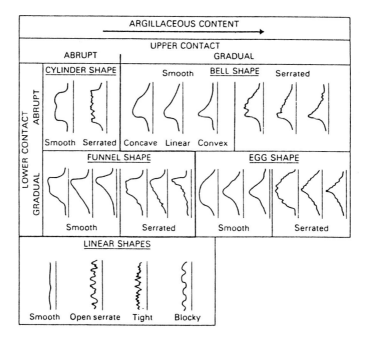

Fig. 32.   Classification of GR/SP log curve shapes based on the clay content. Important is the nature of the upper and lower contact (after Serra and Sulpice 1975).

(Pirson 1983, pp. 97 ff., sandbars in Table 4-1).

For recognition and evaluation of bedding planes and sedimentary structures resistivity-measurement principles are used. Modelling demonstrates that sedimentological bedding features also influence pore geometry and, thus, the electrical conductivity of rocks (Fülöp 1991).

2.4.2.6. Interpretation of Log Curve Shapes

Shape analysis of continuous log curves provides additional information on (1) homogeneity or heterogeneity of composition and grain-size distribution within major beds, (2) number of major beds and their thickness, (3) sequential variations as well as direction and rapidity of changes, and (4) locations of abrupt changes.

Any change of response in a continuous well log corresponds to a change of one or more important physical rock properties. A gradual change of the curve reflects a gradual change of physical rock properties. A break indicates an abrupt change of physical rock properties. Smooth curves denote persistence of the most important physical rock properties and, thus, of the sedimentary environment.

These principles apply to all types of sedimentary rocks. First in 1958, the idea has

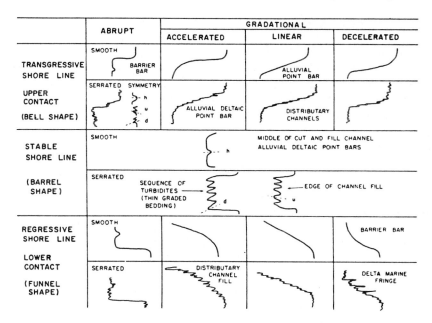

Fig. 33. Classification of SP curve shapes in terms of sedimentary environment (after Pirson 1970, p. 37).

been developed to characterize sand accumulations from the appearance of the SP curve. By analyzing top and bottom contacts of major beds and the shape of the curve, Serra and Sulpice (1975) established a general classification (Figure 32). Certain sedimentary environments, rated within transgressive or regressive shoreline cycles, can be recognized by their SP log curve shapes: barrier bar, alluvial point bars, distributary channels, delta marine fringe, and turbidites (Figure 33). The intensity of regressive and transgressive cycles can be even quantified. Theoretical sedimentation and SP log curve shape pattern can be designed (Pirson 1983). Schematic SP and gamma-ray log curve shapes typical of common reservoir sand types are included in the summary chart on reservoir sandstones (Figure 52).

Transgressions, dominated by erosion and redeposition through wave action, produce only minor sedimentary volumes with thin sands as compared with regressive or deltaic sequences. The higher the wave energy of the transgressive sea, the thinner the sediments deposited.

Graded-bedded turbidites are easily recognized in the SP and resistivity curves by their serrate log shapes with high SP at bottom and low SP at top as well as by thin sand fingers of straight parallel symmetry.

Other log types used for log curve shape analysis are the short normal, induction, conductivity, gamma ray, and for fine correlation the dipmeter curves.

2.4.3. INTEGRATED LOG ANALYSIS

Whenever possible, the use of well logs for sedimentological evaluation is advocated regardless of rock composition. Such an evaluation requires plotting well logs and dipmeter data on one single chart, i.e. the fundamental composite log. Its scale depends on the type of study and degree of accuracy desired. If it has to cover the entire sequence traversed by a borehole, the use of a scale of 1:1 000 or preferably 1:2 000 is recommended. Such a compilation permits better evaluation of well logs and of the facies traversed, leading to the identification of facies and of sequential variations. A scale of 1:500 or preferably 1:200 may be, however, necessary in order to determine details of an elementary sequence. This procedure also includes the analysis of shape and amplitude of the different log curves.

Integrated log evaluation must be supplemented by qualitative and quantitative analysis of porosity and determination of major mineral constituents in the matrix such as quartz, calcite, dolomite, and clay. Moreover, data of petrographic and sedimentological analyses of cuttings, sidewall cores, and conventional cores may be included. Such a combined log and sedimentological evaluation is the only means to calibrate the so-called 'electrofacies'.

Originally nine major electrofacies types have been identified in a study of the ancient Niger delta. This 'faciolog' case history is an example of ancient delta deposition. Petrographic details, however, such as the presence of considerable amounts of radioactive minerals (e.g. zircon or thorite) seem to be rather unique for the Niger delta.

A geological facies can be defined in terms of sedimentary properties (color, texture, grain size, hardness, etc.). Analogously, the entirety of all log readings at any particular depth represents physical properties in terms of resistivity, sonic transit time, level of natural gamma radiation, etc.

The definition of facies in terms of log properties is called an electrofacies. Geological facies evaluation and electrofacies evaluation differ widely, but aim at the same goal, namely at the description of physical rock parameters. They should not be seen separately as they supplement each other.

The proper analysis of sedimentary sequences normally includes the subdivision into layers or zones with constant petrophysical properties or of the same facies type. Since well logs are normally recorded over the total interval drilled they are well suited for such zoning. Most logs are recorded with one reading taken every six inches of the borehole measured. When even finer resolution is required, the high-resolution dipmeter provides a means of detecting phenomena of only a few centimeters in size.

The problem of zonation of well logs is tackled through multivariate analysis. Advantages of automatic log zoning are: (1) production of squared logs substantially free of measurement errors, (2) identification of layers for reservoir modelling, (3) improved determination of porosity, saturations, etc., and (4) dividing the logged intervals into facies units.

Electrofacies was identified first by the use of the 'cobweb' and 'ladder' diagrams. Diagrams can be drawn for all zones identified and their shapes matched. Zones of the same basic 'cobweb' shape belong to the same electrofacies. In this development stage no more than fifteen electrofacies types have been discriminated. This method of zoning has been considerably improved by computer application. Dividing the logged interval into electrofacies zones is a problem for mathematical clustering. This is done by the following steps: depth matching and log corrections, high-resolution dipmeter processing, principal component analysis, cluster analysis, and level by level attribution.

A recent step was the introduction of the faciolog (Figure 34) just a further pace in the development of the electrofacies concept. The faciolog plots the responses for all the logs – including the dipmeter log – on a multidimensional set of axes. Clusters may be recognized and all points in a given cluster are considered to show comparable log response. High-vertical-resolution dipmeter permits the recognition of thin permeability barriers or of fracture zones. Previous electrofacies zonation has been improved; now it is possible to distinguish objectively as many as 24 clearly distinct electrofacies types in a borehole on the basis of their log responses. Even lithological descriptions and petrographic analyses of cuttings and cores or stratigraphic information from paleontological analysis can be integrated into the format.

The data reduction given by averaged logs can aid log interpretation and may simplify cross-plot analysis. The new tools like the high-resolution dipmeter, natural gamma ray spectrometry, and litho-density tool provide a basis for constructing an interpretational model.

The electrofacies zones are more rigorously defined in the context of an improved concept of facies description. The electrofacies zonation furnishes a sound basis for extrapolating geological information obtained from cores through the complete well section. The electrofacies approach allows log data to be integrated with other geological aspects of an oil-field or basin study.

2.4.4. FUTURE PROSPECTS OF SEDIMENTOLOGICAL WELL LOGGING

In the future, well logging in general will gain even greater importance. New log types will be developed as oil and gas exploration becomes more expensive and the need for more sophisticated methods will be greater. Calibration of well logs by direct sedimentological examination of cuttings, sidewall cores, and conventional cores will be required also in the future to develop new plays and to search for new exploration areas on the globe. After such a calibration including a thorough sample analysis in a key well, most other sedimentological data will be extrapolated to other wells in a basin. This trend will also develop in any other search for new energies.

2.5. Sedimentological Evaluation of Seismic Data

This section is not intended to replace an introduction to seismology such as Dohr (1981). It is aimed at making a sedimentologist aware of the potentials of seismics, especially in petroleum sedimentology.

Seismic stratigraphy is an expanding discipline in geoscience. The basic concepts of seismic response to thin transitional beds and the synthesis of seismograms from stratigraphic sequences have been explained long ago, but routine use of these concepts

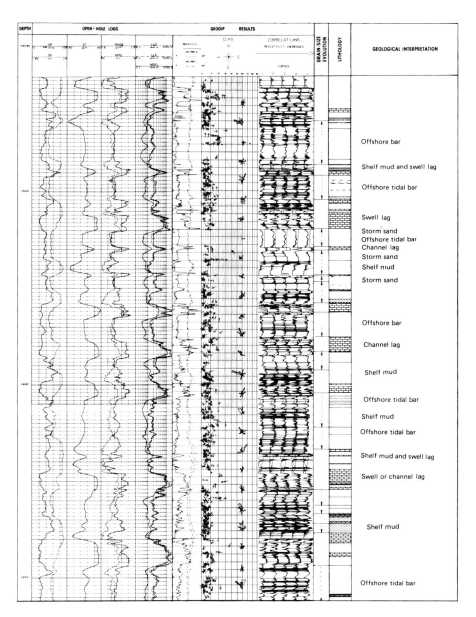

Fig. 34.   Facies log consisting of original open-hole logs (left), GEODIP results (center), grain-size evolution, lithology, and geological interpretation (courtesy Schlumberger, Well Evaluation Conference, India, 1983).

had to await modern electronic technology. Only recently quality of seismic data has been adequate to accurately interpret reservoir conditions and depositional facies.

The traditional objective of interpreting reflection-seismic sections has been the mapping of geologic structures without concluding much about the stratigraphy disclosed by seismic sections. Seismic events could be traced over some distances with reasonable certainty, and could be interpreted as representing the same geologic horizon. Seismic events could be observed which are loosing their continuity, occur or disappear gently or abruptly, change their character.

Since the introduction of digital recording and CDP techniques, seismic processing has become more sophisticated and the analysis of seismic features, such as wave trains and reflection frequencies, has become more quantitative such as:

–   Analysis of travel times allow calculation of seismic velocities that are interpreted in terms of gross lithology.
–   Angular relations between seismic events commonly are interpreted in terms of geologic history (e.g. unconformities, periods of erosion or structural growth, the direction to the source of the sediments, etc.).
–   Data patterns interpreted in terms of depositonal environment to distinguish sediments deposited in calm marine environments from sediments laid down in high-energy environments such as beaches or deltas.
–   Extraction of the seismic wavelets and analysis of the reflection amplitudes enable the study on the acoustic impedance, its change at interfaces and correlation with stratigraphic occurrences (pseudo-acoustic logs).

Modern accounts of sedimentological evaluation of seismic data are by Payton (1977), the classic AAPG memoir 26 entitled 'Seismic Stratigraphy – Applications to Hydrocarbon Exploration' with the contributions by Sangree and Widmier (1977), Vail, Mitchum and Thomson (1977), and Vail, Todd and Sangree (1977), Anstey (1980), Sheriff (1980a+b), and Hardage (1987). Data on principles and methods of applied geophysics in general are compiled by Parasnis (1986) and Sharma (1986).

2.5.1. SEISMIC STRATIGRAPHY AND DEPOSITIONAL SEQUENCES

Inferring stratigraphy in unknown areas without geological control is a challenging, but difficult task. Under such conditions it is not possible to obtain details or accuracy when using seismic data calibrated by borehole control. In such areas even hints at the stratigraphy of the area are helpful.

Seismic stratigraphy is applied on local as well as on regional scales. On a local scale it is used for the development of oil and gas fields, e.g. to track the lateral extent of lithologic units such as porous reservoir rocks. On a regional scale seismic stratigraphy serves to identify and delimit stratigraphic/sedimentologic rock units and to determine the source rock and reservoir rock potential within a basin.

Seismic elements useful for stratigraphic interpretation are listed in Table 20. Interference between successive reflections, which depend on the thickness of rock units, will

Table 20. Seismic facies associated with clastic depositional units (after Sheriff 1980b, Table 1).

Regional setting	Basis of distinction	Subdivisions	Interpretation	Other characteristics
Shelf	Reflection character	High continuity, high amplitude	Interbedded high- and low-energy deposits	Usually neritic marine, occasionally marsh clays and coals.
		Low amplitude	Uniform energy deposits, predominantly one lithological type	Generally shale-prone if seaward from high-amplitude, high-continuity facies; sand-prone if seaward from low-continuity, variable-amplitude facies.
		Low continuity, variable amplitude	Variable energy: generally non-marine	Occasionally coal members will give high amplitude and continuity.
	Broad, low-relief mound		Variable energy, delta complex	Internal reflections gently sigmoid to divergent.
Shelf margin, prograded slope	Internal reflection pattern	Oblique progradational in dip direction	High energy in updip portions	Toplap (offlap) pattern at top of unit. Fairly steep depositional dips (up to 10°). Small channels common. On shelf margin: fluvial delta; lower portions may contain turbidite deposits. Often fan-shaped or multiple fans. Variable amplitude, variable continuity.
		Sigmoid progradational in dip direction	Low energy	Gently S-shaped at top (thinning of members may make cycles disappear because of resolution limitations). Moderate amplitude, high continuity.
Basin slope, basin floor	Overall unit shape	Sheet drape	Low energy; hemipelagic clays	Parallel reflections of relatively constant thickness and character drape over pre-existing topography.
		Slope-front fill	Low energy; deep marine clays and silts	Onlap at top of unit, downlap in lower parts. Often fan-shaped.
		Onlapping fill	Predominantly low energy	Deposited by gravity-controlled flows, mostly low-velocity turbidity currents.
		Mounded	Variable energy	Fan-shaped. Complex pile of sediments by gravity transport through submarine canyons.
		Mounded onlapping fill	High energy, turbidity current deposits	Discontinuous, variable-amplitude reflections.
		Chaotic fill	Variable energy	Overall mound in topographic low.

affect the amplitude and character (or frequency content) of the composite reflection. Significant lateral frequency and phase variations sometimes occur where reefs are present or where other lateral changes in the section occur such as at a major fault.

Continuous reflections normally indicate widespread uniformity of depositional environments. They may be characteristic of marine deposition, although unconformities may be camouflaged by continuous reflections. Abundant reflections point to repeated changes in lithology with depth. In clastic sequences they may be a sign of sandstone and shale intercalations. Lack of distinctive reflections may mean that massive rock units are present such as thick, relatively uniform shale sections, thick carbonate sequences or crystalline basement.

Reflection configuration refers to patterns of non-parallel reflections. Long regional seismic lines, not complicated by structural disturbance, are best suited for detecting features of stratigraphic significance, but the behavior in structured areas also furnishes lithologic information. Provided a seismic section has been properly processed and multiples are removed, non-parallel seismic events indicate thickening or thinning as well as intercalating or disappearing of lithologic units. Reflection patterns of this kind commonly reflect a change in depositional processes or in the environment of deposition. Thickening or thinning may also point to the direction of the source of the sediments. Irregular thickening or thinning may mark irregular sedimentation as in a delta, the occurrence of sediments capable of 'flowing' such as shale or salt, or of soluble sediments such as salt. Thickness variations in folds may indicate competent and plastic rock units. They are also associated with uncompactible rocks such as reef limestone in contrast to more compactible shale surrounding the reef. Reflections from horizontal gas-oil-water contacts ('flat spots') where the reflections from lithologic interfaces are not flat are evidences of hydrocarbon accumulations. The interpretation of some reflection configuration patterns is depicted in Figure 35.

A depositional sequence is – according to the definition by Mitchum, Vail and Thompson (1977) – a stratigraphic unit 'composed of a relatively conformable succession of genetically related strata and bound at its top and base by unconformities or their correlative conformities'. This concept of a 'sequence' is modified from Sloss (1963). Figure 36 illustrates the basic concepts of a depositional sequence. Depositional sequences may range in thickness from a few centimeters to hundreds of meters.

A depositional sequence is determined by a single objective criterion, the physical relations of the strata themselves. The combination of objective determination of sequence boundaries and the systematic patterns of deposition of the genetically related strata within sequences makes the sequence concept a fundamental and extremely practical basis for the interpretation of stratigraphy and depositional facies.

To delimit and correlate a depositional sequence accurately, the sequence boundaries must be defined and traced precisely. Usually the boundaries are set at unconformities and traced to their correlative conformities.

Discordance of strata is the principal criterion used in determining sequence boundaries, and the type of discordant relation is the best indicator of whether an unconformity results from erosion or non-deposition.

Onlap, downlap, and toplap indicate non-depositional hiatuses; truncation indicates an

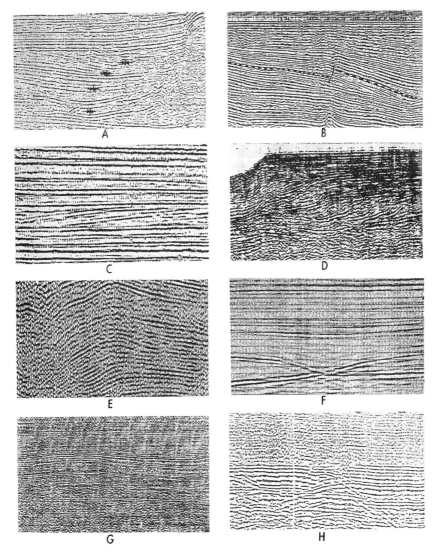

Fig. 35.   Interpretation of seismic reflection configurations (from Sheriff 1976, p. 539, reprinted with permission): A. Migration of syncline with depth; B. Reversal in direction to source of sediments; C. Progradation of ancient delta; D. Outbuilding and upbuilding of shelf; E. Response of competent versus incompetent rock units to folding; F. Two salt lenses; G. Patterns characteristic of salt; H. Massive carbonate overlain by clastics.

erosional hiatus unless the truncation is a result of structural disruption.

Depositional sequences may be recognized on seismic sections, well-log sections, and surface outcrops. Depositional sequences presented on seismic sections depend primarily

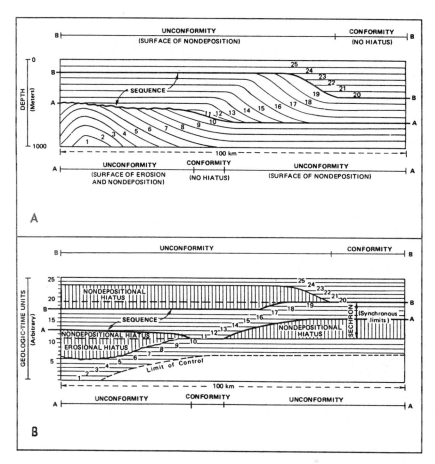

Fig. 36. Basic concept of depositional sequences. A depositional sequence is a stratigraphic unit composed of relatively conformal successions (after Mitchum *et al.* 1977, p. 54, reprinted with permission).

on correlation of physical stratigraphic surfaces for identification of the unconformities bounding the sequences, and on biostratigraphic zonation for determination of the geologic ages of the sequences, if well control is given. Their reliable analysis is an essential step towards seismic facies analysis.

2.5.2. SEISMIC FACIES ANALYSIS

Seismic facies analysis is usually based on recognizing patterns in the seismic data. Among obvious patterns are those attributed to progradation, reefs, erosion (such as channels), etc.

Depositional sequences show a seismic response that is recognizable on a seismic section. According to the property of the seismic events differences can be distinguished,

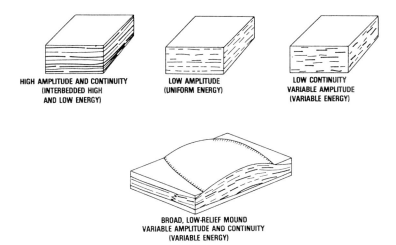

HIGH AMPLITUDE AND CONTINUITY
(INTERBEDDED HIGH
AND LOW ENERGY)

LOW AMPLITUDE
(UNIFORM ENERGY)

LOW CONTINUITY
VARIABLE AMPLITUDE
(VARIABLE ENERGY)

BROAD, LOW-RELIEF MOUND
VARIABLE AMPLITUDE AND CONTINUITY
(VARIABLE ENERGY)

Fig. 37.   Shelf seismic facies types (after Sangree and Widmier 1977, p. 169, reprinted with permission).

grouped, and qualified as seismic facies units which are mappable. 3-D seismic units are composed of groups of reflections whose parameters differ from those of adjacent facies units. Where the internal reflection parameters, the external form, and the three-dimensional associations of these seismic facies units are delineated, the units can then be interpreted in terms of environmental setting, depositional processes, and estimates of lithology.

2.5.2.1. Clastic Facies Patterns

Clastic facies patterns can be recognized in a rather systematic way (Table 20).   A first distinction is based on the paleogeographic setting (Sangree and Widmier 1977): shelf, shelf-margin and prograded-slope, and basin-slope floor. Four sub-groups can be recognized within the shelf clastics group, based on reflection continuity and amplitude and overall unit shape (Figure 37). Shelf-margin or prograded-slope units are separated into two classes based on toplap characteristics. The distinction between oblique and sigmoid patterns (Figure 38) is often not clear, and the disappearance of reflection detail further complicates an already difficult distinction because of resolution limitations. Basin-slope and basin-floor patterns are distinguished both on the basis of the overall three-dimension form of the unit and on the attitude of reflections within the unit (Figure 39).

Shelf environments range from neritic to non-marine. Sediments deposited in these environments tend to generate parallel to gently divergent reflection configurations having a widespread sheet – or wedge-shaped – external form. One facies, characterized by a broad low-relief mound, composed of gently sigmoidal to downlapping reflections, is an exception of these generalizations.

Reflections are normally concordant at the top and range from concordant to gently onlapping and occasionally downlapping at the base. Prediction of depositional energy and

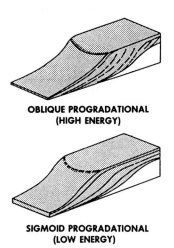

**OBLIQUE PROGRADATIONAL
(HIGH ENERGY)**

**SIGMOID PROGRADATIONAL
(LOW ENERGY)**

Fig. 38.   Shelf-margin and prograded-slope seismic facies types (after Sangree and Widmier 1977, p. 173, reprinted with permission).

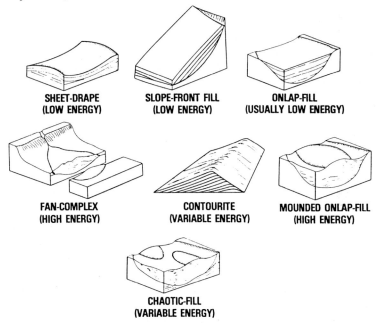

**SHEET-DRAPE
(LOW ENERGY)**

**SLOPE-FRONT FILL
(LOW ENERGY)**

**ONLAP-FILL
(USUALLY LOW ENERGY)**

**FAN-COMPLEX
(HIGH ENERGY)**

**CONTOURITE
(VARIABLE ENERGY)**

**MOUNDED ONLAP-FILL
(HIGH ENERGY)**

**CHAOTIC-FILL
(VARIABLE ENERGY)**

Fig. 39.   Basin-slope and basin-floor seismic facies types (after Sangree and Widmier 1977, p. 177, reprinted with permission).

sand content in shelf seismic facies units must rely on analysis of variations in reflection amplitude, continuity, cycle width, interval velocity, and areal relations with other units.

The next major group of clastic seismic facies units is associated with shelf-margin and prograded-slope deposits. Two principal facies, defined exclusively on reflection configuration, are recognized in this environment (Figure 38): the oblique-progradational facies and the sigmoid-progradational facies. Both facies are characerized by downlapping reflections at their base. Downlap reflects the prograding of sediments from relatively shallow into relatively deep water with thinning of the outer toes of individual beds, usually below seismic resolution. Normally, the upper parts of these patterns represent sediments deposited in fluvial to neritic environments. However, examples have been documented where these patterns originate from sediments deposited in bathyal water, presumably through the action of deep-water currents.

The groups of seismic facies units (Figure 39) that dominate the basin-slope and basin-floor are: (1) sheet-drape seismic facies, (2) slope-front fill seismic facies, (3) onlapping-fill seismic facies, (4) mounded-fan seismic facies, (5) mounded-contourite seismic facies, (6) mounded onlapping-fill seismic facies, and (7) chaotic-fill facies. Some facies types are still poorly documented. Thus, this classification will be refined and probably extended in the future.

These facies commonly overlap from the basin floor onto the slope. For example, submarine fan complexes usually extend from the slope at their apex out into the basin. Chaotic and onlapping-fill seismic facies patterns are observed in topographic lows on the slope and on the basin floor, including the basin-floor plain, local basins, channels or troughs, or areas of prominent topographic flattening in the configuration of the slope. These various facies are likely to be deposited by high-density turbidity currents and other mass transport processes. Displaced, relatively shallow-water faunas are common in cores of these chaotic facies units.

Indicative of high-energy deposition and consequently of sand-dominated units are increasing irregularity of reflection pattern and character and the mounding of external form. Thus, fan complexes, mounded onlapping, and chaotic units are interpreted as higher-energy deposits. The final controlling factor of the lithology is the sediment source area; no process can transport sand if sand is absent.

2.5.2.2. *Carbonate Buildups*

Carbonate buildups including banks, bioherms, and reefs are well suited for stratigraphic interpretation of reflection seismic data because depositional features or bedding character-istics differ markedly between the buildups and the surrounding strata. Seismic recognition of buildups can be either direct, i.e. seismic parameters directly outline buildups such as reflections from the boundaries of the buildups, onlap of overlying cycles, or seismic facies changes between the buildups and the surrounding strata; or indirect, i.e. seismic parameters indirectly outline or indicate the presence of buildups such as draping patterns, velocity anomalies, and location on a shelf edge or fault block. All available geologic and geophysical data (e.g. seismic stratigraphic and seismic facies analysis) should be used. Modern computer programs support the graphic displays of these parameters.

For seismic analysis, the wide variety of carbonate buildups can be grouped into four major types: (1) linear barrier buildups with relatively deep water on both sides during deposition, (2) equidimensional pinnacle buildups surrounded by deep water during deposition, (3) linear shelf-margin buildups with deep water on one side, and shallow water on the other, and (4) patchy buildups formed in shallow water, either in close proximity to the shelf margins, or over wide shallow seas.

Recognition of the various microfacies units composing a so-called buildup is, however, beyond the resolution of conventional seismic data.

Seismic criteria (Figure 40) for directly outlining carbonate buildups include reflections from top and sides of buildups and onlap of overlying reflections onto buildups (I-A), internal patterns of seismic facies change between buildup and surrounding strata (I-B). Criteria that outline indirectly or infer the presence of buildups comprise draping, velocity anomalies, and abrupt changes in internal bedding (II-A) as well as the determination of optimal position within a basin for buildups (II-B).

Well-documented examples of the seismic expression of carbonate buildups, which form important hydrocarbon reservoirs or show reservoir potential, have been compiled by Bubb and Hatlelid (1977) from North Africa, offshore West Africa, the Gulf of Papua, New Mexico, and Texas. Figure 41 depicts a seismic profil through platform-like banks and pinnacle carbonate buildups of Early Tertiary age in North Africa. These carbonate buildups are interpreted based on:

—   The seismic facies change from continuous parallel reflectors into chiefly reflection-free to discontinuous reflection zones.
—   One to two cycles of onlap of overlying units onto buildups.

A borehole south of the seismic line confirmed the buildup seen on right. Figure 42 shows a seismic section through pinnacle carbonate buildups of the Middle Devonian in Alberta, Canada. The presence of two small pinnacle reefs is indicated by small amounts of drape in reflectors above the reefs and by one cycle of onlap (?). The drape is likely to be due to differential compaction and subsequent removal of salt deposited in interreef areas.

On the basis of seismostratigraphy, Pavlow *et al.* (1988) reconstructed a geological-geophysical model of the famous Tengiz paleo-atoll reservoir.

2.5.3. SEISMIC METHODS IN OIL-FIELD DEVELOPMENT

Seismic methods are increasingly used in oil- and gas-field development. In fact, most 3-D work is done with the objective of optimizing field developments. Seismic methods, especially seismic stratigraphy, are applied to delineate the stratigraphic and lithologic characteristics of reservoirs and their aquifers with as much detail as possible in order to minimize the cost of development drilling and to support primary, secondary, and tertiary recovery operations (Rice *et al.* 1981). To meet the goal, close cooperation between geophysicists, geologists, and reservoir engineers is required.

Because interpretation of seismic data in respect to geology, rock properties, and well data in oil- and gas-field development is new and critical, few papers are published up to now (compare Bortfeld 1983).

The first step is to combine seismic reflection measurements at the surface with corre-

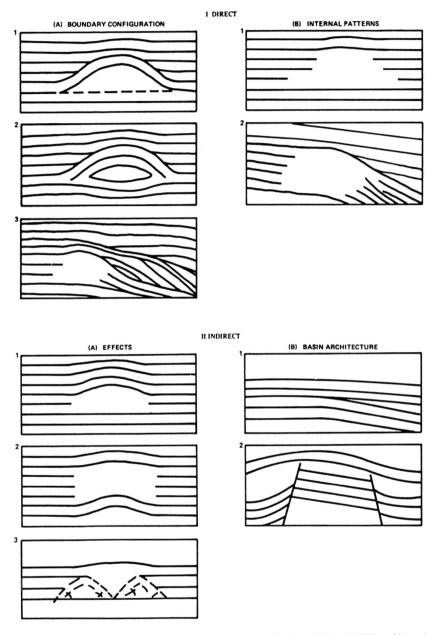

Fig. 40.   Seismic criteria for recognizing carbonate buildups (after Bubb and Hatlelid 1977, p. 88, reprinted with permission). Boundary configurations such as A1, A2 and A3 and internal patterns such as B1 and B2 furnish direct criteria (I). Drape (A1), velocity anomalies (A2), and spurious events (A3) as well as optimum basin positions for buildup (B1 and B2) are indirect criteria (II).

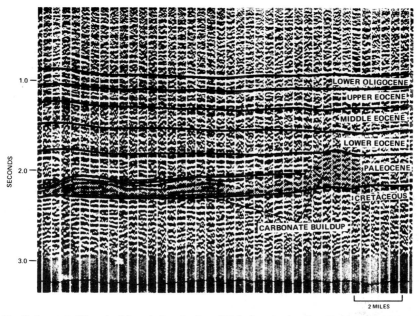

Fig. 41. Carbonate buildups, both low platform banks and high pinnacles (stratigraphy interpreted from 12-fold CDP thumper data), N Africa (after Bubb and Hatlelid 1977, p. 93, reprinted with permission).

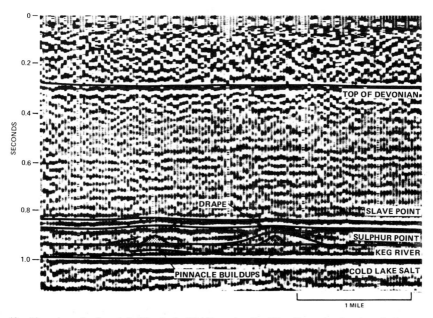

Fig. 42. Pinnacle reef carbonate buildup in the Mid-Devonian Keg River Formation (stratigraphy interpretated from 6-fold CDP data) Rainbow-Zama area, Canada (after Bubb and Hatlelid 1977, p. 195, reprinted with permission).

sponding measurements in boreholes (synthetic seismograms). The synthetic seismogram is obtained by convolving the sequence of acoustic impedances $\rho \cdot v$ (where ρ denotes the rock density and v its velocity) with the seismic wavelet. Both, ρ and v are obtained from borehole measurements (density log and sonic log). They are a guide to what to look for on a seismic section as evidence of lateral changes to be expected (e.g. pinch-outs), development of channel sands or even less conspicuous facies changes, changes in porosity and/or permeability induced by changes in lithology or diagenetic cementation rate are most important in reservoir evaluation for both sandstones and carbonate reservoirs. The sometimes incomplete input data (velocity and density) limits the use of synthetic seismograms.

Another modern method, namely vertical seismic profiling (VSP), allows to match seismic sections with boreholes. Its fundamental goal is to detect boundaries (ρ- and v-anomalies) below the borehole with a better accuracy than surface seismic. VSP can be applied in wildcats to exactly define the position of a known strong reflector below total depth (TD). Multi-Offset VSP, seismic profiles at right angles in the borehole area, even allows to locate faults below TD.

Proper calibration of the seismic trace at the well site in terms of travel time, amplitude, and phase is a crucial problem for the lack of the computation of the true seismic wavelet delimiting the results in seismic facies interpretation. It has become modern practice to extract the seismic wavelet from borehole measurements and to apply inverse filtering processes to the seismic data in order to convert the seismic section into a sequence of pseudo-impedance or pseudo-acoustic logs to be compared with those measured in the borehole. Pseudo-acoustic logs (Figure 43), displayed trace by trace similar to a seismic section, do not show the distinctive changes like a sonic log due to the sample rate applied to the seismic data in digital recording. In fact, sample rate limits the resolution to a few meters. Thus, variations in lithology and fluid content may be inferred from the changes of the pseudologs in the vicinity of the borehole. For instance, gas was trapped in Miocene sands (Figure 44).

Moreover, the introduction of areal seismic surveys, the so-called 3-D seismic, facilitated the three-dimensional evaluation of seismograms. It led to an easier evaluation and more complete definition of the structural and stratigraphic framework of an area, normally in the dimension of an oil field. Geological as well as sedimentological interpretation became easier; processing costs, however, are escalating for genuine 3-D migration. Much development work is left to provide smooth running procedures for 3-D velocity analysis and for application of the 3-D velocity field in 3-D migration.

3-D processing allows to create synthetic cross sections within the survey area and to display them. Interpretation is aided by the possibility of obtaining, after 3-D migration, sections at any position and direction. For instance, horizontal seismic sections from 3-D seismics display much better the spatial extent of subsurface features. These procedures allow reliable identification of very small structural features in sedimentary sequences, such as hydrocarbon flat spots, peculiar carbonate features, stream channels, and sand bars.

Detailed seismic analysis using the techniques of lithologic modelling and reflection character analysis allows even the determination of reservoir qualities such as thickness

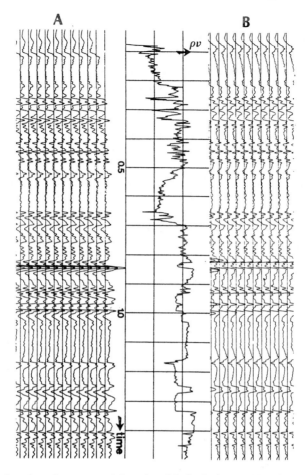

Fig. 43. Pseudo-impedance logs computed from the original seismic trace (A), after transformation of the seismic trace (B) and compared with the actual acoustic-impedance log measured (after Marschall 1982).

and porosity of sandstone reservoirs. Budny (1991) demonstrated the porosity distribution in potential Rotliegende sandstones of NW Germany (Figure 45). Thus, the integration of seismic data allows the extention of geological reservoir description by conventional methods and the more efficient prediction of reservoir distribution, quality, and variability in three dimensions.

2.5.4. THE 'BRIGHT SPOT' EXPLORATION CONCEPT

The 'Bright Spot' exploration concept is a fairly recent development in geophysics (1968, Gulf of Mexico). It is used to predict the existence of hydrocarbons prior to drilling. The name is derived from the fact that high-amplitude reflections appear 'brighter' than low-

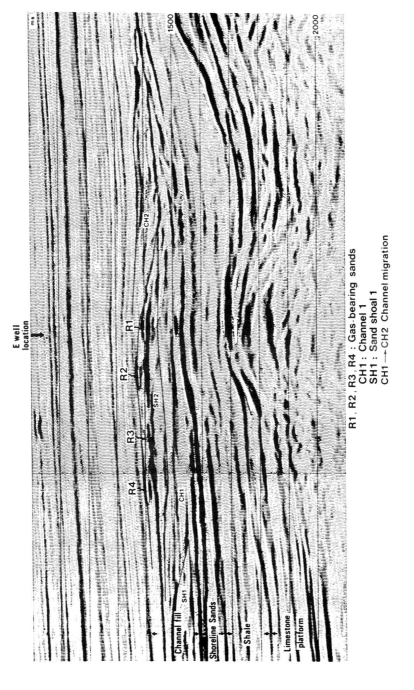

R1, R2, R3, R4 : Gas-bearing sands
CH1 : Channel 1
SH1 : Sand shoal 1
CH1 → CH2 Channel migration

Fig. 44.   Seismic profile showing sedimentary facies and distribution of gas-bearing sands (Miocene), Gulf of Cadiz (after Institut Français du Pétrole 1986).

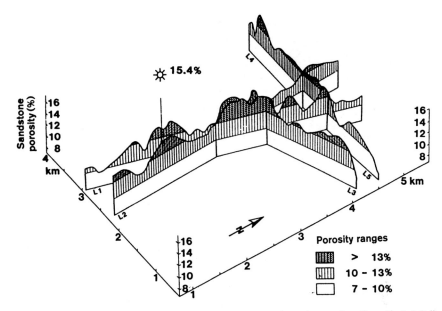

Fig. 45. Porosity distribution and prediction by seismics: Rotliegende sandstones along five seismic 2-D lines, NW German Basin (modified from Budny 1991).

amplitude reflections when observed on a seismic section. This 'brightness' is important because high-amplitude events may indicate the presence of hydrocarbons, especially of gas (Figure 46). 'Bright spots', in common use in petroleum industry since 1972, resulted in some 'spectacular' discoveries of gas.

Velocities and densities of sedimentary rocks vary according to mineral composition, degree of compaction, porosity, pressure, and fluid/gas content in the pores of the rock matrix. The change in acoustic impedance at an interface between two different rock types causes seismic energy to be reflected from the interface. Acoustic impedance is a convenient form of expressing the velocity-density relationship of an individual rock. The formula for acoustic impedance is

$$I = \rho \cdot v$$

(I = acoustic impedance, ρ = density, and v = velocity).

The standard procedure is to scrutinize seismic reflections on a profile for lateral variations in relative amplitude, and to assume that any pronounced amplitude changes of reflections coming from any suspected reservoir are directly related to hydrocarbons in the reservoir. Any observed high amplitudes are termed 'bright spots', low amplitudes 'dim spots'.

The assumption of relating changes in amplitudes exclusively to changes in pore content ignores, however, the other possible causes for such occurrences. Potential contributors to amplitude changes are changes in lithology, tuning effects of varying bed thickness,

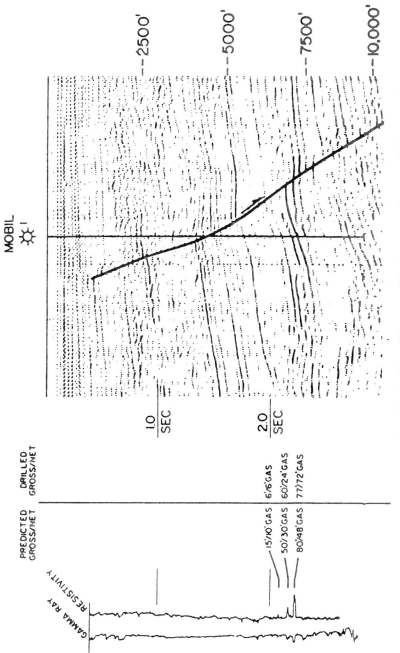

Fig. 46. 'Bright spot' in gas reservoirs of the Louisiana offshore area. The three strong reflectors or bright spots at about -7,000 ft of the seismic line indicate three gas sands (after Poh-Hsi Pan 1981).

focusing of seismic waves, multiples arriving in phase, variations in the weathering layer, common depth-point-stacking techniques applied.

The discovery of 'bright spots' accelerated efforts to analyze more closely other features which are frequently observed on seismic profiles. Consequently, several other geophysical indicators of hydrocarbon have been found: polarity inversions, time sags, absorption of high frequencies, flat spot reflections, velocity slow-downs, dim spots, diffraction patterns at the edges of reservoirs, as well as constructive and destructive interference patterns. Thus, 'bright spots' high-amplitude reflections are only one type of events classified as geophysical hydrocarbon indicators.

Geophysical hydrocarbon indicators are concentrated on gas exploration. Velocities and densities of gases differ radically from those of formation waters which they displace in a reservoir. These substantial contrasts produce observable changes. On the contrary, oil has essentially the same seismic velocities and densities as water. Thus, oil creates no appreciable contrast when displacing the formation waters. Many dry holes in gas exploration drilled on 'bright spots' may be due to the fact that the same anomalies occur even at 95% water saturation.

The art of recognition of 'bright spots' and other potential hydyrocarbon indicators can be used in its simplistic form if risks are clearly considered. Future emphasis will be laid upon the development of techniques to analyze geophysical hydrocarbon indicators in order to predict reservoir qualities and quantities.

Chapter 3

SEDIMENTOLOGY OF ROCKS IMPORTANT TO HYDRO-
CARBON EXPLORATION

Commonly sedimentary rocks form the framework for hydrocarbons to be generated, to migrate, to be stored, to be sealed, and to be trapped. In the following chapters sedimentary rocks are regarded under the viewpoint of hydrocarbon exploration.

3.1. Source Rocks of Hydrocarbons

Source rocks are sedimentary rocks, usually shales or limestones, in which organic matter was transformed to liquid or gaseous hydrocarbons under the influence of pressure, heat, and time. Source rock parameters are (1) amount of organic matter, (2) type of organic matter, and (3) maturity of organic matter. An active source rock can be defined as a rock containing a sufficient amount of organic matter of a proper type and of sufficient maturity. A potential source rock is an immature source rock.

Generation of hydrocarbons from source beds is related to temperature and, to a lesser degree, to pressure and geologic time. Oil is generated and primary migration occurs when the source rocks attain an optimum temperature range. The temperature gradient is variable in different basins. The term 'oil window' has been applied to the temperature range in which hydrocarbons are generated. The present depth of burial may be markedly less than in previous geological epochs because of erosion of large volumes of overburden since oil generation took place.

3.1.1. SOURCE ROCK TYPES

The first step in petroleum formation is the subaqueous deposition of an organic-rich sediment containing abundant lipid-rich organic debris, together with minute clay and/or carbonate particles, and their preservation in a non-oxidizing low-energy environment in which the supply of organic matter exceeds that of oxygen.

In ordinary aquatic environments, such as in open marine environments, the remains of algae and bacteria (phytoplankton) as well as other organic debris are oxidized both abiogenically and bacterially in the aerobic zone. In high-energy environments this zone may extend down to the sediment-water interface and even into the upper few centimeters of the sediment. The organic matter that survives bacterial oxidation in the aerobic zone is anaerobically degraded in the anoxic zones.

The following conditions favor the formation of organic-rich sediments as potential source beds for hydrocarbons: abundant supply of lipid-rich organic particles, sedimentation in a low-energy environment, protection from abiogenic oxidation by rapid transport to the depositional surface, stagnation (enhanced by consumption of oxygen during aerobic oxidation of a portion of the organic matter, and by low water current), low pH acidic environment (e.g. peat bogs), and optimum sedimentation rate which is slow enough to avoid dispersive dilution and rapid enough to prevent oxidation.

The dominant type of organic matter is kerogen. It is fossilized insoluble organic material found in sedimentary rocks, usually in shales, which can be converted by distillation to petroleum products. Kerogen designates organic constituents that are neither soluble in aqueous alkaline solvents nor in the common organic solvents. On the other hand, the fraction extractable with organic solvents is called bitumen. Thus, the term 'kerogen' does not include soluble bitumen. Kerogen is a macromolecule made of condensed cyclic nuclei linked by heteroatomic bonds or aliphatic chains.

The so-called Van Krevelen diagram (Tissot and Welte 1984, p. 152) is a plot widely used for classifying kerogens and for demonstrating their evolution paths. It was first used by Van Krevelen (1961) in order to characterize coals and their coalification paths. In this diagram the atomic compositions of the three major elements (C, H, O) are plotted as atomic H/C versus O/C ratios (Figure 47). These ratios serve for classifying the various types of kerogens.

Three types of kerogen characterized by their evolution paths can be recognized by optical examination and physico-chemical analysis.

Type-I kerogen contains many aliphatic chains and few aromatic nuclei. The H/C ratio and the potential for oil and gas generation are high. This kerogen is either derived from algal lipids or from organic matter enriched in lipids by microbial activity.

Type-II kerogen contains more aromatic and naphthenic rings. The H/C ratio and the oil and gas potential are lower than those observed for type-I kerogen. This type of kerogen with a medium to high sulfur content is related to marine organic matter deposited under reducing conditions.

Type-III kerogen contains mostly condensed polyaromatics and oxygenated functional groups with minor aliphatic chains. The H/C ratio is low and the oil potential is moderate only. This type of kerogen may still generate abundant gas at greater depth. The O/C ratio is higher than that in the other two types of kerogen. The organic matter is mainly derived from terrestrial higher plants.

The major part of kerogen is amorphous, probably due to microbial alteration during sedimenation. Amorphous organic material, commonly referred to as sapropelic matter, may be associated with minerals. Only a minor part of kerogen is made of recognizable organic remnants. These organic remnants may include planktonic organisms (microscopic algae, copepods, ostracods) and micro-organisms normally living in the fresh sediment (bacteria, algae, etc.).

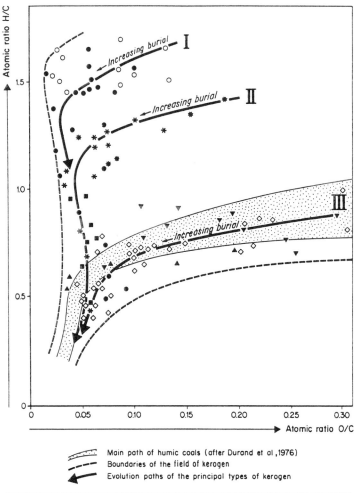

Fig. 47.   Van Krevelen diagram showing the elemental composition of oil-shale kerogens. The evolution paths of three main kerogen types and the field of humic coals are shown for comparison. Letters indicate sample locations (after Tissot and Welte 1984).

Geochemical 'fossils' are molecules synthesized by plants or animals and incorporated in sediments with only minor changes. In particular, the carbon skeleton of hydrocarbons or other lipids is preserved. These molecules represent only a minor fraction of crude oils. However, they provide a hint at the origin of the organic material.

Alkanes, fatty acids, terpenes, steroids, and porphyrins are the major groups of geochemical 'fossils'. They can be traced from recent to ancient sediments where they progressively suffer thermal degradation and/or dilution by other hydrocarbons generated at greater depths.

Kerogen is a possible additional source of geochemical 'fossils'. Some lipids may be trapped in the kerogen network, or alternatively bound to kerogen by chemical bonds. These molecules are released from kerogen with increasing depth and temperature.

The following sedimentary rock types can become source rocks for hydrocarbons under the above specified conditions: shales, carbonate rocks, anhydrite, and coal.

3.1.1.1. Shales As Source Rocks

Organic-rich shales represent the most widespread potential source rock for petroleum. In thin section they are characterized by abundance of yellow to red to brown organic matter and commonly by the presence of microlaminated textures as a result of sedimentation undisturbed by the activity of aerobic organisms. Geochemically, they show a high content of total organic carbon, which provides an approximate measure of the source-rock potential to generate hydrocarbons upon thermal maturation.

One of the most complete work on shales as source rocks for hydrocarbons is that by Bitterli (1963) on bituminous rocks of western Europe.

Empirical data from hydrocarbon-producing and non-producing areas suggests that potential source beds in sand-shale sequences must normally contain at least 0.5% organic carbon in order to yield significant amounts of hydrocarbons. This threshold value, which reflects the source-rock potential as a function of organic carbon content, is widely accepted.

Significant oil production is associated with shales containing more than 0.5% organic carbon. However, it is unwise to preclude any shale section from being considered a potential source bed only on the basis of the above minimum value of the organic carbon content. Measurements of the organic carbon content are only one step to define areas of best source-shale potential.

The interaction between clay minerals and organic matter is of fundamental importance, because clay minerals may catalyze, to varying extents, hydrocarbon-forming reactions. Moreover, dehydration of clay minerals during burial diagenesis may trigger primary migration of hydrocarbons from argillaceous source rocks.

3.1.1.2. Carbonate Rocks As Source Rocks

Normally carbonate rocks are considered not to possess great hydrocarbon source rock potential, possibly because of their light colors. In fact, pure carbonate rocks contain little organic matter, normally less than 0.1 C_{org} (Seemann 1980). Nevertheless, carbonate sequences may include carbonate-bearing source rocks such as marls or argillaceous limestones rich in organic matter. The clay content of such carbonate rocks commonly ranges between 10–30%. The minimum values of organic carbon for potential carbonate source rocks are 0.3% (Tissot and Welte 1984, p. 497).

Seemann (1980) determined the C_{org} content in selected carbonate rock samples of Permian to Jurassic age from outcrops in Central Europe. Samples from basins are characterized by relatively high organic carbon contents (about 0.3% C_{org}) and relatively low bitumen contents (up to 100 ppm extractable organic matter). Reef carbonates and most of the lagoonal carbonates show only low amounts of C_{org} and extractable organic matter. Exceptionally, lagoonal carbonates such as bituminous, laminated carbonate rocks of the Seefeld Shale type of the Grenzbitumen horizon (Triassic) and the so-called bituminous facies of the Upper Jurassic (Kimmeridgian), southwest of Geneve are rich in C_{org} and extractable organic matter. Most of the organic matter from the carbonates investigated, however, was immature. The distribution of n-alkanes indicates the presence of high amounts of micro-organisms such as bacteria. If mature, two thirds of this microbial organic matter can be transformed in petroleum-like hydrocarbons.

Carbonate-rich source rocks are also described from the Jurassic of Saudia Arabia (Ayres *et al.* 1982), the Cretaceous of Venezuela (Hedberg 1964), and from the Devonian of Alberta.

In order to assess the hydrocarbon potential of carbonate source rocks the great thickness of many carbonate rock series and their ease to release hydrocarbons has to be taken into consideration.

3.1.1.3. Evaporites As Source Rocks

In many petroleum provinces, evaporites are coeval with and spatially related to known source rocks. The occurrence of hydrocarbons in sediments associated with evaporites is too common as though this association should be purely coincidental. Also Brongersma-Sanders (1971) stressed the common association of black shales and evaporites; Busson (1988) and Warren (1989) the significance of evaporites for hydrocarbon accumulations.

In fact, evaporation produces dense brines which normally sink and create density stratification in a standing water body, resulting in eutrophication and anoxic bottom conditions. Organic matter could be preserved in such sediments making them potential source beds for petroleum.

Study of modern saline lakes, solar-evaporation ponds, and hypersaline lagoons (Kirkland and Evans 1981) shows that evaporitic environments can be productive of organic matter. Few species of organisms such as phytoplankton (e.g. blue-green algae) or fauna (e.g. shrimps or flies) survive in the brines. Prolific growth of phytoplankton may be similar to that in areas of upwelling in modern oceans.

The prolific productivity within the mesosaline environment results in a nearly continuous rain of dead lipid-rich organisms into the bottom waters of the basin. Brine stratification often causes anoxia, even in very shallow waters. Thus, much of the organic matter is probably preserved. In the Middle East mesosaline conditions are known to have occurred repeatedly from the Triassic to Cretaceous. These conditions might have been responsible for the vast petroleum reserves in that area. In the Pennsylvanian Paradox Basin, Utah, U.S.A., black carbonate-rich shales interstratified with evaporites contain up to 15% organic matter. Most limestones associated with evaporites are dark and highly bituminous with relatively considerable amounts of organic matter.

3.1.1.4. *Particulate Coaly Matter And Coal As Source Rocks*

Pelitic sediments with particulate coaly matter which is disseminated in marked quantities may act as source rock in a similar way as coal proper.

The significance of coal as a source for gas was recognized by Karweil (1969). Patijn (1964a) stressed the role of Upper Carboniferous coals forming the northwestern Europe gas fields. Additional evidence that Carboniferous coals are a primary gas source was presented by Stahl (1968), Bartenstein and Teichmüller (1974), and Lutz *et al.* (1975). Commercial oil accumulations, however, have not yet been traced to coal as a source, though small amounts of liquid hydrocarbons are probably generated in coal during catagenesis. Nevertheless, oil shows, small oil seeps, and oil-impregnated sands were observed in association with coal world-wide.

Crude oils associated with coal normally belong to the high-wax type, such as the oils from the Midland coal measures in Great Britain (Hedberg 1968). Large oil fields associated with coal are not known. The potential of coals to produce higher hydrocarbons is limited. It is similar to type-III kerogen which yields gas rather than oil. The limited primary migration of heavier hydrocarbons out of coal is probably caused by the high absorption capacity of coal, the sizeable microporosity, and the fact that coal normally occurs as a massive, continuous, solid organic accumulation.

3.1.2. SEDIMENTARY ENVIRONMENTS FAVORABLE FOR HYDROCARBON SOURCE ROCK FORMATION

Hydrocarbon source rocks of marine origin are widespread and well studied. However, oil has also been produced from non-marine strata, e.g. in the Uinta basin of Utah, U.S.A. Such occurrences were initially considered to be the results of migration from marine source rocks. Recently, however, investigations on oil-producing basins in China gave evidence of large-scale hydrocarbon generation in cratonic basins from lacustrine or fluviatile source rocks. This observation stimulated search for oil in basins in the interior of Africa. Moreover, large gas fields sourced from coal measures are known from northwestern Europe, northwestern Siberia, and New Zealand.

Specific environments favorable for the deposition of organic-rich sediments, i.e. potential source beds, are (Demaison and Moore 1980, Moorkens 1991):

1. Shelves of open oceans and silled seas (lipid-rich phytoplankton) and deltas and estuaries on shelves of open oceans and silled seas (lipid-rich phytoplankton/coal and carbonaceous shales) in the marine realm with nutrient supply and preservation as controlling factors.
2. Peat bogs (coal and carbonaceous shales) and freshwater lakes (lipid-rich phyto-

plankton) on the continent with rainfall and natural fertilizers as controlling factors. Lacustrine petroleum source rocks are discussed by Fleet, Kelts and Talbot (1988); they compile data on their geological and tectonic framework, geochemistry, and biology, as well as diagnostic paleoenvironment indicators, and illustrate their discussion with case studies.

Episodic volcanism can favor the formation of organic-rich sediments and hydrocarbon-source rocks (Zimmerle 1985) as well as the generation of anoxic conditions, preservation of organic matter and alteration of volcanic ash to highly surface-active smectites.

Normal open marine basins such as most present-day marine basins are not favorable for the deposition of oil-source rocks. Only small amounts of organic matter are finally preserved.

Exceptions are deltaic areas where rapid deposition rates can overcome the effects of oxidation and biological destruction of organic matter. Good source rocks were identified in the oxygen-poor depositional environment of the upper continental slope associated with deltaic sedimentation (prodelta shales). Prodelta source shales are observed in the Gulf of Mexico and Niger Delta. In deltas good source shales may exist in deep-water environments as a result of slumping from the up-slope, prodelta environment.

Gas sources may be found in both deltaic and non-deltaic settings where nearshore, marginal marine sediments are characterized by a high content of terrigenous plant material.

Anoxic bottom conditions can also occur locally in more open oceans with highly fertile and productive surface waters usually associated with upwelling.

Basins with restricted water circulation (stagnant basins) are favorable for source-rock deposition. Many of the best source rocks such as pyritic black shales were deposited in this type of basin.

In such land-locked basins (barred basins), separated from the ocean by narrow and shallow sills, the water tends to be stratified as a result of restrictions to circulation combined with the development of temperature or salinity gradients with depth. Should stratification persist for extended periods of time, the water beneath the density boundary will become anoxic.

The Black Sea, the Carioca Trench, and certain Norwegian fjords are modern-day examples of this type of basin; the Los Angeles and other organic-rich basins in California are paleobasins of this type.

3.1.3. METHODS TO DETERMINE SOURCE ROCK QUALITY

Source-rock analyses are usually performed by mineralogical, organo-chemical, and coal-petrographical methods.

The amount of organic matter is determined as a total organic carbon percentage. The Total Carbon (TC) and Total Organic Carbon (TOC) are usually determined by the Leco carbon analysator.

The type of organic matter is determined by optical and chemical methods. Optically the organic matter is rated qualitatively in reflected light supplemented with blue-light

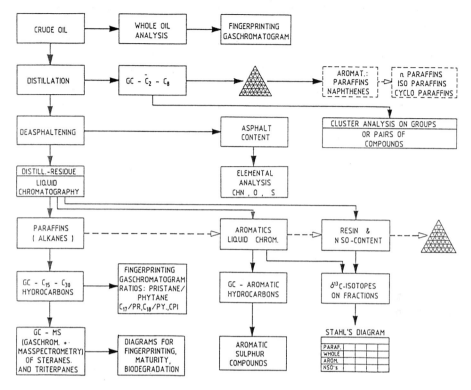

Fig. 48. Flow diagram showing main steps of organic-chemical oil and gas analysis (courtesy R. Gedenk).

induced fluorescence.

The maturity of the organic matter is determined by organo-chemical parameters (extrability, pristane/n-C17 ratio, CPI values, alifate/hydrocarbon ratio, and composition of alifate fraction from gas chromatography) and by vitrinite reflectance.

Soluble Organic Matter (SOM) is determined from the extracts from a Soxhlet extraction of the crushed sample with methylenchloride in 24 hours. The separation of SOM is performed by column-chromatography with hexane, methylenchloride, and methanol as efluents. Gas Liquid Chromatography (GLC) of the alifate fraction is performed on an OV 1 capillary column. Pyrolysis is carried out by the Rock-Eval apparatus. Figure 48 depicts a flow diagram showing the main steps of organic-chemical oil and gas analysis.

Viewing and measurements of vitrinite reflectance are made with a reflected-light photomicroscope and photometer. Measuring procedures should follow the outlines in Stach *et al.* (1982), rating the organic matter as vitrinite, liptinite, and inertinite. Approximate threshold values for the maturity levels corresponding to the onset of oil generation, expulsion of oil, and the peak zone of oil generation, are found in Hood *et al.* (1975) and Tissot and Welte (1984).

The mineralogical analysis determines parameters as to the sedimentation and diagenetic alteration of the host rock (degree of diagenesis of carbonates, iron compounds, and clay minerals; catalytic effect of clay minerals) and, thus, contributes to the integrated analysis of the source rock.

Semiquantitative mineralogy and clay mineralogy are determined by X-ray diffraction on powdered bulk samples and on pretreated oriented clay samples. Qualitative and for carbonates and sulfides quantitative mineralogy is determined by differential thermal analysis (DTA) on powdered bulk samples with detection of CO_2, H_2O, and SO_2. The oxidation state and mineralogical positions of Fe can be determined by Mössbauer spectroscopy on powdered bulk samples.

3.1.4. REGIONAL SOURCE ROCK OCCURRENCES

Each source rock is unique in composition, structure, and geological setting. Thus, the spectrum of source rock occurrences worldwide and in Europe is exemplified in the following.

3.1.4.1. Best Known Source Rocks Worldwide

Most attention in respect to analytical work has been paid to dark-colored shales, the so-called black shales, as potential or actual source rocks (Bitterli 1963, Welte and Tissot 1984, Huc 1990).

In the Van Krevelen Diagram (Figure 47) the kerogen composition of numerous oil-shale samples worldwide is demonstrated. Other well studied source rocks are: Silurian source shales of the Hassi Messaoud field, Algeria; Tertiary-Cretaceous shales of the Quiriquiri field, Venezuela; Devonian-Mississippian shales of the Williston Basin, North Dakota; Jurassic thinly laminated carbonate shales, Saudi Arabia (Ayres *et al.* 1982); Upper Jurassic shales, Bazhenov Formation, West Siberia (Krylov and Korzh 1984); thin-bedded, carbonaceous-bituminous limestones, Cretaceous La Luna formation, western Venezuela, and dark-colored aphanitic limestones between reefs and calcarenites, Devonian Swan Hills formation, Alberta, Canada.

There are certainly much more source rocks in all formations and many localities but they are less conspicuous in regard to color, high organic carbon content, and thickness.

3.1.4.2. Source Rocks in Europe

In Europe, possible source rocks of Early Paleozoic age have probably been overprinted by Caledonian metamorphism to a considerable extent. However, in the course of deep-gas exploration in northwest Germany potential source rocks for gas were postulated to occur in the Silurian, Early Devonian, and Early Carboniferous of the Variscan foreland as well as of the Caledonian foreland in NW Europe (Pratsch 1983).

Carboniferous oil source rocks are of minor importance, but Lower Carboniferous oil shales are believed to be the source rocks of the oil discoveries in the East-Midlands area of England. Upper Carboniferous coal measures source the gas in the Upper Carboniferous,

Rotliegende, Zechstein, and Buntsandstein reservoirs of the southern North Sea, the Netherlands, and northwestern Germany. The nature of the Upper Carboniferous coal seams and carbonaceous shales as major source rocks for gas was recognized since early 1960. The coal-bearing Upper Carboniferous, however, is limited to the basins south of the Mid-North Sea High and Rinkøbing-Fyn High.

Source rocks of Rotliegende age may occur locally. In the Zechstein Kupferschiefer, Stinkschiefer, and Stinkkalk are considered to be possible source rocks, both north and south of the Rinkøbing-Fyn High.

The majority of Triassic deposits, which consists of sandstones and evaporites in red-bed facies, show no source-rock potential. Only parts of the Winterton Formation, present in wells of the Danish Central Graben, are rich in organic carbon. The material may be a potential source rock for gas (Michelsen 1982). Early Lower Jurassic shales are considered to be possible source rocks in the North Sea area and northwest German Basin.

The Late Lower Jurassic Posidonia Shale (Toarcian) is regarded to be the most important source rock for oil in northwestern Europe. The pyritic, bituminous shales of an average thickness of 8.5 m extend from France, through Switzerland, southern Germany to northern Germany. The middle part of the shale is richest in oil. The organic matter amounts to a maximum of 20%, but the average oil content is 4–6%, decreasing from south to north. The caloric value of the shale lies in general between 3.780 and 5.040 kJ/kg. The shale contains higher amounts of heavy metals such as chromium, cobalt, vanadium, molybdenum, nickel, and uranium. New activities to exploit the shale have been started at Schandelah near Braunschweig. The resources accessible by various technologies are estimated at 2×10^9 t of oil shale in southern and northern Germany each; 100×10^6 t of oil could be extracted.

The Middle Jurassic, when represented by coal-bearing fluvio-deltaic deposits, is regarded to be poor source rocks for oil due to the dominance of vitrinite and inertinite. However, sapropelic deposits found in this depositional environment may possibly source oil (Barnard and Cooper 1981).

Pyritic, bituminous claystones and marlstones as well as dark limestones in the Upper Jurassic presumably source some oil reservoirs of northwest Germany and gas reservoirs in the Molasse.

The Late Jurassic Kimmeridge Clay is considered to be the major source of hydrocarbons in the North Sea oil province (Bjørlykke, Dypvik and Finestad 1975, Johnson 1975, Barnard and Cooper 1981). It is present in most of the North Sea. Only on salt structures and on structural highs it may be absent. Locally it might exceed two thousand meters. The upper boundary is the Late Cimmerian unconformity, normally overlain by Early Cretaceous sediments of varying age. The Kimmeridge Clay is generally characterized by high gamma radiation, the so-called 'Hot Shale', and a low sonic velocity. It is a dark grey, laminated , slightly silty, partly calcareous claystone with mica, micro-lignite, and pyrite. The organic carbon content is normally high and of marine origin. Numerous thin limestone and dolomite beds are intercalated. Thin oil-shale horizons occur throughout most of the Kimmeridge Clay. The hydrocarbons produced by pyrolysis of Kimmeridge Clay oil shales are mostly viscous oils and tars with a high sulphur content. Trace elements are not appreciably enriched in the oil shales. The oil shales formed from algal (largely dinoflagellate) blooms. The fine grain size, the flora and fauna, and the high organic content indicate a marine environment with high organic productivity and restricted bottom circulation, probably deep shelf.

The bituminous paper shales (Blättertonstein) of the Berriasian, previously exploited as oil shales, are hydrocarbon source rocks for the Lower Cretaceous sandstone reservoirs of northwestern Germany (Emsland) due to their considerable thickness and high organic carbon content (up to 30%). The elemental composition of the insoluble organic matter and the atomic H/C and O/C ratio indicate a highly bituminous substance of sapropelic origin (kerogen type 1). The organic matter of the bituminous paper shales is largely identical with that of the adjacent albertite veins, which are filled with solid bitumen.

The Barremian paper shales and Aptian fish shales of northwestern Germany are immature hydrocarbon source rocks of minor importance due to their reduced thickness (only a few meters) and low organic carbon content. The organic matter is terrigenous, gas-prone, and commonly reworked.

Anoxic rocks occur also in the Lower Cretaceous of the northern North Sea (Johnson 1975, pp. 391–392). Dark shales were observed in the Barremian.

Upper Cretaceous and Tertiary sediments of the North Sea area did not generate hydrocarbons in any significant quantities (Weismann 1979).

In the Rhine Graben Mesozoic as well as Tertiary source rocks are known since long. Paleozoic sourcing from Upper Carboniferous and even Rotliegende (Kusel and Lebach Beds) was debated, but not proven yet. Mesozoic source rocks (Lettenkohlenkeuper, Liassic alpha, Posidonia Shale, and Dogger alpha) are of minor importance. Tertiary source rocks are the Eocene Lymnaea Marls and the Eocene-Oligocene Pechelbronn Beds, the Oligocene Foraminifera Marls, Fish Shales, Meletta Beds, Cyrena Marls, and Corbicula Beds for oil, and the Oligo-Miocene Hydrobia Strata and Miocene-Pliocene limnic marls and clays for gas. The organic matter in the Rhine Graben was mainly derived from terrestrial sources.

Sourcing for the reservoirs in the Subalpine Molasse Trough is still much debated. Potential source rocks appear to be widespread, especially in the eastern portion of the trough such as marine shales of the Lower Meeresmolasse, the Fish Shale, and Lower Oligocene marlstones. However, they did not reach maturity where they are found. Source rock quality show also dark-colored, pyritic claystones and marlstones of the Upper Cretaceous, Lower Cretaceous bituminous shales, and bituminous marlstones of the Upper Jurassic. In the western Molasse Trough liquid hydrocarbons might have been derived from the Helvetic Upper Jurassic carbonate facies as isotope data indicates. The methane-rich gas in the eastern Molasse Trough (Rupelian, Chattian, and Aquitanian) is derived by bacterial decay.

3.1.5. OIL AND GAS MIGRATION

After an organic-rich shale has been buried to a sufficient depth hydrocarbon generation initiates. Hydrocarbons migrate from the source to the porous reservoir rock (primary migration) and then through the porous reservoir rock into a suitable hydrocarbon trap (secondary migration) where the hydrocarbons are collected. It has been postulated that during primary migration hydrocarbons move either as discrete globules, as colloidal suspensions (micelles), or in solution. All three modes contribute under different circumstances. Hydrocarbons are presumably transported within an aqueous phase in response to compaction. It was suggested that redistribution of hydrocarbons along fluid-potential gradients reaches a maximum during the release of interlayer water from mixed-layer clay minerals. Any porous interbed within a compacting claystone sequence will act as a conduit for hydrocarbons present in moving connate fluids.

Three factors influence secondary migration: (1) the tendency for buoyant rise of oil and gas in water-saturated rock pores, (2) capillary pressure that determines multiphase flow, and (3) the hydrodynamics of pore-fluid flow (Tissot and Welte 1984). Oil globules or gas bubbles larger than any given pore throat diameter must undergo deformation before they pass the pore throat. The interfacial tension between oil, gas, and water must overcome the capillary pressure. Petroleum trapped in a porous rock, thus, represents an equilibrium between buoyant/hydrodynamic driving forces and capillary resisting forces.

The direction of hydrocarbon movement is determined largely by local or regional pressure gradients; upward, lateral, downward migrations are all possible depending on local conditions. Hydrocarbon migration will continue as long as driving forces are active. Some studies postulate hydrocarbon migrations for hundred kilometers or more. Liquid migration in the subsurface is facilitated by the low viscosities of crude oil due to elevated temperatures and by effects of gas carried in solution.

Magara (1986) compiled data for models of petroleum generation and primary hydrocarbon migration.

3.1.6. OIL AND GAS SHALES

Kerogen-bearing, mostly brown or black shales subcropping in shallow depths and yielding oil in commercial amount upon pyrolysis are termed oil shales. Russell (1990) summarized their world-wide distribution, origin, and exploitation. Oil shales and petroleum source rocks contrast each other to a certain extent. An oil shale preserves normally an immature stage of kerogen developed prior to deep burial; it yields oil upon pyrolysis only. A petroleum source rock, on the other hand, requires considerable burial and a certain degree of catagenic alteration in order to transform kerogen into oil. Shales rich in organic matter which have been buried deeply are not prolific oil shales, even if subsequent folding and erosion has brought the shale back to the surface.

Oil shales contain large quantities of insoluble organic matter (kerogen) which yields oil upon pyrolysis at temperatures of about 500°C. The shales themselves do not contain producible oil, and normally little extractable bitumen. The shale matrix consists of clay minerals, carbonates, and quartz. The organic matter is usually composed of marine or freshwater algae, but other planktonic organisms and bacterias may contribute significantly. The kerogen of the oil shales belongs to type I or II. Shale oil contains a characteristic proportion of unsaturated olefins which are absent in natural crude oils. Sulfur and nitrogen are abundant in shale oil too.

Oil shales are normally characterized by thin lamination of alternating laminae, rich in organic matter or clay matrix. The lamination indicates quiet sedimentation and absence of benthos; influence of periodic events is inferred. In this restricted environment, the decaying organic matter consumes oxygen to produce CO_2. Thus, the organic matter is protected from being reworked by benthic fauna and from aerobic microbial degradation. Well logging permits in-situ appraisal of the oil shale potential. The density log allows prediction of oil yield, the sonic log and the natural gamma log of the kerogen content.

Environments in which oil shales have been normally deposited are rather identical with those of hydrocarbon source rocks. Duncan (1967) distinguished between oil shales deposited (1) in shallow seas, (2) in large lake basins, and (3) in small lakes, bogs, and lagoons.

Oil shales deposited in shallow seas are thin (a few meters to a few tens of meters thick) and of large extent (up to hundreds of thousands of square kilometers).

Oil shales deposited in large lake basins, particularly in those formed during orogenies, consist of marls or argillaceous limestones. Associated sediments may include volcanic tuffs and evaporite deposits. Fossil examples are Triassic beds in the Stanleyville Basin, Zaire; and the Mississippian Albert Shales, New Brunswick, Canada.

Numerous oil shale occurrences are known. In the Precambrian oil shales are rare, though shungite is probably a metamorphic equivalent of Lower Proterozoic oil shales. Lower Paleozoic oil shales were deposited in shallow-marine environments bordering the Scandinavian-East Canadian Precambrian shield (central-eastern U.S.A., Canada, Sweden, and C.I.S.). Cambrian oil shales occur in Siberia. The kukersite, found in the Ordovician of Estonia, is among the richest oil shales. It consists of thin (2.5–3 m) calcareous oil shale horizons, occurs at a shallow depth (5–100 m), and contains up to 40% organic matter.

Silurian and Devonian oil shales were deposited on wide shelf areas rimming North Africa (Libya and Algeria) and in the central-eastern U.S.A. Late Paleozoic oil shales occur widely on the Gondwana continent. The Permian Irati oil shale in southern Brazil, the world's second largest oil shale deposit, consists of two horizons (respectively 3.20 m thick with an oil yield of 9%, and 6.50 m thick with an oil yield of 6.4%), separated by 8.6 m of barren sediments. Stratigraphic equivalents are known from Uruguay and Argentina. Rich Permo-Carboniferous oil shales are known from Australia (kerosene shales, tasmanite) and South Africa (Ermelo). The tasmanite and the Ermelo shale contain mostly yellow aggregates of organic material, considered to be remains of algal colonies. Minor oil-shale deposits, commonly associated with coal, occur in tectonic or post-orogenic basins of Carboniferous or Permian age along the Hercynian belt of western Europe and of the Appalachians. Small lakes, bogs, and swampy lagoons are also suited for oil-shale deposits. They are associated with coal seams. Fossil examples are Permian oil shales in western Europe (e.g. St. Hilaire, France; Saar-Nahe Basin) and Tertiary beds in China (Fushun, Manchuria).

Triassic lacustrine oil shales with a high oil yield and associated with limestones and volcanics occur in Zaire. Triassic oil shales are known from Switzerland, Austria, and Italy. Lower to Upper Jurassic black shales (e.g. Posidonia Shale) of varying richness are widespread in western Europe. The oil content is rather low (4–6%), but the large areal distribution makes it an important potential hydrocarbon reserve.

Other Jurassic-Cretaceous oil shales are known in northern and eastern Asia, where they are associated with coal, and in Alaska, where marine deposits include tasmanite, channel coal, and a shaly coal named 'whale blubber' rock. Low-grade Cretaceous oil shales occur in southern-central Canada and western U.S.A. Late Cretaceous marine black shales with phosphorite and chert extend over several Middle East countries (Syria, Israel, Jordan).

The major Tertiary oil shales are the lacustrine Eocene Green River shales in Colorado, Utah, and Wyoming, U.S.A., which constitute the world's largest oil reserve (circa 300 billion t). Lacustrine deposits of Tertiary age also include the Paraiba Valley shales in Brazil, the Aleksinac shales in Yugoslavia, and oil shales associated with coal beds in China (Fushun, Manchuria). Small deposits occur in Tertiary basins in Europe, South America, and western U.S.A. Thus, marine black shales of Tertiary age are known from Sicily, Italy (circa 5 billion t of potential oil), California, and the southern C.I.S.

Middle Eocene (Lutetian) oil shales crop out at Messel, 9 km northeast of Darmstadt, F.R.G. (Kubanek *et al.* 1988). The bituminous shale was deposited in an area of 0.8 km^2 as a lacustrine sapropelic, partly coaly shale, up to 180 m thick. It is composed of 35% ash, 25% organic matter, and 40% water. The oil content ranges from 6 to 14% (average 8%). The oil yield amounts to a total of 28×10^6 t of oil in place contained in 35×10^6 t of oil shale. The shale was deposited in the Rhine Graben at the southern side of the crossing Saar-Nahe Basin. It is famous for its richness of well preserved fossils, which indicate a subtropical, low-oxygen environment. The fossils comprise leaves, seeds, fruits, pollen, silicified wood, freshwater snails, sponges, robbery fishes, crocodiles, turtles, frogs, insects, and even birds and mammals. Commercial utilization of the openpit mine started in 1884; it was in operation until 1962. A total of approximately 1×10^6 t of oil and 0.5×10^6 t of by-products such as electrode coke, ammonium sulfate, paraffins, phenols, and pitch was produced.

Lacustrine oil shales of Miocene age, about 40 m thick, occur in the Noerdlingen Ries impact crater (20 km in diameter), 110 km NW of Munich. The laminated oil shales, composed of smectite and illite, are characterized by a high C_{org} content (up to 22%). The oil content (4–22%) is estimated at 100×10^6 t (Lemcke 1977).

A very minor, but genetically interesting occurrence of oil shale (i.e. bituminous laminites) was reported also from a lacustrine environment in an Eocene maar in the Eifel, F.R.G. by Negendank *et al.* (1982).

At the moment, only Estonia and China (Manchuria) continue exploitation of oil shales by mining. The U.S.A. and Brazil developed pilot plants on the world's largest oil-shale deposits: the Green River shales and the Irati oil shales. With energy becoming increasingly short, other countries are again investigating their oil-shale resources.

Gas shales gain increasingly importance, especially Paleozoic shales. The best known example are the Devonian gas shales in the western U.S.A. (see Section 4.2.3). The Eastern Gas Shales Project is a multidisciplinary effort to better exploit these gas shales using existing technical expertise from U.S. government agencies, state geological surveys, universities, research institutes, the U.S. Geological Survey, national laboratories, and private industry.

3.2. Reservoir Rocks for Hydrocarbons

Hydrocarbon reservoirs are rocks that have sufficient porosity and permeability to permit the accumulation of crude oil or natural gas under adequate trapping conditions. In most instances the rock is a sediment, generally a sandstone or a carbonate rock; exceptionally, shales, pyroclastics, tuffs, or fractured igneous rocks serve as reservoir rocks.

3.2.1. PETROPHYSICAL PARAMETERS OF RESERVOIRS

The study of the pore space, i.e. the physical properties of pores and entire pore systems in solid rocks, is termed petrophysics. Measurements of petrophysical parameters in hydrocarbon reservoirs, regardless of their structure and mineral composition, are important both in search for oil and gas, and for planning subsurface liquid waste disposal and gas storage.

Pores are minute to small openings in a rock, either evenly disseminated in a sponge-like fashion as interstices bound to vugs and druses, and/or as plane-linear fractures, not occupied by the solid rock frame mostly by grains, matrix or cement. Pores may contain gases (e.g. N_2, CO_2), or gaseous hydrocarbons such as methane, and/or liquids like brine and oil. The study of pore liquids and gases and their behavior in a rock is part of petroleum engineering.

Explorationists should know origin, configuration, and filling of porous systems in order to predict actual and potential reservoir behavior.

The porosity of a rock is the ratio of its total pore space to its total volume

$$\text{porosity} = \frac{\text{volume of total pore space}}{\text{volume of rock sample}} \times 100.$$

Porosities in sedimentary rocks commonly range between 5–25%. Porosities of > 25% are regarded as excellent, 15–25% as good, 5–15% as fair, and < 5% as low (Table 21).

Important is to distinguish between total porosity of a rock and its effective porosity, i.e. the amount of interconnected pore space present in a rock. The effective porosity gives a rock the property of permeability. Permeability is the capacity of a porous rock for transmitting gas or fluids. It is a measure of the relative ease of flow under unequal pressure.

Permeability is determined conventionally based on Darcy's law using the equation:

$$Q = \frac{K \Delta A}{\mu \cdot L}$$

Q = rate of flow in cm^3/s, Δ = pressure gradient in Pa, A = cross-sectional area in cm^2, μ = fluid viscosity in Pa.s, L = length in cm, and K = permeability.

The conventional units of measurements are darcy (D) or millidarcy (mD). One darcy is the equivalent to the passage of one cubic centimeter of fluid of one centipoise viscosity flowing in one second under a pressure differential of one atmosphere through a porous medium having a cross-sectional area of 1 cm^2 and a length of 1 cm. Because most rocks

Table 21. Generalized texture classification of reservoir rocks (after Robinson 1966, reprinted with permission).

Type	Visual Characteristics		Reservoir Properties			
	Major	Minor	Pore-Size Distribution*	Porosity Permeability	Examples	Remarks
Carbonate						
I. Limestone, partly dolomitized	Dense	Few pin-point pore spaces, surface luster	$C=P$, $Pd=H$, $Sm=H$	$\phi=F$, $K=L$	Fig. 2	Poor reservoir quality
	Less dense (than above)	More pin-point pore spaces, reduced luster	$C=M$, $Pd=H\text{-}M$, $Sm=M$	$\phi=M$, $K=F$	Fig. 3	Solution can enlarge and increase pore spaces
II. Dolomite	Saccharoidal (microgranular)	Sugary, usually brownish in color	$C=G$, $Pd=M$, $Sm=L$	$\phi=E$, $K=G$	Fig. 4	Comprises better carbonate reservoirs; found in association with cal. and anhy. xls.
	Granular	Denser, visible pore spaces	$C=G$, $Pd=L$, $Sm=M\text{-}L$	$\phi=G$, $K=G\text{-}E$	Fig. 5	
III. Limestones: Bioclastic Oolitic Algal Fine matrix	Few large vugs	Dense rock framework (matrix)	$C=P$, $Pd=VL$, $Sm=H$	$\phi=L$, $K=L\text{-}E$	Fig. 6	Difficult to evaluate permeability from cuttings and small cores
	Many small visible pore spaces	Rock framework breaking up (matrix)	$C=P$, $Pd=VL$, $Sm=M\text{-}H$	$\phi=F$, $K=F\text{-}G$	Fig. 7	Smaller, more numerous pore spaces; usually higher porosity
IV. Limestone and dolomite, dense	Smooth—dense	No visible pore spaces, assoc. cal. and anhy. xls.	$C=P$, $Pd=H$, $Sm=H$	$\phi=L$, $K=L$	Fig. 8	Usually very dense; important as reservoir seal
Sandstone						
I. Slightly altered	Granular	Open surface texture appearance	$C=G$, $Pd=L$, $Sm=L$	$\phi=E$, $K=E$	Fig. 9	Slightly consolidated; clean to handle
II. Compacted	Granular (but denser than above)	Closed surface texture appearance	$C=G$, $Pd=M$, $Sm=M$	$\phi=F\text{-}G$, $K=G$	Fig. 10	Can become well consolidated and still be of fair reservoir quality
III. Pores filled	Granular	Clogged surface texture appearance	$C=M$, $Pd=M$, $Sm=M\text{-}H$	$\phi=G$, $K=F\text{-}G$	Fig. 11	Difficult to evaluate permeability; lge. amt. "fines;" spl. dirty to handle
	Granular (but dense)	Closed surface texture appearance	$C=G$, $Pd=M$, $Sm=M$	$\phi=F\text{-}G$, $K=G$		Similar in appearance to Fig. 10
IV. Highly altered	Dense to dirty	Smooth to gritty surface texture appearance	$C=P$, $Pd=H$, $Sm=H$	$\phi=L$, $K=L$	Fig. 12	Ranges from friable to highly consolidated; imp. as res. seal

* Parameters taken from Pc Curve.

C—Measure of Pore Sorting.
.1-.5 Good (G)
.6-1.0 Medium (M)
>1.0 Poor (P)

Pd—Entry Pressure (PSIA)—Indication of Largest Pore Size (microns)
<10 Very Low (VL)   >24µ
10-25 Low (L)   24-8µ
25-100 Medium (M)   8-2µ
>100 High (H)   <2µ

Sm—Unsaturated Pore Volume in Per Cent
<10 Low (L)
10-20 Medium (M)
>20 High (H)

ϕ—Total Porosity in Per Cent
< 5 Low (L)
5-15 Fair (F)
15-25 Good (G)
>25 Excellent (E)

K—Air Permeability in Millidarcys
<1 Low (L)
1-10 Fair (F)
10-100 Good (G)
>100 Excellent (E)

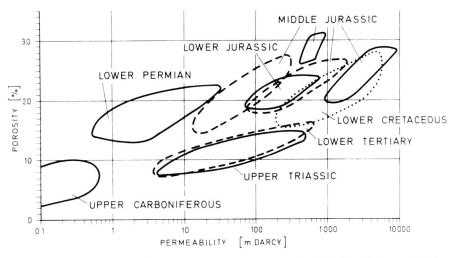

Fig. 49.   Porosity vs. permeability plot of prospective sandstones in the F.R.G. (after Gaida *et al.* 1973).

have a permeability considerably less than one darcy, the millidarcy, one thousandth of a darcy, is mostly used.

The permeability of sedimentary rocks is variable from rock to rock, even in the same rock sample, depending on the direction of measurement, i.e. whether perpendicular or parallel to the bedding plane. The vertical permeability, measured perpendicular to the bedding, is normally lower than that parallel to the bedding. The usually slightly elongate detrital grains of a sandstone are commonly aligned parallel to the current direction. This arrangement can lead to a slightly increased permeability in the direction of preferred grain elongation. Permeabilities above one darcy are exceptionally high, between 100–1 000 millidarcies excellent, 10–100 millidarcies good, 1–10 millidarcies fair, below 1 millidarcy low. Table 21 lists porosity, effective porosity, and permeability in various rock types.

Another measure defining properties of the pore space in a reservoir are the capillary pressure curves. These curves are obtained by measuring the pressure required to displace a wetting fluid (e.g. mercury) from the pores of various sizes. They normally furnish data in pore-size distribution and distribution of pore-throat sizes. For capillary pressure curves of selected reservoir carbonate rocks from northwestern Germany see Füchtbauer (1974, pp. 342–345).

The internal surface area of a porous medium, determined by absorption methods, is another petrophysical parameter. Rocks with a small internal surface area release oil easier than rocks with a large internal surface area.

Petrophysical parameters such as porosity, permeability, distribution of pore radii, and internal surface areas of selected reservoir sandstone samples in the F.R.G. were investigated (Gaida *et al.* 1973). The interrelation between geological age, porosity, and permeability corresponds closely to that between depth of burial, porosity, and permeabil-

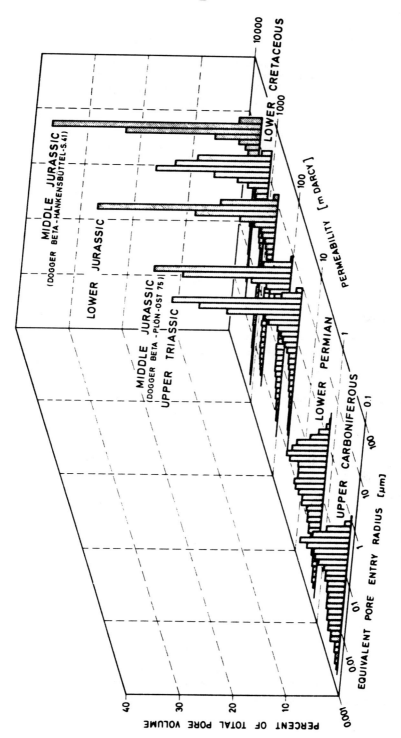

Fig. 50. Pore radii distribution in prospective sandstones of the F.R.G. (after Gaida *et al.* 1973).

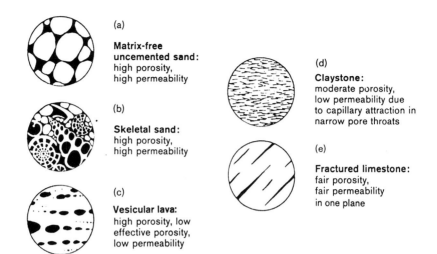

Fig. 51.   Simple classification of porosity types as seen in thin sections (compiled after Selley 1982, pp. 24 and 28).

ity. This is demonstrated best in a porosity versus permeability plot of sandstone samples from formations of different age (Figure 49). The distribution of the pore radii of the same sandstone samples as a function of permeability (in mD) likewise stresses this relationship (Figure 50).

Different methods can be used to measure porosity, permeability, directional permeability, and capillary pressure of rocks. Direct measurements of porosity or permeability require rock samples, namely small plugs taken either from hand specimens or conventional cores. In boreholes, the porosity of a rock can be approximated by indirect geophysical techniques, especially by the so-called porosity logs (e.g. sonic log, density log). However, no indirect methods are available so far to determine the permeability in rock sequences by geophysical techniques. The permeability of a given reservoir can be evaluated from drill-stem and production tests. Attempts to discern a general quantitative relationship between porosity and permeability have been futile up to now. Diagrams of laboratory data from individual formations on the other hand, show not rarely convincing linear trends when porosity and permeability values are plotted on semilogarithmic coordinates.

For details see standard texts (Harris and Hewitt 1977, Dickey 1986, Kobranova 1989, Cushmann 1990).

Since porosity and permeability are the most important parameters of any reservoir rock, petrophysical studies of reservoir rocks have to include a concise description of amount, type, and origin of porosity. Various attempts have been made to classify porosity types. Selley (1976) presented a simple classification of porosity types (Figure 51) applicable to both sandstones and carbonate rocks, Choquette and Pray (1970) a widely used scheme for carbonate rocks.

The empty, partly or completely filled pore space can be studied in various ways: (1) examination of rough or polished rock surfaces by hand lens or binocular microscope, (2) in thin sections, and (3) under the scanning electron microscope (see Photoplates 6–14). The microscopic geometry sets limits to porosity, permeability, and hydrocarbon saturation. Another technique of studying pore configuration and their three-dimensional appearance is to impregnate the porous rock with a suitable plastic resin and then to dissolve the rock with an appropriate solvent. The examination of the cast left gives hints at pore size and shape and at the kind of pore throats. Size of pore throats and tortuosity of pore systems influence the permeability of the rock.

Not all sandstones are porous and permeable. The pore space may be filled with primary clay matrix and/or diagenetic mineral cements which can block fluid flow. The primary matrix usually consists of clay-sized particles. Diagenetic mineral cements comprise in general newly formed clay minerals, silica, carbonate, and sulfate cements. For instance, in the Rotliegende sandstones of northwestern Germany minute flakes and fibers of diagenetic clay encrust sand grains, and reduce porosity and even more so permeability considerably. The permeability of reservoir sandstones depends on numerous parameters such as grain-size distribution, grain shape, sorting, pore-size distribution, and pore configuration.

Clay minerals in the form of minute flakes and fibers, the so-called fines, are easily displaced by drilling mud or during enhanced oil recovery. They may clog the pore throats.

3.2.2. SANDSTONES AND CONGLOMERATES AS RESERVOIR ROCKS

More than 60% of the reservoirs in the world's giant fields are sandstones. The following parameters define the quality of a sandstone as a hydrocarbon reservoir: sand body morphology, reservoir extent, reservoir anisotropy, mineral composition, and petrophysical properties.

Modern texts on sandstones as reservoir rocks are those by Berg (1986), Tillman and Weber (1987), and Barwis, McPherson and Studlick (1990).

Sand bodies, traceable in the subsurface of mature exploration area, have been classified in terms of their plan shapes (Pettijohn, Potter and Siever 1987): sheets, more or less equidimensional and more than about 5 km in extent; pods with length to breadth ratios of less than 3:1; and elongate bodies, having length to breadth ratios of more than 3:1. Elongate bodies are further subdivided into ribbons, dendroids (having tributaries), and belts (developed by fusion of several dendroids).

Most sandstone reservoirs extend in the lateral direction much further than in the vertical direction, thus contrasting most carbonate reservoirs. The dimensions of sand bodies show striking similarities to the wide range of shallow-marine depositional environments. Individual sand bodies as terrestrial or marine channels, bars, shoreline sands, and barriers are between 5 and 75 m thick, but the averages of each environmental group fall between 16 and 36 m; most lie between 20 and 30 m. Widths range from one to several hundred kilometers in extreme cases, the average in each group is close to 4 km. Lengths vary from less than 5 km to 300 km, the average is around 50 km. Massive sands, several hundreds of meters thick, are known to occur; however, no major hydrocarbon accumulations were found.

The initial porosity of freshly deposited sands is about 40% regardless of their sedimentary environment. However, most reservoir sandstones have porosities lower than 40%, mainly due to cementation rather than compaction. Porosity values between 20% and 30% are common in Mesozoic oil fields of northwestern Germany or of the North Sea.

Reservoir sandstones are generally water-wet. The water is held as a thin film around grains, in greater volumes in collars at grain contacts, and in pores bounded by smaller than usual throats. Minimum water saturations of producing sands are seldom less than 10%; commonly they range between 15–40%. Small amounts of clay can considerably increase the amount of pore water. Thus, shaly sands can produce water-free oil when their water saturation is as high as 65%. Small-scale packing heterogeneities are probably the chief factor responsible for the levels of irreducible water saturation in clean sands.

In deep sandstone reservoirs compressibility of the pore system needs to be taken into account.

Lateral extent and thickness of sand accumulations, their shape in plan, the internal structures, and the nature of their contact with the enclosing sediments are largely controlled by the environment of deposition. However, beware of over-prediction in environmental analysis, especially to predicting sand-body geometry.

Isotropic sandstone reservoirs are rare, as sedimentation tends to segregate different size particles into separate laminae. Grain-size lamination, presence of clay or mica partings, layer-cake mineral cementation, and/or preferred grain orientation may induce variations in vertical and even horizontal permeability. These inhomogeneities effect rate and completeness with which a reservoir can be drained. Reservoir anisotropies have to be taken into account when deciding on well spacing. In reservoirs with fining-up sands hydrocarbons tend to accumulate in the poor part of the reservoir. This fact has to be considered when assessing reservoir potential of sands from different environments.

Each reservoir sandstone is characterized by an individual assemblage of matrix and mineral cements due to conditions of primary environment and diagenetic development. Quartz arenites make good reservoirs, except when pores are filled with secondary silica or calcite. Examples are the Rhaetian quartz arenites of the Hardesse field, northwestern Germany, the clean permeable Simpson sand of the Oklahoma City field, U.S.A. The pore space of graywackes, on the other hand, is filled with clay and fine sand particles, causing rather reduced porosity and permeability. Many important producing sands, however, are graywackes, such as the Upper Carboniferous sandstones of northwestern Germany, the Bartlesville sand of Oklahoma, U.S.A., and the Bradford sand of Pennsylvania, U.S.A.

Arkoses, normally found only within a hundred kilometers of granitic outcrops as 'granite wash' are not widespread as reservoirs, e.g. they produce oil in southwestern Oklahoma.

Significant features of common reservoir sand types from selected depositional environments are compiled in Figure 52. The data refers to grain size, log profile, typical grain-size curve, dipmeter expression, plan geometry, primary porosity, and permeability isotropy. The environmental analysis has to be supplemented by observation on the diagenesis, because type and degree of mineral cementation, especially of early diagenetic origin, are often related to the depositional environment.

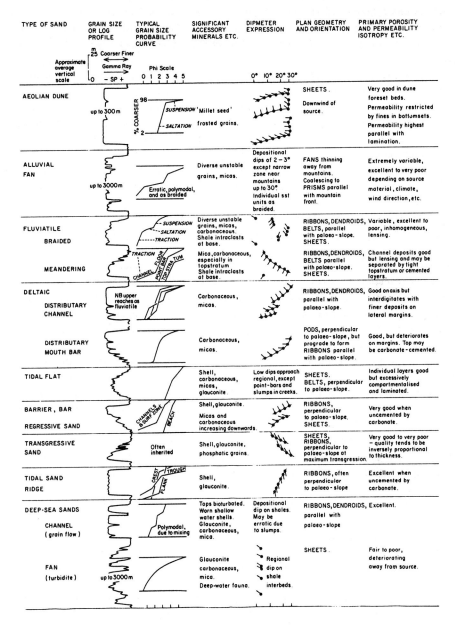

Fig. 52.   Common reservoir sand types characterized by log profiles, grain-size distributions, significant accessories, dipmeter expression, geometry and orientation relative to paleo-slope and reservoir data (after Taylor 1977, Fig. 1).

Eolian sandstones are in general very good reservoirs, provided adequate seals, favorable trapping geometry, and prolific source rocks are available. Alluvial fan deposits show a good reservoir potential, if the coarse debris consists of non-carbonate components and the fans are limited in size. Because primary porosity and permeability may be very high in sand and gravel lenses, alluvial fan deposits are particularly sensitive to compaction and burial diagenesis. Early ferruginous and carbonate cementations are common. Sandstone reservoirs, formed in alluvial channels and point bars, are usually layered by poorly permeable horizons. These horizons are either tops of individual fining-upward sequences, or carbonate-cemented horizons of early diagenetic origin in the basal part of each sequence. Lateral and vertical inhomogeneities are likely to be greater in braided than in meandering river deposits.

Fluviatile sheet sands might have gone unrecognized because of the lack of obvious channel morphology. Tidal sands in some estuaries are fairly clean and well sorted; they can provide fair hydrocarbon reservoirs. Complex inhomogeneities due to depositional processes in tidal flats, however, are adverse to efficient oil or gas drainage of a reservoir. Beaches, barriers, and bars form potent reservoir sands. No generalizations about early mineral cementation can be made yet. Calcite cement, likely to be formed in a warm and dry paleoclimate, is common. Lag concentrates of aragonitic shells may provide carbonate cement for adjacent beds.

Essential for optimum oil extraction of sandstone reservoirs are the following parameters: (1) large reservoir size, (2) high porosity and low minimum water saturation, (3) relatively high permeability, and (4) continuity and regularity of the pore system.

Search for sandstone reservoirs in a new exploration area includes three steps: (1) evaluation of the regional geology on a basin-wide scale, (2) reinterpretation and prediction of facies from reconnaissance geophysics, especially from seismic, and (3) more detailed delineation and trend prediction of the sandstone reservoir following drilling. The first step involves stratigraphic analysis, using evidence from pertinent literature, aerial photographs, outcrops, and boreholes. Well location, depositional environments, and even sandstone petrology have to be evaluated under the aspect of plate-tectonic processes. The second step essentially consists of seismic interpretation. Increasing velocity with depth observed in sandstones can be attributed to cementation and loss of porosity rather than to compression.

In general, environments of deposition can be inferred from considering the geometry of strata drawn from seismic reflections. For instance, the fore-sets of Tertiary deltas in the North Sea area can be well recognized. The third step of evaluating and further predicting sandstone reservoir trends is initiated subsequently to drilling, evaluating data from the borehole, regardless whether a potential sandstone reservoir has been proven or not. This phase includes direct measurements of petrophysical parameters on cores, indirect determinations of petrophysical data by means of well logs, direct recognition of the environment of deposition in cores, supplementary interpretation of environments using cuttings and well logs, and finally determining direction trends in the sandstone and the regional paleoslope evaluating dipmeter logs.

Sedimentological characteristics of some major sandstone reservoirs for oil and gas in the F.R.G. and northwestern Europe are briefly described.

Fine to coarse, rarely conglomeratic greywackes of Later Carboniferous age (Photoplate 6) are gas reservoirs in northwestern Germany in a depth between 2 500 and 4 500 m. They have been deposited in a coal-bearing, limnic-fluviodeltaic environment. Siderite, kaolinite, quartz, dolomite/ankerite, calcite, chlorite, and illite/sericite are the predominant diagenetic mineral cements. Hematite and sericite are detrital as well as authigenic. Platy and fibrous illite/sericite is widespread as neoformation in the pore space. This illite/sericite phenotype has a relatively large internal surface. The variety of all mineral cements reduces – as a rule – porosity and permeability considerably. Average porosities range between 5 and 15% and permeabilities between 0.1–10 mD. The primary environment of acid coal swamps notably determined extent and type of mineral cementation (diagenetic siderite-kaolinite paragenesis). Tectonic deformation, time, depth of burial and elevated temperatures negatively influenced the reservoir parameters of the sandstone.

The fine to coarse, partly conglomeratic sandstones of Rotliegende age are the major gas reservoirs in northwestern Europe. In northwestern Germany they occur in depths between 1 500 and 4 500 m. Most sandstones have been deposited as cross-bedded dune sands (Photoplate 7) and poorly bedded wadi sands in the continental realm. Diagenesis was severe; it is characterized by the neoformation of iron oxide, feldspar, illite, quartz, chlorite, kaolinite, anhydrite, and carbonate. Illite forms tangential rims, radial rims, mesh-work, and dense homogeneous masses. The primary hypersaline and alkaline environment influenced the diagenetic development from the beginning (diagenetic dolomite-silica-anhydrite paragenesis). Hydrothermal influence on mineral cementation by the underlying volcanic rocks is of minor importance. The problem in many Rotliegende reservoirs of northwestern Germany is that good porosities (up to 20%) are often associated with poor permeabilities (< 1 mD). Sandstones cemented mainly by meshy illite show good porosities, but poor permeabilities; those cemented by kaolinite have both, good porosities and good permeabilities. Different particle size and shape of illite and kaolinite are responsible for this marked contrast in poroperm relations. The sedimentological aspects of Rotliegende deposition are discussed in more detail in Section 4.2.1.

Jurassic sandstones form prolific reservoirs in the North Sea area and in northern Germany. Non-fossiliferous, well to moderately sorted, fine to medium, very rarely coarse or conglomeratic quartz sandstones exemplify Middle Jurassic sandstones occurring in northern Germany, in the Gifhorn and East Holstein Jura Troughs. The oil-bearing sandstones are found in depths between 1 500 and 3 300 m. Up to eight sandstone horizons are stacked one above the other as the source area is approached. The environment of deposition is marine to restricted marine. A gradual increase of diagenetic alterations from sandstones in outcrops into the deeper buried sandstones can be well observed. The sandstones in sandpits are friable and poorly cemented (porosity 35%, permeability 7 000 mD). Incipient overgrowths of secondary silica can be seen under the scanning electron microscope only. At a moderate depth of about 1 500 m the original porous framework of loosely cemented sand grains still exists (Photoplate 8). With augmenting depth of burial (down to 3 200 m), mineral cements such as secondary silica, carbonate minerals, and various types of clay minerals (kaolinite and chamosite) increasingly fill the pores. Quartz overgrowths increase with depth. The intensity of secondary quartz overgrowth is thought to depend mainly on the overburden pressure; thus, it indicates the maximum depth of burial before oil accumulation. Close to salt domes, detrital feldspar is albitized by hypersaline sodium-rich brines. Also authigenesis of titanium minerals is notable. Reservoir sandstones as well as enclosing shales of the Gifhorn Trough have been thoroughly studied under the aspect of hydrocarbon-migration history (Philipp *et al.* 1963). The time of oil accumulation has been determined by evaluating the grade of quartz diagenesis and by means of subsidence-time plots. Oil migration in the Gifhorn Trough was confined to a subsidence interval from 270 to 2 000 m, corresponding to a shale porosity ranging between 10 and 35%. Additional compaction did not cause further migration.

Lower Cretaceous sandstones, chiefly Valanginian in age, produce most of the oil in the F.R.G., namely the Bentheim Sandstone in the Emsland area and Valanginian sandstones in the Gifhorn Trough. They are found in rather shallow depths, between 500–2 000 m. These sandstones were not buried deeper before; thus, their diagenesis was rather weak. The Bentheim Sandstone, about 40 m thick, is the prime oil horizon in the Rühler Moor, Bramberge, and Georgsdorf fields. The clean, well to fairly sorted, fine to medium quartz sandstone (Photoplate 9) which is characterized by a variety of sedimentary structures and ichnofossils was deposited in a coastal barrier-bar system. The petrophysical data, porosity about 25% and permeability 500–10 000 mD, is favorable. The pore space is partly filled by clay minerals such as cryptocrystalline illite clay and vermicular kaolinite; secondary silica and carbonate minerals are of minor importance. Locally, calcareous sandstones and clay ironstones are intercalated.

The Valanginian sandstones of the Gifhorn Trough have been deposited on a shallow-marine shelf, including barrier-bar sands. They are mainly cemented – in the relative order of frequency – by calcite, dolomite, phosphate minerals, glauconite, pyrite, siderite, quartz, kaolinite, potassium feldspar, and chamosite. Calcite and dolomite locally reduce porosity and permeability to a large extent; other authigenic minerals contributed only slightly to the pore reduction.

3.2.3. CARBONATE ROCKS AS RESERVOIRS

Carbonate rocks from the world's giant fields produce 38% of oil and gas. The greatest oil fields in the world are found in Jurassic limestones in Saudi Arabia. As the Middle East fields will reach maximum development, about 65% of hydrocarbon production outside of the C.I.S. and China will come from carbonates rocks.

The methods exploring carbonate-petroleum reservoirs are described and illustrated by case histories in Reeckmann and Friedman (1982) and Roehl and Choquette (1985). Bathurst (1975) and Moore (1989) summarize data on carbonate-rock diagenesis.

Carbonate reservoirs are found mostly in marine sediments from various environments: tidal flat muds, shallow-water carbonates, reefs, pelagic chalks, calcareous debris flows, and turbidites. Lacustrine carbonates, however, are not appropriate reservoirs.

The following peculiarities contrast carbonate reservoirs from sandstone reservoirs. Carbonate rocks are normally more heterogeneous than sandstones, in both lateral and vertical direction. This heterogenity is caused by the low-permeability matrix on the one side, and openings of super-capillary size such as vugs, fractures, fissures, and solution channels on the other side. Thus, the development of porosity can be very erratic. Some wells in carbonate reservoirs may produce thousands of m^3 a day, while nearby wells are dry. A considerable portion of porosity can be formed by fractures, vugs, and in some cases by large caverns.

Diagenetic alteration can be even more complex in carbonte rocks than in sandstones. Early diagenetic leaching, dolomitization, and leaching at unconformities are diagenetic processes common in carbonate rocks creating permeability and porosity, particularly in carbonate buildups. Carbonate sediments show initially high porosity. The world's greatest carbonate reservoir is in lime sand with primary porosity preserved and a simple diagenetic history. Consequently, search is important for conditions and areas in which sediments have been protected from cementation, solution-compaction, and replacement.

Carbonate reservoirs have generally a smaller ultimate recovery than sandstone reservoirs. Moreover, carbonate reservoirs are characterized by regular and directional patterns of natural fracture systems and a particular way of water invasion into fractured reservoirs.

The following parameters define the quality of a carbonate rock as a hydrocarbon reservoir: carbonate-rock body morphology, reservoir extent, reservoir anisotropy, lithology, petrophysical properties, fracturing, and trapping mechanisms.

Carbonate-rock bodies mainly consist of two basic genetic types: autochthonous bodies like carbonate buildups and reefs and allochthonous bodies with indications of lateral transport. The autochthonous reef bodies can be often circumscribed by seismic profiling (see Section 2.5). The allochthonous carbonate-rock bodies largely resemble sandstone bodies. The extent of economic carbonate reservoirs fluctuates between a few hundreds of meters to tens of kilometers in reefs and within the same range in detrital carbonate reservoirs.

Appraisal of anisotropies in carbonate reservoirs is as essential for proper exploration and exploitation as it is in sandstones. There are primary anisotropies caused by specific sedimentation features of a given environment of deposition and/or secondary anisotropies induced by diagenetic processes. Anisotropies can be examined on macroscopic scale as well as in microscopic dimensions. Macroscopic anisotropies can be produced by sedimentation and/or tectonic fracturing. Many carbonate-rock reservoirs are anisotropic also in respect to permeability. Such a permeability anisotropy is commonly caused by anisotropies in fracture systems due to tectonic stress. Evaluating this anisotropy properly the spacing for primary recovery as well as for subsequent fluid-injection wells can be optimized. Injection wells are usually lined up with fracture trends. Microscopic anisotropies are observed in fabrics, leaching, and cementation patterns. Depositional carbonate fabrics deserve careful examination because the presence of primary porosity enhances diagenetic processes such as leaching, cementation, and dolomitization. In addition, carbonate particles and sedimentary structures in carbonate rocks may give clues as to type, shape, and trend of carbonate bodies which contain the reservoir. Thus, well formed multicoated ooids in grainstones generally indicate belts of tidal bars of predictable size and trend. Or mounds composed of phylloid algae are commonly oriented parallel to the shelf margin.

Carbonate-rock reservoirs are built up of aragonitic, calcitic, and/or dolomitic rock bodies. Their primary porosity, diagenetic development, and reservoir performance are strongly influenced by this particular mineral composition.

About 30% of the world's carbonate reservoirs are found in dolomite. 90% of Precambrian carbonates are dolomite and many carbonate complexes, older than Middle Paleozoic, are of dolomite rather than calcite. Dolomite rocks, essentially composed of the mineral dolomite and also called dolostone, are important as potential carbonate reservoirs for the following reasons:

Dolomites may be coarse- and even-grained and their intercrystalline porosity as well as their permeability may be more uniform and, thus, more predictable than in limestones.

Our knowledge on dolomite is based on studies of Holocene dolomite formation. Applying this knowledge to ancient examples, studies of ancient dolomite must also contain data on the stratigraphic dolomite distribution, its association with tectonic elements, unconformities, and other rock types, and a thorough petrographic examination of all diagenetic textures.

The formation time of dolomitization differs from sequence to sequence. In many instances dolomitization is early. In such cases dolomite rhombs, generally between 20 and 100 μm in size, but of equal size in each layer, form preferentially in micritic lime mud, avoiding consolidated peloids and skeletal grains. With advanced dolomitization (up to 40–50%) open pore space develops between the dolomite rhombs forming a sucrose texture. Complete dolomitization, in which original dolomite rhombs show overgrowths or fine dolomite crystals have replaced grains and matrix regardless of original permeability, is the final stage. Microstructures of shells are preserved. Porosity, however, is nil. This completely dense, commonly very fine-crystalline dolomite may be associated with anhydrite, which may destroy permeability in sucrose dolomite reservoirs, but it is susceptible to later leaching and secondary porosification. Replacement dolomites, commonly occurring beneath regional unconformities or in brecciated zones along old fractures or faults, are generally coarse-crystalline with zoned rhombs from 100 to 500 μm in size. Vein dolomite may also occur in form of large twisted white limpid crystals.

Dedolomitization is another diagenetic process associated with unconformities and leaching.

Dolomite is comparatively brittle and easily fractured compared with calcite under overburden pressure greater than 3 000 meters. This property is advantageous to deep-seated carbonate reservoirs.

Porosity, permeability, capillary pressure, fracture porosity, and whole-core analysis are petrophysical parameters of carbonate reservoirs with specific peculiarities. Porosity types in carbonate rocks are shown in Figure 53. Primary porosity is found in a wide variety of depositional settings ranging from reefs covering less than one km^2 to carbonate banks that often extend over thousands of km^2. With decreasing energy in the depositional environment, average grain size, pore size, and permeability decrease. Subsequent diagenetic processes normally modify the original pore size and distribution to more complex pore systems. Leaching generally affects reservoirs favorably because it improves porosity and enhances permeability.

Dolomitization can play a dual role: it can improve a reservoir by increasing the pore size or it can destroy porosity by advanced dolomitization creating a dense, interlocking crystal fabric.

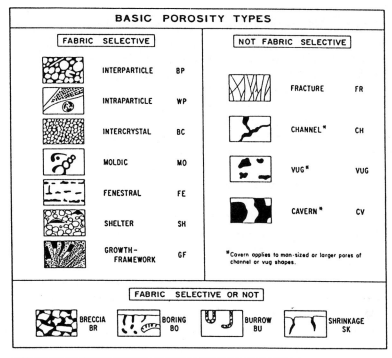

Fig. 53.   Geological classification of pores and pore systems in carbonate rocks (after Choquette and Pray 1970, Fig. 2, p. 224, reprinted with permission).

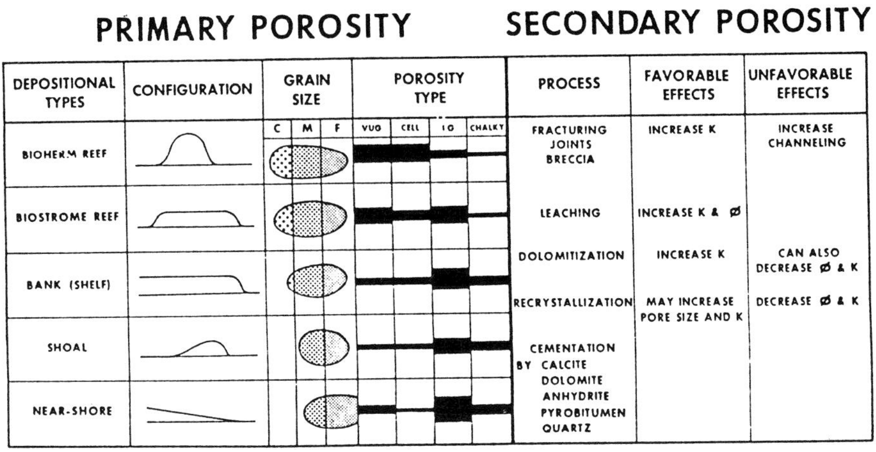

PRIMARY POROSITY   SECONDARY POROSITY

DEPOSITIONAL TYPES	CONFIGURATION	GRAIN SIZE			POROSITY TYPE				PROCESS	FAVORABLE EFFECTS	UNFAVORABLE EFFECTS
		C	M	F	VUG	CELL	IO	CHALKY			
BIOHERM REEF									FRACTURING JOINTS BRECCIA	INCREASE K	INCREASE CHANNELING
BIOSTROME REEF									LEACHING	INCREASE K & ∅	
BANK (SHELF)									DOLOMITIZATION	INCREASE K	CAN ALSO DECREASE ∅ & K
									RECRYSTALLIZATION	MAY INCREASE PORE SIZE AND K	DECREASE ∅ & K
SHOAL									CEMENTATION BY CALCITE DOLOMITE ANHYDRITE PYROBITUMEN QUARTZ		
NEAR-SHORE											

Fig. 54.   Primary porosity and secondary porosity in carbonate rocks (after Jardine *et al.* 1977, p. 874).

Also recrystallization or micrite enlargement can be reservoir-enhancing processes producing larger crystals and chalky textures. Carbonates with chalky texture can have high porosity as seen on well logs, but very low permeability. Major processes of porosity alteration leading to secondary porosity are shown in Figure 54.

Mineral cementing, on the other hand, destroys porosity in carbonate rocks. Even small amounts of certain cements such as pyrobitumen can severely reduce porosity and permeability by clogging the narrow pore throats.

Porosity and permeability in carbonate reservoirs may also reflect petrophysical peculiarities of the primary environment of deposition. Biohermal reefs are characterized by limited size, a positive relief above the sea-floor, steep flanks, and abundance of skeletal debris. A lagoonal facies of laminated carbonate mud is usually present in the center. Porosity and permeability are highest in the coarse rim sediments rich in skeletal debris. The center is less porous; it shows poor vertical permeability due to fine lamination.

Sheet-like deposits such as biostromal reefs can extend over hundreds of km². Skeletal and coarse debris is normally abundant. Porosity is highest near the seaward edge but some porosity may be preserved throughout the deposit. Horizontal permeability is normally higher than vertical permeability due to marked stratification.

Carbonate banks are tabular deposits of considerable thickness and wide extent formed by carbonate buildups. Because bank carbonates are normally composed of lime mud; low porosity and permeability predominate. Good porosities are confined to a narrow seaward rim along the margin or to patches within the bank.

Shoal carbonates are mound-like deposits of abundant skeletal debris or of other coarse detritus. They are deposited in varying water depths and in various geographic positions such as reefs, banks, and more basinal areas. The porosity is mainly intergranular.

Nearshore carbonate deposits include beach, tidal flat, and supratidal sediments which are subject to minor sea-level fluctuations causing cyclic sequences of evaporites and fine carbonates. These sequences are thin, well stratified, and restricted in areal extent. Porosity can be high in some beds, but permeability is generally very low (less than 1 mD).

Fracturing plays an important role in carbonate reservoirs. It can create permeability in carbonate rocks where none existed before and form additional pathways for leaching or cementing solutions. The permeability of fractures is very high. It increases as the square of the fracture width; e.g. a fracture only 0.1 mm wide has a permeability of 833 darcies.

On the contrary, the permeability of a limestone matrix may be 0.001 darcies (1 mD) or less. Fractured carbonate reservoirs are characterized by high initial production rates. Huge fields such as Kirkuk (Iraq) and Gach Saran (Iran) show daily production rates above 3.000 m^3 per well. Fractures are interconnected for kilometers and pressure changes are transmitted fast throughout the reservoir. Marked discrepancies between conventional analyses of small core plugs and pressure-buildup analysis are commonly indicative of fractures present. Fractures are essential for oil production from carbonate rocks with low matrix permeability (e.g. Upper Cretaceous chalk in the North Sea such as Ekofisk, or some Middle East carbonate reservoirs).

Fracture systems in carbonate rocks are recognized by comparing neutron-derived porosities with core plug porosities and in the borehole by the microlaterolog and proximity log, the micro-seismogram fracture-finder log, borehole televiewer log, and/or Formation MicroScanner.[1] Surface traces of fracture systems may be detected and interpreted by means of aerial photography, airborne radar imagery, and thermal infrared imagery.

Fractures, appreciated in primary recovery, may cause problems during secondary recovery. Fractures remain to be pathways for the driving fluids to bypass large volumes of oil retained in the carbonate matrix of the reservoir. Thus, the presence of fracture systems may lead to failures of conventional water injection projects. Conventional water flood prediction methods normally give too optimistic predictions for carbonate reservoirs with fractures and solution channels.

There are various trapping mechanisms in carbonate rocks. Stratigraphic traps are common; they are caused by the primary limitation of coarse carbonate debris accumulations such as in reefs or oolite barriers. Thus, the stratigraphic framework surrounding carbonate reservoirs must be carefully examined. Some carbonate shelf reservoirs are due to updip pinch-out of the porous facies. Also, the regional extent of unconformities and indications of unconformities in cores and samples must be thoroughly studied when searching for carbonate reservoirs. Evaporites may serve as updip and overlying seals. Leaching of secondary anhydrite may occasionally lead to the formation of porous and permeable collapse breccias.

Carbonate reservoirs are the result of a favorable combination of depositional features, diagenetic history, specific rock associations, and proper timing of tectonic events. A few reservoir types account for the majority of the oil and gas found in carbonates. Wilson (1980) classified carbonate reservoirs into six major groups:

1. Grainstone reservoirs with preserved primary porosity;
2. Organic carbonate buildup reservoirs at shelf margins;

---

[1] Mark of Schlumberger

3. Downslope carbonate debris reservoirs with preserved void space and some enhanced secondary porosity;
4. Stratigraphic trap in reservoirs of shelf cycles;
5. Reservoirs with porosity-permeability, developed beneath a regional unconformity through leaching and dolomitization;
6. Chalk-textured reservoirs.

These reservoir types do not precisely follow the facies classification of Figure 55. Several of the oil fields and oil provinces mentioned below as examples include more than one of the six reservoir types.

Grainstone reservoirs with preserved primary porosity normally form widespread pro-grading sheets, elongate bars, or barchan-shaped dunes of lime sand. Ooids and coated bioclastic grains are common. Initial porosities range between 40 and 75%. Dolomitization is uncommon in grainstones, but occurs in interbedded wackestones. This reservoir type is common in the Tertiary and Mesozoic of North Africa and the Middle East (e.g. Jurassic Arab C-D Zones, Paleocene Zelten field, Libya).

Organic carbonate buildup reservoirs at shelf margins vary in size from long and thick composite buildups at platform margins to individual loaf-shaped mounds of low relief. Such accumulations contain a reef framework, debris piles, flank beds, and trapped finer sediments. Sessile benthonic colonial animals and plants are abundant. The original porosity is high. Additional porosity and permeability is caused by decaying of and boring in reefs. Original porosity is easy subject to diagenesis by meteoric vadose and phreatic waters. Low-Mg calcite cement may be precipitated in phreatic zones in the reefs at the water table. Faulting, folding or compaction induces fracturing to the massive reef. Dolomitization can enhance permeability of the matrix. Giant reef fields like Kirkuk (Iraq), the fields in the Golden Lane trend (Mexico), and those in the Leduc and Swan Hill trends (Canada) belong to this reservoir type.

Downslope carbonate debris reservoirs with preserved void space and some enhanced secondary porosity occur as irregular patches with linear trends beyond the shelf margin. Outcrops of this reservoir type have been studied in the Permian of Texas, Devonian of Alberta, Cretaceous of Mexico, and Triassic of northern Italy. Recent examples are the Belize reef margin and the Bahama Banks. Sedimentary processes involved are mass flows triggered by gravity and slumping as well as turbidity flows. The sediments are highly fossiliferous. Angular lithoclasts of slope sediments and blocks of cemented reef boundstone may be admixed. Coarse particle size is responsible for the high initial porosity. The Poza Rica reservoirs below the Golden Lane (Mexico) (see Section 4.2.5) and the fore-reef shoals in the Oligo-Miocene Main limestone of Kirkuk (Iraq) are examples of this reservoir type.

Stratigraphic traps in reservoirs of shelf cycles are caused by the wide spectra of carbonate facies across the commonly cyclic and progradational shelves. Carbonate shelf deposits grade into tidal flat sediments and, in arid climates, into supratidal sabkha evaporites, which form the updip seal. Peloidal and oolitic lime sands are subject to splash-zone cementation. Bioclastic lime wackestones are usually dolomitized and contain high solution porosity. Intertidal algal laminites and swamp deposits, also usually dolomitized, are

DIAGRAMMATIC CROSS SECTION — labels: Oxygenation level; Storm wave base; Normal wave base; Increasing salinity; 37–45‰; >45‰

	1	2	3	4	5	6	7	8	9
FACIES AND GENERAL ENVIRONMENT	Basin (euxinic or evaporitic) (a) Fine clastics (b) Carbonates (c) Evaporites	Open marine neritic (a) Carbonates (b) Shale	Toe of slope carbonates	Foreslope (a) Bedded fine grained sediments with slumps (b) Foreset debris and lime sands (c) Lime mud masses	Organic build up ('reef') (a) Boundstone (b) Encrusting masses (c) Bafflestone	Sands on edge of platform (a) Shoal lime sands (b) Islands with dune sands	Open platform (a) Lime sand bodies (b) Wackestone mudstone areas, biotherms (c) Areas of terrigenous clastics	Restricted platform (a) Bioclastic wackestone, lagoons and bays (b) Litho-bioclastic sand in tidal channels (c) Lime mud on tidal flats (d) Fine grained terrigenous clastic interbeds	Platform evaporites (a) Nodular anhydrite and dolomite on salt flats (b) Laminated evaporites in desiccated ponds
LITHOLOGY	Dark shale or silt, then limestones (starved basin). Evaporites fill basin if desiccation occurs.	Very fossiliferous limestone with marl interbeds	Fine grained limestone, locally cherty	Variable depending upon water turbulence upslope, sedimentary breccias and lime sands	Massive limestone, dolomite	Calcarenitic, oolitic lime sand or dolomite	Variable carbonates and terrigenous clastics	Often dolomite and dolomitic limestone.	Irregularly laminated dolomite and anhydrite locally may grade into red beds
COLOUR	Dark brown, black and red	Grey, green, red, brown	Dark to light	Dark to light	Light	Light	Dark to light	Light	Red, yellow, brown
GRAIN TYPE AND DEPOSITIONAL TEXTURE	Lime mudstones, fine calcilutites.	Bioclastic and whole fossil wackestones, some calcilutites.	Dominantly lime mudstone with some calcilutites	Limesilt and bioclastic wackestone, packstone, lithoclasts.	Boundstones and pockets of grainstone-packstones	Grainstones, well sorted rounded	Variable textures in grainstone and mudstone. Bioturbation	Clotted pelleted mudstone and grainstone, laminated mudstone, coarser wacke stones in channels.	Anhydrite after gypsum 'nodular rosettes,' 'chicken wire,' and blades, irregular lamination, caliche
BEDDING AND SEDIMENTARY STRUCTURES	Very even lamination on mm scale. Rhythmic bedding, occasional ripple cross lamination	Bioturbated, then to medium bedded with nodular layers	Minor lamination. Often massive beds, lenses of graded sediment. Lithoclasts and exotic blocks.	Slumps, foreset bedding, slope build ups, exotic blocks	Massive organic structure or open framework with roofed cavities. Injection dykes. Sometimes stromatactis.	Medium to large scale cross bedding	Intense bioturbation	Birdseye, stromatolites, fine laminations, dolomite crusts. Cross-bedded sand in channels	Aeolianites and terrigenous interbeds may be important
TERRIGENOUS CLASTIC COMPONENT	Quartz silt and shale, fine grained siltstone, often cherty	Quartz silt and shale in well segregated beds	Some shales, silt and fine grained sandstone.	Some shales and silts	None	Local quartz sand	Terrigenous and calcareous beds, well segregated	Interbedded terrigenous and calcareous beds possible.	
BIOTA	Planktonic and nektonic only. Occasional mass mortality deposits.	Diverse. Shelly fauna and trace fossils represent both infauna and epifauna	Bioclastic debris derived mostly from upslope	Colonies of whole fossil organisms and bioclastic debris	Major frame building colonies and communities associated with them	Few indigenous organisms. Specialised community. Mostly abraded shell debris from other platform environments	Fauna dominated by more tolerant groups (e.g. bivalves, gastropods, sponges, forams, some algae); less tolerant groups (e.g. cephalopods, brachiopods, echinoderms) often segregated	Limited fauna. Mostly grazing gastropods, algae and some forams (e.g. miliolids) and ostracodes.	Stromatolitic algae almost the only indigenous biota

Fig. 55. Standard facies spectrum in carbonate rocks recommended by Wilson (1975) (after Reading 1986, p. 302).

reservoir rocks of minor quality. Dolomitic reservoirs of this reservoir type, occurring in a supratidal environment, are known from the Permian in Texas and from the Ordovician-Silurian of the Williston Basin, U.S.A. Presumably 5% of giant oilfield production falls within this reservoir type.

Reservoirs with porosity-permeability, developed beneath a regional unconformity through leaching and dolomitization, are specific cases. Porosity is developed on irregular surfaces of subaerially exposed unconformities. The hydrocarbon distribution in the reservoir is controlled by facies variations and by karstic relief. Porosity and permeability are mainly caused by leaching and dolomitization, to a lesser extent by collapse-brecciation and dedolomitization. Mosaic calcite cement occurs below the water table. Coarse-crystalline dolomitization beneath regional unconformities is common. This reservoir type is well known from examples in the U.S.A., e.g. top surface Arbuckle, top Hunton, and top Mississippian in the Andarko Basin. Production underneath unconformities also includes the Silurian of the Michigan Basin, the Devonian of the Williston Basin, and the Devonian of the Zama-Rainbow trend in Alberta. In the giant Cretaceous fields of the Golden Lane trend (Mexico) leaching and karstification are widespread on top of the rudist reefs. The Cretaceous of Oman produces from muddy limestones leached beneath an unconformity.

Chalk-textured reservoirs consist of micritic carbonates with pores of microscopic scale only. The matrix consists of low-Mg calcite coccoliths and their fragments, 2–20 μm in size. Chalky-textured carbonates from subtidal to tidal flat environments are widespread in Cretaceous and Tertiary strata of North Africa and the Middle East. Reservoirs of pelagic chalks form vast sheets, including shallow-water and deep-water environments. The initial porosity of a micritic chalk is very high. Porosity remains high (20–35%), even after burial and diagenesis. Permeability is low due to the minute particle size. Shallow burial appears to preserve the chalky porosity. Dolomitization is rare in chalk reservoirs. Fracturing induced by deep burial might be important for improving the permeability of the chalk. Orientation of micrite grains may cause directional permeability.

The Late Cretaceous Chalk of the North Sea area is representative of this reservoir type. Other chalk-textured reservoirs of Mesozoic and Tertiary age occur in the Persian-Arabian Gulf in Sirte Basin of Libya, e.g. in the Lower Cretaceous Thamama, in the Middle Cretaceous Wasia, and in the Austin Chalk of Texas. But only 5% of the giant oil fields' production comes from chalk-textured carbonates.

Exploration for carbonate reservoirs requires thorough examination of all geological data available. The complexities of carbonate reservoirs require a multidisciplinary approach, with strong emphasis on geology and sedimentology. Exploitation should be likewise planned carefully since carbonate reservoirs are normally characterized by erratic porosity and permeability distribution ranging from massive, vuggy, and fractured reservoirs of an organic-reef facies to laminated, vertically discontinuous reservoirs of a back-reef and shoal facies, often within a single oil field. Depletion plans and preparation of enhanced recovery operations require integrated studies by geologists and engineers. For instance, the geological description of carbonate reservoirs such as the subdivision of a reef complex into different zones like reef, fore-reef, and back-reef is essential to define flow patterns in both the aquifer and the oil zones.

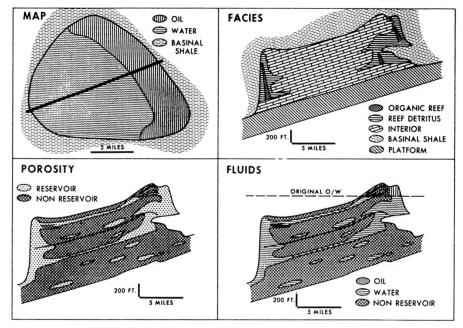

Fig. 56. Map, facies, porosity, and fluid distribution in the Redwater Oil Field, central Alberta (after Jardine *et al.* 1977).

Especially in fractured reservoirs fluid behavior, total oil in place, and recoverable oil are difficult to predict. Fluid flow is enhanced by conductive stylolites, solution channels, fissures, and hairline fractures. Whole-core analysis assists to better evaluate fracture, vug, or solution porosity and permeability. Petroleum engineers show special interest in the total permeability of a fracture-matrix system.

Trace element and stable isotope studies are supplementary means to reconstruct past water movements and to distinguish conditions of cementation (cement stratigraphy). Reservoir studies comprising lithofacies descriptions, cross-sections for correlation, and evaluation of fluid-saturation studies are used to develop reservoir models. Continuous monitoring of operation performance is indispensable to test and upgrade the geological model.

Each carbonate reservoir has to be considered as an individual case.

The Redwater Field of central Alberta (Figure 56), the second largest field in Canada, serves as an example. The oil field produces from a large limestone reef resting on a carbonate platform, but it is entirely enclosed by shale. The reef covers 518 km^2 and is 250 m thick. In cross-section, it resembles an atoll with an exterior rim of organic reef debris and a lagoonal to supratidal algal limestone facies in the center. The best reservoir is in the organic facies of the rim; the porosity was enhanced by leaching. The laminites of the center show low and discontinuous porosity. Oil is present in both the rim and the center. Only about 5 percent of the gross volume of the reef contains oil.

The major carbonate reservoirs are found in the Near East and in Mexico; they are described in detail in Sections 4.2.2 and 4.2.5 as case histories. But also a broad spectrum of potential or actual carbonate-rock reservoirs of different origin is known in Europe including the F.R.G.: Middle/Upper Devonian reef limestones, Zechstein biostromes and reefs, Upper Jurassic carbonates, the Upper Cretaceous chalk, and the Lower Tertiary *Lithothamnion* Limestone. Areal trends and developments of diagenesis are quite different from those in sandstones. The diagenesis is isochemical in most instances.

Middle/Upper Devonian reef carbonates occur in the Harz Montains, in the Rhenish Schiefergebirge, and in the North Sea along a NE–SW striking belt. Their potential as oil and gas reservoir in the subsurface of northwestern Germany was discussed by Krebs (1969, 1975). The reef limestones had been favorable reservoir rocks in respect to primary porosity and permeability. However, severe diagenetic alteration and mineralization tightened the known limestones completely (Schneider 1977). The occurrence of impsonite, a black solid bitumen, gives evidence of previous hydrocarbon presence in the reef limestones (Photoplate 11). Only early oil migration combined with continuous low-grade coalification would have preserved such a reservoir. Secondary porosity due to thorough dolomizitation or to secondary solution did not develop during the subsequent geological history. The geotectonic setting of the European Devonian reefs in the mobile Hercynian belt is the main difference to the cratonic Devonian reefs of Canada.

The Upper Permian biostromes in the subsurface of northwestern Germany form an east-west trending belt, 180 km long and 30 km wide and about 90 m thick, especially between the Weser and Ems rivers. Except for the minor occurrence of oil in Volkenroda, Thuringia, and in Heide, Schleswig-Holstein, only gas, derived from coalification of Upper Carboniferous coal, is found in the Zechstein carbonates. Gas pools are known to occur in the dolomites of the second and third cycle (Ca 2 = Stassfurt dolomite and Ca 3 = Leine dolomite). The interrelation between paleogeography, lithology, diagenesis, porosity, gas saturation, and structural history of these carbonate rocks has been demonstrated by Füchtbauer (1964a).

The biostromes were deposited in shallow water (water depth less than 20 m) on the rim of a carbonate platform. Dolomites and anhydrites prevail in the landward lagoons; dolomites and limestones with biogenic detritus basinward. Oncolites are the dominant constituents of the biostromes.

Dolomitization was the main diagenetic process. The majority of dolomites is early diagenetic. Dolomitization started in the soft sediment under the influence of connate water which had the same composition as the sea water above. Epigenetic dolomites were formed in consolidated limestones. Secondary silica and anhydrite precipitation reduced porosity only locally.

The following types of porosity are developed: interparticle porosity (between algae, composite algal pellets, or calcite nodules), vuggy porosity (excavated algae), and intercrystalline pores of different size (Photoplate 12). The porosity of the Ca 2 averages between 2 and 20%. The average permeability of samples with prevailing interparticle and vuggy porosity is between 0.1 and 10 mD, in samples with larger intercrystalline pores (sucrose dolomites) permeabilities range from 20 to 200 mD. Locally, permeability is improved considerably by fracturing.

Thin-bedded Upper Jurassic limestones and dolomites in the Lower Saxony Basin are carbonate reservoirs within an arid shoreline and evaporite environment. The fossiliferous limestones were deposited in a hypersaline epeiric sea as basal layers of major evaporitic cycles. Bioclastic and oolitic calcarenites and micrites were formed in shallow water; shales and sulfates in the adjacent basins. Early diagenetic dolomite replaced more than 50% of the limestone. Dolomitization increased as thickness decreased and clay content increased creating biomoldic porosity (Photoplate 13). Dolomites are the main reservoir rocks. Large-scale dedolomitization is related to weathering prior to a transgression. Intensive, late-diagenetic cementation, mostly by calcite, took place prior and subsequent to oil migration.

The Late Cretaceous chalk produces oil and gas in the fields of the greater Ekofisk complex of the central North Sea area (Van den Bark and Thomas 1980). Its sedimentology was described in detail by Feazel and Farrell (1988) and Moore (1989). The chalk was deposited in a pelagic basin throughout northwestern Europe at the end of Cretaceous times. The very pure, micritic limestones are mainly composed of coccoliths and coccolithic debris which are smaller than 10 μm in size. This small primary particle size causes excellent porosities (up to 40%), relatively low permeabilities (up to 20 mD), and large internal surfaces. The average pore size ranges between 1 and 5 μm (maximum 25 μm). The pore throats are extremely small and prevent easy oil migration. Diagenetic alteration only consists of minor mechanical compaction and subordinate mineralization by secondary calcite, dolomite, and silica.

The presence of low-Mg calcite and the high-Mg content of pore water prevented intense cementation. Dolomitization is absent. In northwestern Schleswig-Holstein the Heide Oil Chalk (Maastrichtian), a peculiar carbonate rock of the same type, forms a rather unique oil reservoir close to the surface. The oil, however, cannot be extracted economically, because of the peculiar minute pores. The resources are large, the recoverable reserves low. From 1919 to 1945 the oil chalk was mined. After numerous futile attempts during the last 50 years the problem still remains how to extract the considerable reserves in the Heide Oil Chalk economically. Under different petrophysical circumstances, i.e. higher brittleness and fracturing, this rock type produces considerable oil in Ekofisk.

As compared with size, number, and extent of the Zechstein and Upper Cretaceous carbonate reservoirs in northwestern Europe the Lower Tertiary (Eocene) *Lithothamnion* Limestone gas reservoirs in the Molasse Basin are relatively small. The natural gas reservoirs – locally up to 100 m thick and found in a depth range of 1 400–4 000 m – are increasingly used for underground storage of gas. This limestone is mainly composed of the encrusting or nodular, red calcareous alga *Lithothamnion* (Photoplate 14). It was deposited on a carbonate platform and grades laterally and vertically into the Priabonian Sandstone. The poorly sorted, generally pure biogenic limestone is characterized by fluctuations in porosity and permeability which are due to primary differences in rock composition. The porosity is mainly intergranular; it reaches up to 17%. Permeabilities range between 0.1 and 10 mD. Dolomitization amounts to 10%. Dolomitized portions of the limestone contain commonly more clay. Primary kaolinite content, secondary dolomitization, and dissolution along stylolites favor preservation of the moderate porosity (Blind 1964). Such an erratic distribution of porosity and permeability is characteristic of many carbonate reservoirs.

3.2.4. OTHER RESERVOIR ROCKS

Unconventional hydrocarbon-reservoir rocks include fractured shales, fractured cherts, pyroclastic deposits, and fractured rocks of the crystalline basement.

The best example of fractured shales which act as both source and reservoir rocks are the Devonian gas shales of the eastern U.S.A. (see Section 4.2.3). Another example of fractured shales are the famous Bazhenov shales in western Siberia (see Section 4.2.6). Fractured shales produce commercial oil also in Colorado and California, U.S.A.

Fractured cherts of Franciscan age (Jurassic) are reservoir rocks in the Santa Maria district, California. Other Californian fields have produced oil from the fractured Monterey cherts (Miocene). Best production was found in wells of faulted or intensely folded areas. Oil reserves of a fractured reservoir, however, are difficult to appraise.

Coarse pyroclastic deposits derived from Cenozoic volcanic rocks are potential hydrocarbon reservoirs in Japan. In Tertiary oil and gas fields of Slovakia tuffaceous reservoirs are observed, too. Economic gas production is also found in an Early Tertiary tuffite of the De Wijk gas field, the Netherlands (Gdula 1983). Of course, pyroclastic reservoirs are subject to exceptionally strong diagenetic alterations (e.g. zeolitization and silicification). Other occurrences of volcaniclastics as potential hydrocarbon reservoirs have been compiled by Seemann and Scherer (1984).

Known occurrences of hydrocarbons in crystalline basement rocks are numerous (Chung-Hsiang P'an 1982), though the number of commercial fields (U.S.A., Mexico, and Cuba) is small. Fractured crystalline basements produce oil commercially in Kansas (e.g. fractured pre-Cambrian quartzite and granite) and in California. Part of the production in the El Segundo field, California, comes from fractured schists. In Texas, volcanic rocks contain oil reservoirs. The porosity is apparently in part primary, due to the vesicular character of volcanic flows, and in part secondary, due to the alteration of volcanic rock material. In Mexico, oil comes from porous zones in gabbro and metamorphozed shales. Cuban fields produce from fractured serpentinite.

3.3. Cap Rock (Seals)

Any oil- or gas-bearing reservoir rock or any underground-storage oil or gas cavern is overlain by an impervious stratum, commonly called cap rock or seal. The cap rock prevents further upward migration of oil or gas. Comparatively impervious sedimentary rocks can be of different lithology; they occur in most stratigraphic sections.

A common belief is that a cap rock should be impermeable both to oil and natural gas as well as to water. This belief is not entirely correct. The only true water-impermeable rocks in sedimentary basins are evaporites and rocks under permafrost conditions. The thickness of permafrost, originally developed in Pleistocene glacial climates, can be high, up to 1 500 m in the Arctic zone of the C.I.S. Permafrost is the ideal seal for stopping the dispersion of gas accumulated at shallow depth in the giant gas fields in the northern West Siberian platform. The Urengoy gas field, the largest in the world, has producing horizons at a depth of only 1 035 to 1 200 m.

Shales are the most common cap rocks. The majority of sandstone reservoirs and many limestone reservoirs are capped by shales. Even moderately compacted, they are normally impermeable to gas and oil. It is necessary to reach the displacement pressure to force the droplets of oil or bubbles of gas through the pores of the shale. Generally, a cap rock holds oil better than gas. Thick, smectite-rich, partially compacted shale sections act as very low-permeability rocks and generate overpressured zones. The giant Lacq field, France, and the Malossa field in the Po Valley, Italy, are examples of overpressured gas fields with hydrocarbons trapped below a thick shale cap rock. In both fields, the gas is in a single-phase state.

Other cap rocks are dense limestones and tight sandstones. Sandstones may be impervious because of thorough mineral cementation and/or because of the presence of a fine clay or silt matrix. Less common barriers to upward migration of hydrocarbons include evaporites such as rock salt, anhydrite, or gypsum. Rare are sealing faults or sealing igneous rocks.

In northwestern Europe effective cap rocks for the Paleozoic reservoirs (Upper Carboniferous and Rotliegende) are Zechstein evaporites such as rock salt and anhydrites (Photoplate 15, Figures 1 and 2) similar to the cap rocks in the Panhandle area, Oklahoma, U.S.A. The Mesozoic and Cenozoic reservoirs (Triassic, Jurassic, Cretaceous, and Tertiary) in the F.R.G. are capped mainly by tight shales (Photoplate 15, Figures 3 and 4).

The presence of a suitable source rock, a porous reservoir rock, and a tight cap rock is essential for the accumulation of oil and gas. Tectonic activity with folding and faulting contributes to the formation of the majority of traps. The oil and gas prospectiveness, classified according to various play types, of an oil province such as the Tertiary of Indonesia is depicted in Figure 57.

3.4. Diagenesis

Diagenesis are the chemical and physical changes that sediments undergo during and after their accumulation, but before consolidation (Gümbel 1868). Since Gümbel many more names, synonyms or subdivisions, have been introduced for diagenesis, such as authigenesis, metagenesis, catagenesis, epigenesis, etc. Halmyrolysis, also called submarine weathering, refers to the geochemical reaction of sea water and sediments immediately after deposition.

Diagenetic pathways depend on the initial sediment composition and grain size, depositional environment, temperature and pressure conditions during progressive burial, and depth of burial. Diagenetic processes modify porosity and permeability largely. In general, constructive diagenesis lowers porosity and permeability of sandstones or carbonate rocks, destructive diagenesis creates new porosity and permeability.

Diagenetic studies become increasingly important to oil and gas exploration. Wilson (1977) even proposed a third category of hydrocarbon traps besides 'structural' and 'stratigraphic' traps, namely the 'diagenetic' trap. The ultimate goal of diagenetic studies in petroleum exploration is to predict type and magnitude of diagenetic alterations to be found in subsurface target areas prior to leasing and/or drilling. Thus, diagenetic studies

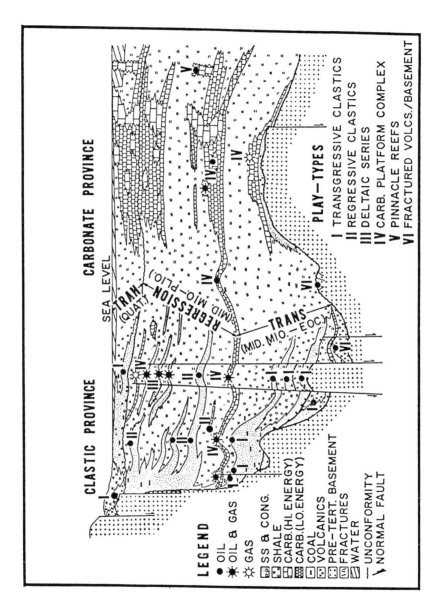

Fig. 57.   Different play-types of a hydrocarbon province as shown in a diagrammatic cross-section of the Tertiary of Indonesia (after Soeparjadi *et al.* 1975).

will enhance better predictions of new plays.

Sandstone diagenesis is referred to in detail by Pettijohn, Potter and Siever (1987) and carbonate rock diagenesis by Bathurst (1975). Chemistry of diagenetic processes is discussed by Berner (1980) and Krauskopf (1985). Krumbein and Garrels (1952), Garrels (1960), and Garrels and Christ (1965) are classic references on chemical diagenesis, stressing the use of Eh-pH diagrams.

3.4.1. DIAGENETIC PARAMETERS AND PROCESSES

All sedimentary rocks are more or less subject to diagenesis. In the frame of petroleum sedimentology one focuses on the diagenesis of source rocks, claystones and shales, and of reservoir rocks, namely sandstones and carbonate rocks. Included is also the diagenesis of organic substances, oils, brines, and pore waters.

Porosity and permeability are the most important parameters in oil and gas exploration and production which are influenced by type and intensity of diagenesis.

Porosity reduction penecontemporaneous with deposition is caused by bioturbation, soil-forming processes (pedogenesis), synsedimentary deformation, and slumping.

Compaction is the mechanical process responsible for the loss of porosity and per-meability in sedimentary rocks after deposition. During compaction grains move closer together by rotation and slippage of grains, by flexible, ductile, and brittle grain defor-mation, and by pressure solution. The porosity of sand can be reduced from 40% initial porosity to 28% by sole grain rearrangement.

Beside the compactional effects porosities of shales, sandstones, and carbonate rocks decrease normally with depth of burial. Thus, porosities of sands range from an initial porosity of 45% to practically zero for compacted and cemented sandstones. However, even in tight sandstones an irreducible porosity of approximately 2% remains. The porosity-depth interrelation, an exploration parameter of utmost significance, differs from borehole to borehole and basin to basin depending upon sandstone composition, geother-mal gradient, geotectonic history, and local factors (Figure 58).

Permeability or hydraulic conductivity measures the capacity of a porous rock to transmit gas or a fluid. Holocene sands from various environments show permeabilities ranging from 0.004 to 500 mD. Permeability is roughly proportional to porosity and inversely proportional to the square of the specific surface.

Primary factors which influence considerably the course of diagenesis in sandstones and carbonate rocks are the initial mineral composition of the detritus and the initial com-position of pore solutions. The mineral composition of detrital grains, i.e. the mechanical and chemical environment of the detrital mineral assemblage, may cause differences in compaction. A rigid sandstone rich in quartz and chert clasts shows less compaction than a soft sandstone rich in shale clasts. The physical chemistry of the environment of deposi-tion, especially the chemical environment of pore solutions, may cause cryptocrystalline ankerite, dolomite, or siderite to be formed soon after deposition, especially in siltstones and sandstones of swamp and organic-rich lagoonal environments.

Main motor of diagenesis are the changing chemical and physical conditions, mainly associated with the increase in temperature and pressure and/or with augmenting depth of

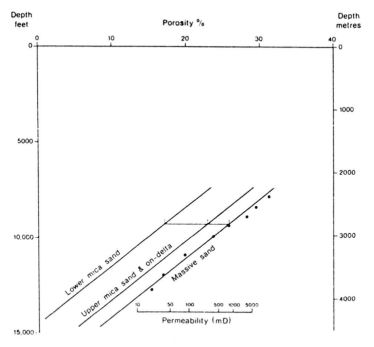

Fig. 58.   Porosity-depth trends in the Brent Sandstone, U.K. Block 211, North Sea (after Selley 1978).

burial.

Temperature plays a major role in initiating and controlling diagenetic reactions. The average geothermal gradient is about 3.3°C/100 m, but it may vary from region to region. Higher geothermal gradients occur in sedimentary sequences surrounding igneous intrusions where short-distance gradients of several hundred degrees/100 m are recorded. In regions of a high geothermal gradient porosities are normally lower than in regions with a low geothermal gradient.

One may distinguish between constructive and destructive diagenesis. Constructive diagenesis mostly tends to fill and, thus, reduce the pore space of sandstones and carbonate rocks by the authigenic growth of one or more isochemical or allochemical minerals such as quartz, carbonates, and/or anhydrite. Also mineral replacement occurs. Replacement is the dissolution of one mineral and precipitation of another mineral in its place on a volume-for-volume or molecule-for-molecule basis. It can occur early as well as late in the diagenetic history. In addition, replacement may generate secondary porosity. When dolomite replaces calcite cement in sandstones, porosity can be increased from 6 to 8%. Pyrite, siderite, and ankerite are minerals that can indiscriminately replace any cement or framework grain to a minor extent. In contrast, feldspar replacement by calcite likely accounts for the largest volume per cent of replacement in sandstones. Replacement of feldspar can be incipient, partial or complete. Ca-plagioclase is most susceptible to replacement followed by Na-plagioclase and finally potassium feldspar. Calcite does normally replace only the margins of quartz grains.

In contrast to constructive diagenesis, destructive diagenesis is directed toward decomposition and solution of minerals in reservoir sandstones and carbonate rocks or in claystones or shales. It may remove components and produce solution porosity. The significance of solution porosity was first recognized in carbonate rocks. Later Schmidt, McDonald and Platt (1977) demonstrated that solution porosity exists in sandstones, too. Many examples of sandstones in which solution of carbonate, anhydrite, and halite cements led to secondary porosity have been reported. Petrographic criteria for recognizing solution porosity in sandstones are partial dissolution, molds, inhomogeneity of packing and 'floating' grains, oversized pores, elongate pores, corroded grain margins, honeycombed grains, and fractured grains.

The fact that mineral compounds such as oxides, carbonates, and sulfates predominate among mineral cements, regardless of space and time, and even in a similar succession of precipitation, leads to the conclusion that principles of physical chemistry and physics govern the diagenesis of sedimentary rocks. Modifications of the succession of precipitation are minor and of local nature only. Thus, Eh and pH are important parameters in diagenetic mineral reactions. Eh-pH diagrams (Krumbein and Garrels 1952, Garrels and Christ 1965, Brookins 1988) are an auxiliary and indispensable tool to study diagenetic processes (Figure 59). They can be drawn for a wide variety of reactions and reactants and help to interpret weathering, depositional, and diagenetic chemical reactions and to determine mineral stabilities. Eh-pH values characteristic of depositional and interstitial waters show normally a positive Eh; slightly alkaline marine depositional waters and of interstitial waters negative Eh and approximate neutrality (Figure 23).

The stable-isotope distribution in minerals and sedimentary rocks is another parameter for deciphering conditions of diagenetic formation and subsequent alteration. The most common isotopic ratios used in diagenetic studies are $^{13}C:^{12}C$ and $^{18}O:^{16}O$. In both cases the magnitude of the ratios is compared with the ratios of some standard sample. This comparison is presented in the delta (δ) terminology. Further details are given by Fuex (1977) and Hoefs (1987). Figure 60 provides sound criteria for the determination of the site of diagenetic precipitation processes of carbonates.

A depth-related zonation of diagenetically formed minerals depending mainly on the decomposition of organic matter was introduced by Curtis (1978, pp. 110 ff.). This zonation applies not only for porous sandstones and carbonate rocks but also for shales and concretions.

Compacting mud sequences provide fluids and ions in solution causing pore-filling cementation in the adjacent porous sandstone horizons. Early diagenesis of marine near-surface muds, well studied in detail, serve as a model. Timing and mechanism of diagenesis of organic material and the production of hydrocarbons are intimately interwoven with mud diagenesis, fluid expulsion, and the origin of diagenetic iron and manganese minerals in mudstones. Table 22 presents a general model for marine-mud diagenesis within depth-related zones.

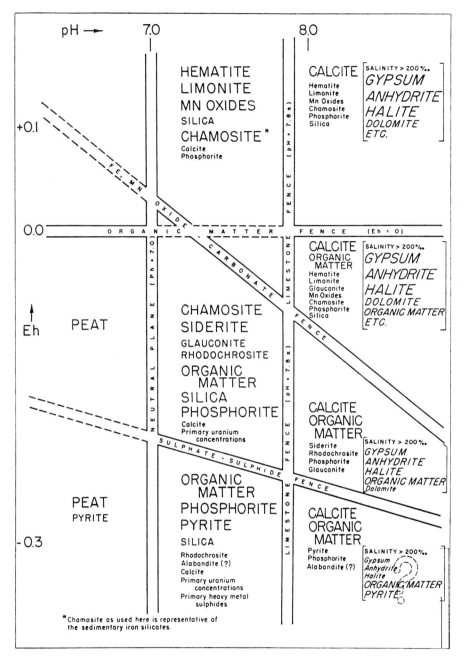

Fig. 59.   'Fence diagram' displaying authigenic mineral formation in the diagenetic realm as a function of Eh and pH (after Krumbein and Garrels 1952).

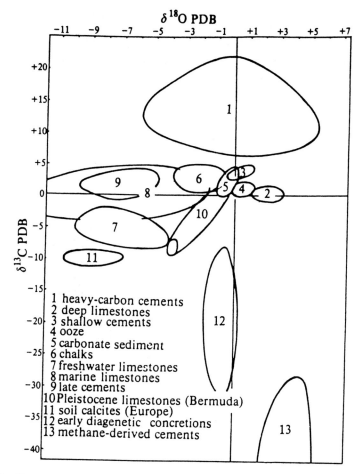

$\delta^{18}O$ PDB

1 heavy-carbon cements
2 deep limestones
3 shallow cements
4 ooze
5 carbonate sediment
6 chalks
7 freshwater limestones
8 marine limestones
9 late cements
10 Pleistocene limestones (Bermuda)
11 soil calcites (Europe)
12 early diagenetic concretions
13 methane-derived cements

Fig. 60. $^{13}C-^{18}O$ isotope grid shows compositional ranges of $CaCO_3$, sediment/rocks and cements (after Hudson 1977).

3.4.2. DIAGENESIS OF SOURCE ROCKS AND ORGANIC MATTER

Methods studying diagenesis of shales as source rocks and organic matter include thin-section petrography; X-ray diffractometry; organic-chemical analysis of organic carbon and other elements (H-N-O) and chemical analysis of trace elements; coal petrography with vitrinite reflectance measurements and fluorescence microscopy; determination of palynomorph colors; pyrolysis (Rock-Eval) and stable carbon isotope analysis.

Detailed data on shale diagenesis is compiled by Potter, Maynard and Pryor (1980). Up-to-date texts on hydrocarbon geology with emphasis on organic matter diagenesis are by Tissot and Welte (1984) and Hunt (1979).

Chemical composition, bulk clay-mineral content, illite crystallinity (Dunoyer de

Table 22.   Diagenetic zones for marine mud successions (after Curtis 1977). $\Delta T^{\circ}C$ increase in temperature with depth below sediment/water interface due to a gradient of $27.5^{\circ}C\ km^{-1}$.

Depth (km)	ΔT (°C)	Porosity (%)	Diagenetic zones (minerals formed)
			1 **oxidation**
0.0005			
			2 **sulphate reduction** pyrite calcite dolomite (low Fe carbonates ^{12}C enriched) kaolinite? phosphates?
0.01	0.2	80	
			3 **fermentation** high Fe-carbonates calcite dolomite ankerite siderite ^{13}C enriched
1.00	28	31	
			4 **decarboxylation** siderite
2.50	69	21	
			5 **hydrocarbon formation** (a) wet – oil (b) dry – methane montmorillonite → illite (a) disordered, (b) ordered
7.00	192	9	
			6 **metamorphism** (a) 200°C chlorite (b) 300°C mica, feldspar, epidote?

Segonzac 1970), and conodont colors are main parameters to appraise the diagenetic grade of shales in their function as source rocks. Increasing attention is also paid to the chemistry of associated brines, because marked salinity variations are found in deeply buried brines and pore waters.

Early diagenesis seems to have little effect on clays. The formation of concretions in clays is a diagenetic process which preserves synsedimentary to early diagenetic features and sedimentary structures. Moreover, heavy minerals are normally better preserved

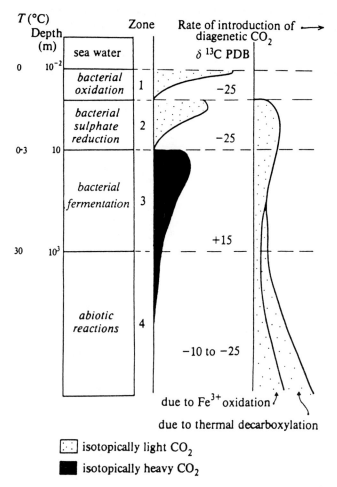

Fig. 61. Diagenetic evolution of organic matter from deposition to incipient metamorphism within the diagenetic zones 1–4 in mudstone sequences (after Irwin *et al.* 1977). The isotope signal of each zone is preserved in carbonate concretions.

in concretions and shales than in associated sands. Both, early concretions and heavy-mineral composition retain information on fabric, depositional environment, initial mineral composition, and/or provenance, now extinguished in the enclosing shale.

Compaction expels large amounts of water. And compaction of thick shale sequences can generate tectonic features such as clay and shale diapirs and mud lumps, commonly associated with natural gas (e.g. methane).

Some clay minerals may also act as catalysts for the transformation of organic matter into hydrocarbons.

Diagenetic processes in argillaceous sediments are difficcult to study because of the

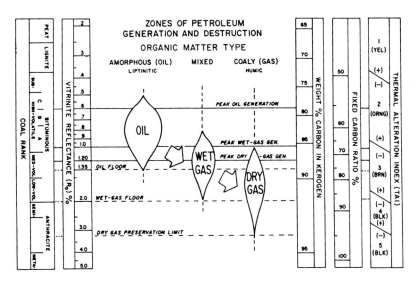

Fig. 62.   Equivalent coal ranks and stages of petroleum generation. Zones of oil and gas generation and destruction with the corresponding maturation parameters: coal rank, vitrinite reflectance, weight-% carbon in kerogen, fixed-carbon ratio, and thermal alteration index. The relative importance of petroleum generation zone depends on the composition of the original kerogen (after Dow 1977).

minute particle size of their rock constituents. Moreover, classical methods such as macroscopic examination and thin-section analysis including the evaluation of ultra-thin sections are not applied to their full extent (Zimmerle 1991).

Diagenesis of organic matter is influenced mainly by biological activity at an early stage as well as by temperature and pressure at a later stage. The diagenetic evolution of organic matter from deposition to incipient metamorphism is shown in Figure 61. Equivalent coal ranks and stages of petroleum generation are listed in Figure 62.

During early diagenesis, microbial activity is the main agent of organic matter transformation. The depth interval of early diagenesis ranges from zero down to 2 000 m. Temperature and pressure increase slightly. Aerobic microorganisms in the uppermost sediment layer consume free oxygen. Anaerobes reduce sulfates to sulfides. The decomposing organic matter is converted into carbon dioxide, ammonia, and water. Simultaneously the Eh decreases abruptly and the pH increases slightly. Biogenic $CaCO_3$ and SiO_2 dissolve and normally re-precipitate together with authigenic sulfides and siderite. During sedimentation and early diagenesis biogenic polymers such as proteins and carbohydrates are transformed by microbial activity into new polycondensed structures ('geopolymers'), the precursors of kerogen. With advancing diagenesis, the organic matter grades mainly into kerogen. When large amounts of organic matter derived from plants are deposited, peat and subsequently brown coals (lignite and subbituminous coal) are formed. Methane is the most important hydrocarbon generated during diagenesis. Additionally, decomposing organic matter releases CO_2, H_2O, and some heavy heteroatomic compounds late in diagenesis.

Diagenesis of organic matter is thought to cease at a level at which extractable humic acids have largely decreased and most carboxyl groups have been removed. This level corresponds to the boundary between brown coal and hard coal and to a vitrinite reflectance of approximately 0.5%.

Increasing burial down to several kilometers leads to an increase in temperature and pressure. Igneous intrusions and tectonic activity may enhance this increase. This zone of catagenesis is characterized by temperatures ranging from about 50 to 150°C and geostatic pressures from 300 to 500 bar. With advancing compaction water continues to be expelled; porosity and permeability decrease markedly. Normally salinities of interstitial waters tend to be higher. Progressive alteration of kerogen leads to the release of liquid petroleum for the first time. 'Wet gas' and condensate are generated at a later stage. Thermal degradation of kerogen leads to the generation of oil and gas. A vitrinite reflectance of about 2.0% (beginning of anthracite stage) marks the base of catagenesis. Aliphatic carbon chains in kerogen disappear and no petroleum is generated anymore. Coal deposits pass through various grades of coalification. They also produce hydrocarbons, mostly methane.

Metagenesis and metamorphism, the late stages of transformation, are normally restricted to deep troughs and geosynclines. Temperature and pressure reach elevated values and the sedimentary rocks may be affected by hydrothermal and igneous activity. The organic matter consists of methane only and a carbon residue. Coals were transformed into anthracite. Clay minerals lose the interlayer water and gain a higher grade of crystallinity. Recrystallization occurs.

During metamorphism of the greenschist facies coal is converted into meta-anthracite with a vitrinite reflectance of above 4%. Residual kerogen is transformed into graphite.

3.4.3. DIAGENESIS OF RESERVOIR ROCKS

The following analytical techniques are applied to study diagenesis of reservoir rocks: thin-section petrography, mineral staining, X-ray photography, autoradiography, X-ray diffractometry, scanning electron microscopy, electron microprobe analysis, fluorescence and cathodoluminescence microscopy, and stable-isotope analysis.

3.4.3.1. Diagenesis of Sandstones

Diagenetic processes affecting sands after deposition modify initial porosity, permeability, entry pressure, and irreducible water saturation. The most important of these processes is cementation apart from grain fracturing and pressure solution. Mineral cementation normally reduces pore dimension and permeability, and increases entry pressure because of enlarging the internal surface area. Introduction of diagenetic clay minerals can also negatively affect entry pressure, permeability, and irreducible water saturation without changing porosity appreciably. This effect may be caused by modifications of pore size, crystal shape, and surface chemistry of different clay minerals. Aspects of reservoir-sandstone diagenesis in relation to the preservation, destruction, and generation of porosity are discussed by various contributors in Chilingarian and Wolf (1988a, b).

Cements are minerals which are chemically precipitated from pore fluids. The most common sandstone cements – in decreasing order of frequency – are calcite, dolomite, quartz, siderite, anhydrite, muscovite, kaolinite, chlorite, potassium feldspar, albite, hematite, halite, barite, celestite, and zeolites. In total, more than 40 minerals are known as cementing agent.

Clay minerals like·kaolinite, illite, chlorite, smectite, and mixed-layer minerals can grow authigenically in sandstones (Figure 146). They occur either as crystal aggregates in pores, or as oriented coatings around detrital sand grains. The particle size is usually less than 5 μm, occasionally it reaches 40 μm. Grain coatings may be arranged with the long crystal dimension either parallel or perpendicular to the detrital grain surface. Continuous coatings of clays, and more rarely of other minerals, inhibit subsequent quartz cementation (Storz 1931). There seems to be an order of decreasing inhibition: chlorite > illite > chert > carbonate > hematite. Pseudohexagonal plates of kaolinite are regularly arranged in vermicular stacks. Chlorite can form fibrous aggregates or radially oriented plates. Illite occurs either as interlocking aggregates of platelets or as fine fibers. Smectite shows a similar habit as illite, but usually an order of magnitude smaller.

Porosity and permeabiltiy may be influenced by clay-mineral composition. Sandstones containing authigenic kaolinite normally show higher permeabilities than sandstones with smaller and more thread-like illite crystals. Effects of clay minerals on the permeability of sandstones are striking, even with a total clay content of only 5%.

It is important, but difficult to separate the different clay types and to distinguish between detrital and authigenic clay in sandstones, particularly under the polarizing microscope. Authigenic clay minerals are in general more homogeneous, better crystallized, and coarser crystalline. In addition, they may display specific authigenic growth forms. The role of authigenic clay minerals influencing reservoir properties, see also Section 4.4.1, was stressed by Wilson (1982) and Wilson and Pittman (1977).

Each chemically defined cement type represents its own phenotype based on crystal shape and size. A schematic view of dominant phenotypes of cements is shown in Figure 63. The precipitation sequence of mineral cements as observed by numerous authors shows mostly a similar succession regardless of age and geographic position.

Cryptocrystalline siderite, chamosite, and pyrite normally are of synsedimentary to early diagenetic origin. Authigenic kaolinite and titanium oxides follow. Quartz and feldspars form another paragenetic 'cluster', followed by sparry carbonates and by sparry sulfates. Repetitions of cementing-mineral sequences or minor deviations may occur.

Various combinations of diagenetic processes lead to diagenetic models such as the model of surface cementation, early cementation, leaching of framework and cement, late cementation, hydrocarbon effect, overpressure–undercompaction, framework collapse, stable framework, and shielded framework (Nagtegaal 1980). Each model is characterized by a specific porosity–depth/age curve (Figure 64).

Exploration aims at predicting porosity prior to drilling, particularly for deep prospects. As shown above, in different parts of the world sandstone porosities diminish with increasing depth to a varying extent. In sedimentary rocks of the same age the relationship

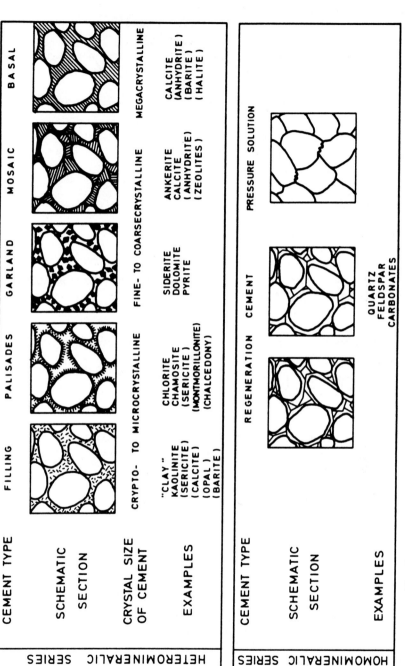

Fig. 63.   Phenotypes of mineral cements in sandstones divided into a hetero-mineralic and homomineralic series.

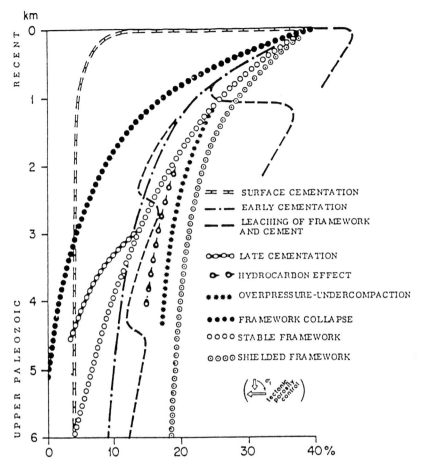

Fig. 64.   Porosity-depth/age curves of diagenetic sandstone models in the 0–6000 m depth range and porosity range 0–40% (after Nagtegaal 1980, p. 7).

is normally linear for a given area: e.g. Miocene and younger sandstones of southern Louisiana with 4.15% porosity reduction per km; Oligocene sands worldwide 3.28% per km; Jurassic sandstones of the North Sea 6.56% per km. In general, gradients of porosity reduction are highest with severe temperature gradients.

Absolute age dating of clay minerals as sandstone cement, especially of illite, by means of K-Ar chronology is now common practice to date neoformation of diagenetic clay minerals and to relate it to corresponding geological events (Aoyagi and Asakawa 1984). Pagel, Walgenwitz and Dubessy (1986) stressed the role of fluid inclusions in oil and gas-bearing sedimentary formations. Fluid inclusion microthermometry on diagenetically formed quartz and carbonate minerals allows to determine the crystallization temperature of these minerals in the course of diagenesis (Rieken 1988, Rieken and Gaupp 1989).

Mineral cementation can be largely inhibited also by the presence of hydrocarbons. In gas-filled sands inhibition is less than in oil-filled sands.

The rather uniform diagenetic gradients result from the interaction of numerous factors. Some factors are introduced by the depositional environment such as depositional porosity, grain-size distribution, shape, mechanical properties, and chemical stability of original constituents. For instance, diagenesis of volcaniclastic Cretaceous sandstones in Montana, U.S.A., was essentially controlled by the environment of deposition. Corrensite, accompanied by calcite and dolomite, is restricted to sediments from delta distributary channels and mouth bars. Smectite plus calcite characterize bay-beach, crevasse-splay, lagoon, barrier island, and shallow subtidal environments (Almon, Fullerton and Davies 1976). The depositional environment may also control the distribution of calcite cement. Abundant shell debris and other primary calcium carbonate can lead to local calcite cementation during early diagenesis in littoral environments, e.g. formation of beachrock. Selley (1977) constructed porosity/depth lines representing different environments. He even ventured to predict permeabilities to be expected.

Sandstones as competent and brittle rocks are also subject to fracturing. The actual porosity of fractured sandstones is much larger than the porosity measured on a core sample. Permeability of sandstones might have been increased in many producing areas by the presence of open fractures or joint planes.

The conditions of some diagenetic reactions in clay minerals are well established. Kaolinization of feldspar requires pore waters poor in dissolved solids, including K^+ and H_4SiO_4, and with a low pH. Such conditions are normally found in meteoric groundwaters. Consequently kaolinization is thought to take place soon after burial, or during later uplift into the weathering zone, or adjacent to faults with circulation of suitable waters. Smectite dehydrates between $110-150°C$. It may take up magnesium or potassium to form mixed-layer clays, chlorite or illite. This reaction has been demonstrated to take place in the Tertiary of the Gulf Coast area, U.S.A., in a depth between 1 500 and 4 000 m, in the Upper Cretaceous of Cameroun below 1 500 m, and in the Mesozoic of the Verkhoyansk area, Siberia, C.I.S., between 3 000 and 4 000 m. Illites and chlorites are relatively stable, except under acid conditions. Increasing temperature augments the degree of crystallinity. Thus, the determination of illite crystallinity by X-ray diffractometry is a method to appraise the metamorphic grade of sedimentary rocks, especially of sandstones.

Taylor (1977) distinguished the following three stages of porosity reduction that normally overlap slightly:

- Early cementation occurring at shallow depth of burial and related to the environment of deposition. The distribution of mineral cements is variable.
- Late cementation taking place during deeper burial. The mineral cements are more uniformly distributed.
- Compaction, mainly by pressure solution, takes place subsequently to the main cementation.

3.4.3.2. Diagenesis of Carbonate Rocks

As shown carbonate reservoirs differ from sandstone reservoirs in several aspects. The porosity is unevenly distributed and more localized, both laterally and vertically. Laterally extended porosity distribution is the exception in carbonate rocks. The porosity of even thick carbonate formations, up to 200 m, is confined to the uppermost portion of 10–15 m. On the other hand, pores may be much larger than in sandstone reservoirs, leading to unusual permeabilities. These facts have to be kept in mind when investigating the diagenesis of carbonate rocks.

Being restricted to the realm of carbonate mineralogy, carbonate-rock diagenesis is more isochemical than sandstone diagenesis. Allochemical diagenesis is not as widespread as in sandstones. Important are diagenetic processes under subaerial freshwater conditions with the influence of vadose and/or phreatic waters. These processes contrast much with those in shallow marine and deep marine environments. The carbonate compensation depth (CCD) plays a role, only in deep-water carbonate diagenesis (Ramsay 1974). The generally constructive, isochemical diagenesis comprises neomorphism, cementation, and stylolitization. In addition to dolomitization and dedolomitization, allochemical diagenesis includes anhydritization, silicification, and neoformation of other authigenic minerals such as silicates, other sulfates, and pyrite. Destructive diagenesis comprehends mainly processes of biological and mechanical erosion and chemical dissolution.

Modern carbonate sediments and reef rocks show an average porosity of about 50%. Reefal boundstones are characterized by porosities between 20 and 50% due to the loose rock frame and boring activity of reef dwellers. Reef debris is just as porous, but more permeable. Oolitic lime sands commonly show porosities up to 40%. Lagoonal lime muds and pelagic chalks reach a water-saturated pore space of 50%. Schroeder and Purser (1986) present work of various authors on reef diagenesis.

Pleistocene carbonate rocks show porosities below 30%. Porosities of chalky lime reservoirs and fine-crystalline dolomites range between 25 and 30%. Paleozoic limestones or dolomites show commonly less than 5% porosity. Early diagenetic cementation stabilizes mechanically the rock frame by solution-compaction. Diagenesis of Holocene sediments is largely controlled by the CO_2 content of circulating waters.

The following diagenetic processes act on original lime muds and consolidated limestones (Wilson 1980): Organic decay, neomorphism, cementation, solution-welding, telogenic dissolution, carbonate-grain rearrangement, pressure solution, stylolitization, and fracturing.

Organic decay causes breakdown and grain-size reduction of biogenic calcite. Organic rims encrust particles and burrowing loosens the sediment. Neomorphism converts unstable marine aragonite and high-Mg calcite into low-Mg calcite or dolomite through submicroscopic replacement processes. Practically all fossil limestones are composed of low-Mg calcite.

Cementation destroys initial porosity within and between the coarser carbonate rock particles. Four cement types can be discriminated based on their appearance: (1) Eogenetic cements characteristic of beach splash or high-energy subtidal zones comprise druse fans of coarse aragonite in reefs, palisade cements between grains, fibrous druses interpreted as original aragonite cement, and cements incorporating marine microfossils. (2) Vadose cements include microstalactites or pendent cements, eroded tops of grains, meniscus cement, irregular whisker crystal cement, association of cement with altered particles, solution-enlarged pores, and microspar matrix. (3) Phreatic freshwater cements are blocky cements with sharp crystal boundaries, bladed calcite crystals lining solution pores, dolomite crystals with zonal overgrowth, and ferroan calcite cement. (4) Connate water cements of the subsurface resemble those from the phreatic freshwater zone. Both cements consist mainly of blocky mosaic ferroan calcite.

Solution-welding takes place in lime sands and lime muds. Contacts of skeletal and non-skeletal grains are modified by dissolution under the influence of freshwater near the surface or at shallow depths of burial. Solution-welding is thought to be involved in the fabric transformation from lime mud to microcrystalline calcite such as in partly cemented chalk.

Telogenetic dissolution, a common process in carbonate rocks, causes secondary porosity cutting across grains and cements and enlarging intergranular pores. Even major caverns may form along joints and fractures.

Carbonate-grain rearrangement prior to burial commonly forces grains into planar contact early in diagenesis. During mechanical compaction grains alter position through dissolution and further crushing or through the formation of megabreccias caused by dissolution of intercalated evaporites. Metasomatic replacement, common in some limestones, involves the following minerals in order of their abundance: dolomite, anhydrite or gypsum, several modifications of silica from chalcedony to microquartz, siderite, ankerite, manganese oxides, and phosphates.

Pressure solution, stylolitization, fracturing, and vein formation are late diagenetic processes subsequent to lithification.

The sequential reconstruction of diagenetic processes in carbonate rocks is essential to explain preservation and formation of various porosity types (Figure 65).

Dolomites and dolomitized limestones, common in the geological column, are important as reservoir rocks. Precipitation of primary dolomite, however, is only known from two locations, Deep Spring Lake (U.S.A.) and Coorong Lagoon (Australia). Since modern seawater is supersaturated in regard to dolomite, dolomite should precipitate widely. Dolomite is even more stable in seawater than calcite. The fact that dolomite does not precipitate from seawater is presumably explained by kinetic factors arising from the extreme regularity of the dolomite lattice. Experiments to precipitate dolomite at room temperature produced disordered Mg calcite only. Under certain conditions, however, protodolomite, a metastable modification of dolomite, may be precipitated. Protodolomite, in effect a calcian dolomite with an excess of Ca^{2+} in its lattice, seems to preceed dolomite in respect to replacement reactions. High temperatures ($> 200°C$), however, are required to experimentally precipitate pure dolomite. The discrepancy between theoretical and observed behavior of dolomite in seawater represents the so-called 'dolomite problem'.

Three major dolomitization mechanisms are being discussed these days: the evaporite brine-residue model, the groundwater mixing model, and the formation-water model.

The evaporite brine-residue model is based on chemical and mineralogical processes below the surface of supratidal sabkhas as observed around the Arabian Gulf. Aragonitic

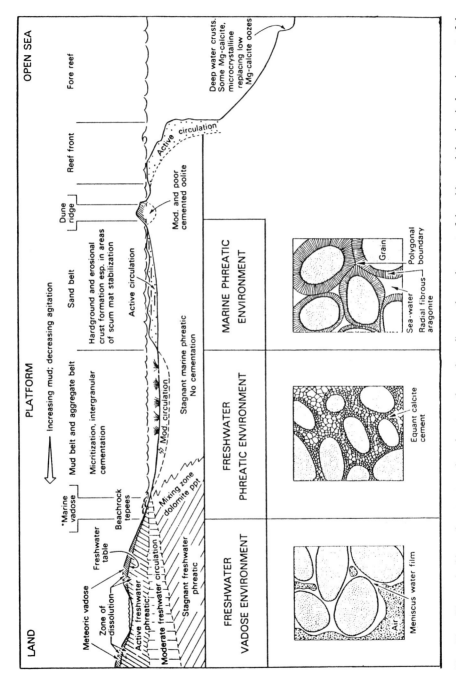

Fig. 65.   Diagenetic thin-section structures in carbonate rocks as depending on the environment of deposition and the chemical environment of the circulating waters (modified from Reading 1986, Fig. 10.1, p. 287).

sediments from former lagoons have been replaced by fine-crystalline protodolomitic mudrocks. Massive dolomitization followed. Nodular gypsum is precipitated within the dolomitized carbonates. Examples of sabkha brine dolomitization are common in the past. Reflux of high Mg:Ca ratio brines in the subsurface can dolomitize wide areas of the vadose and phreatic zones. The presence of evaporitic bands in thick dolomite sequences of regional extent is one criterion for this model.

The groundwater mixing model is used to explain examples of ancient dolomite occurrences of regional extent for which the sabkha model does not apply. This model accounts for the low Mg:Ca ratios (1:2 to 1:4) found in formation brines. Large-scale dolomitization takes place at the boundary of fresh phreatic and marine groundwater. Modern examples are known from aquifers in Jamaica and Florida. Fossil examples are likely to occur at horizons related to regressions and/or sea-level fluctuations. The presence of clear and euhedral dolomite crystals is thought to be indicative of such a mixing process. The dolomite crystals show perfect stoichiometric composition because of slow precipitation.

The formation-water model applies mainly to deeply buried limestones with pore water deriving from shale compaction. Release of Mg^{2+} and Fe^{2+} from the smectite-illite transformation may cause dolomitization of precursory carbonate minerals. The ferroan dolomite produced displays a fabric diagnostic of late diagenetic precipitation or replacement. Thin interbeds of dolomite and calcite in ancient stromatolites represent this dolomitization type. Dolomite is concentrated in dark, algal-rich layers. Dolomite precipitation is favored by high alkalinity.

An integrated effort should be made to decipher dolomitization of a given case including megascopic sample examination, evaluation of petrographic and petrophysical parameters, and geochemical analysis. Investigating dolomitization in thin sections, replacement features such as 'ghosts' or replaced allochems have to be looked for. If dolomite formed in a closed system an intergranular porosity of about 10% should be recognized. Multistage dolomite growth is indicated by zoned dolomite crystals. Early diagenetic ferroan dolomite may indicate temporary exposure to oxidizing conditions.

Dedolomitization refers to dolomite dissolution and replacement by calcite. It takes place when dolomites are penetrated by vadose meteoric waters rich in SO_4^{2-} ions. Such a process occurs commonly below dissolving evaporite sequences. Instructive examples are known from the Jurassic of central Arabia and from the Zechstein of Europe. Sulfate ions seem to favor dolomite dissolution and calcite precipitation.

Case histories on the diagenesis of potential and actual carbonate reservoir rocks are those of Paleozoic carbonate rocks with local remnants of solid hydrocarbons, the so-called impsonite (Schneider 1977), and the productive Zechstein carbonates (Quester 1964; Füchtbauer 1964a, 1968).

3.5. Sedimentary Traps

Hydrocarbon traps are barriers to the upward movement of oil or gas. A trap includes a porous reservoir rock and an overlying or updip impermeable seal or cap rock. The seal commonly consists of sediments whose pore diameters are sufficiently small to exert a

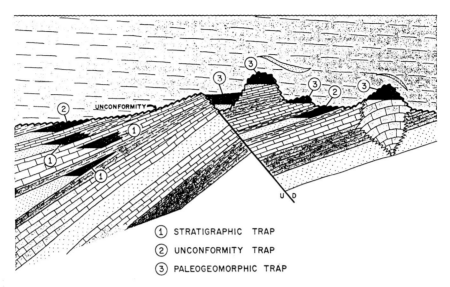

Fig. 66. Non-structural hydrocarbon traps: Stratigraphic traps, unconformity traps, and paleogeomorphic traps (after Halbouty 1981, p. 1218, reprinted with permission).

capillary pressure in excess of the hydrodynamic driving force.

Various types of traps can be distinguished: structural traps, stratigraphic traps, diagenetic traps, and combination traps. Structural traps are most important and widespread. The structural trap is detectable by surface mapping or geophysical surveys; it is formed by folding, faulting, or salt doming. Tectonic hydrocarbon traps, especially those associated with grabens, are discussed by Harding (1983, 1984). It is not discussed any further in the context with petroleum sedimentology.

Non-structural traps, also called sedimentary traps or subtle traps (Halbouty 1981), are trap types recently discovered. They can be subdivided into stratigraphic traps s.str., unconformity traps, and paleogeomorphic traps (Figure 66). Later, the concept of diagenetic traps was introduced. Sedimentary traps can be found in a wide stratigraphic range and in all basins. Some of them are preferentially developed in areas adjacent to ancient shorelines. Remote-sensing techniques may assist in discovering certain types of non-structural traps. The morphology and development of oil and gas traps and their seismostratigraphic detection is discussed by Jenyon (1990).

Stratigraphic traps are the result of primary lithologic changes (e.g. shale-out, pinch-out). They are laterally limited reservoirs sealed by impermeable strata. For instance, deposits of alluvial or submarine fans, fluvial sands, channel and bar sands, carbonate shoals, or carbonate reefs can form stratigraphic traps. Extensive pinch-out traps, for instance, are found in western Siberia. The seal may either be formed by impermeable strata conformably deposited on the reservoir, or the reservoir can be truncated and unconformably overlain by the seal.

Only 6% of the giant oil and gas fields are formed by stratigraphic traps proper; another 11% are a combination of stratigraphic and structural traps. The geology of and prospection methods for stratigraphic traps were discussed by King (1972).

In conformable stratigraphic traps the reservoir is conformably overlain by the seal. Trap size and geometry are delineated by the original upper depositional surface of the reservoir and possibly also by synsedimentary faults, e.g. growth faults. In unconformable stratigraphic traps, i.e. in unconformity traps, the reservoir is truncated and unconformably overlain by the seal. Trap size and geometry are defined by the topography of the erosional surface of the reservoir.

Stratigraphic traps may retain hydrocarbons in flat-lying, undeformed sedimentary sequences. Subsequent deformation may improve or reduce the hydrocarbon trapping capability. Stratigraphic traps are preferentially found along structural highs, e.g. the Paleocene submarine fans in the northern Viking Graben. They might be even improved by differential compaction, when flank sediments undergo stronger compaction, thus increasing the vertical closure of the stratigraphic trap.

Paleogeomorphic traps are reservoirs connected with fossil erosion or weathering surfaces and topographic features now concealed below the surface, e.g. ancient grabens in the Rotliegende or paleokarst in the Argyll field, U.K. North Sea. A paleogeomorphic trap of Paleocene age with the reservoir sand truncated by a clay-filled submarine channel forms the East Brentwood gas field, California, U.S.A.

Diagenetic traps (Wilson 1977) are generated by diagenetic alteration processes, e.g. sealing mineralization or secondary porosification. Sealing mineralization refers to hydrocarbon-filled reservoirs sealed off by ongoing diagenesis in the enclosing sediments. Such a trap may be displaced by tectonics later into a position in which trapping is difficult to be recognized. For sealing of hydrocarbons by diagenetic cementation see also Section 4.2.4.

Search for diagenetic traps is complicated. The following steps are recommended: (1) identify paleostructure, (2) date oil migration precisely, and (3) relate migration to tectonic movements. In diagenetic traps of Middle East carbonate reservoirs oxidized bitumen and stylolites locally interact with the porous reservoir near the oil/water contact making exploration and production more difficult.

Combination traps include structural, stratigraphic, and/or diagenetic trap elements.

Examination of cuttings, conventional cores, and fossil content, isopach mapping based on available logs from producing wells and dry holes, and seismic survey must be fully integrated in the search for sedimentary traps.

The first step to search for stratigraphic traps in cross-sections by means of well logs is to recognize a stratigraphic interval without discontinuities in sedimentation. Secondly, this stratigraphic interval has to be confined between reliable marker horizons. Thirdly, in the correlation cross-sections one of the upper marker horizons has to be marked as a horizontal line = datum horizon. And finally the present structural position of a projected sand pinch-out has to be defined.

Pirson (1983) exemplified the following external geometries of potential sandstone reservoirs from the U.S.A. recognizable in cross-sections and on maps: strike valley sands, offshore bar and barrier-island sands, channel-fill sands, onlap of shoreline sands, chenier sands, delta sands, and turbidite sands.

It is important to future exploration strategy to keep in mind that more petroleum is preserved in stratigraphic plus diagenetic than in structural traps. As the number of large structural traps that remain untested decreases, stratigraphic traps will increasingly become exploration targets. Stratigraphic traps, however, are difficult to be discovered by known methods. Thus, geological research has to be focused on improving known and developing new methods of stratigraphic trap finding. Especially seismic methods as well as supplementary high-precision gravimetry, magnetics, and electric surveys have a great potential for stratigraphic trap exploration. Halbouty (1981) stressed the urgent need for searching sedimentary traps, the so-called subtle traps, because future hydrocarbon reserves will be mostly found in subtle traps. And a considerable portion of them will be discovered in mature producing basins.

Type and frequency of sedimentary traps are related to various stages of the so-called tectonic-sedimentary cycles. Most favorable conditions for sedimentary traps exist apparently at the initial and late stages of tectonic-sedimentary cycles. At the initial stage of a tectonic-sedimentary cycle on platforms, narrow horst and graben structures, only a few kilometer wide and filled by terrigenous debris, develop. Individual silt and sand layers pinch-out towards the graben faults. This trap type is widespread in the Upper Devonian of the Russian Platform. Paleogeomorphic traps associated with buried reliefs (paleoreliefs) characterize the late stage of tectonic-sedimentary cycles. In western Siberia Late Jurassic reservoirs are locally limited by shore cliffs about 10–15 m high (Ovanesov *et al.* 1979).

3.6. Overpressured Shales

Shales adjacent to permeable sands can contain fluids at high pressure. Then they are more porous than shales at comparable depths. This can be attributed to both stratigraphic and structural causes. Such a constellation leads to the formation of 'overpressured' (or 'abnormally high-pressure') reservoirs, whereas the associated shales are termed 'under-compacted' or 'overpressured'.

High abnormal pressure in oil or gas wells usually leads to difficulties during drilling operations. In oil fields detection of abnormal formation pressures has become imperative. Geologists and petroleum engineers need to discover a way to detect and predict high fluid pressures in advance.

Frequently, wells either have been abandoned before reaching the desired formation or have blown out. The methods employed to detect and evaluate high-pressure zones are summarized by Fertl and Timko (1970).

The development of deep-drilling techniques and hydrocarbon production from reservoir rocks associated with thick sequences of argillaceous sediments (e.g. Tertiary basins) have become increasingly pressing in oil industry. Drilling to depths greater than 6 000 m and subsequent production depend greatly on the knowledge of the basic mechanical properties and deformation characteristics of the formations penetrated, especially shales, and on the chemistry of the fluids associated. Interstitial water in undercompacted shales, however, should be more saline than that in well-compacted shales, because in the former case a smaller portion of the more saline fluid present in the center of capillaries is squeezed into adjacent sandstones.

In the world-wide search for oil and gas, fluids at pressures much higher than normal hydrostatic pressures have been encountered in many countries: in the United States (Gulf Coast Basin, Anadarko Basin, Williston Basin, and Ventura Basin), in the Arctic Islands, Africa, Europe (Austria, Carpathians, Romania, France, Germany, The Netherlands, Ukraine, Urals, and Caucasus), Far East, Middle East (Iran, Iraq, and Pakistan), and South America.

A comprehensive definition of abnormal pressure is a fluid pressure that materially exceeds the weight of a column of interstitial fluid. Abnormally high fluid pressures are also defined as fluid pressure which exceeds the hydrostatic pressure of a column of brine containing 80 000 mg/l total solids. Deviations from this generalization occur.

Prediction of fluid pressures is important in successful planning of exploration and drilling programs. The magnitude of fluid pressure must be known in order to: select drilling fluids to prevent blow-outs; optimize penetration rates; recognize hydrocarbon shows from productive formations; lessen damaging fluid entry into potential pay sands; and permit more reliable interpretation of formation log data.

The formation of abnormal pressures in sedimentary basins is ordinarily attributed to the following causes: continuous loading and incomplete gravitational compaction of sediments, faulting, phase changes in minerals during compaction, salt and shale diapirism, tectonic compression, osmotic and diffusion pressures, and/or geothermal temperature changes creating fluid-volume expansion or contraction. Other mechanisms might also conceivably produce high pore pressures: existence of fossil pressures that correspond to a previous greater depth of burial, invasion of water derived from magmatic intrusions, and infiltration of gas.

Pressure seals are the low-permeability envelopes that enclose abnormally pressured reservoirs. Three types of seals are observed: basal, lateral, and top seals. Basal seals form the bottom of abnormal pressure compartments and normally follow stratigraphic horizons. Lateral seals are generally faults. Top seals, developed in any lithology, may parallel or cut across time-stratigraphic boundaries.

Chapter 4

APPLIED PETROLEUM SEDIMENTOLOGY

This fourth part of the book, called applied petroleum sedimentology, comprises the following sections: (1) environments of deposition of modern as well as ancient sediments, (2) case histories in petroleum sedimentology from various exploration areas world-wide, and (3) applied sedimentology to well treatments, hydrocarbon exploitation, and enhanced oil recovery (EOR) operations.

4.1. Environments of Deposition

Sedimentary environments are characterized by physical, chemical, and biological processes which are distinct from those of adjacent depositional areas. The sedimentary properties of the environment include climate, fauna and flora, lithology, geomorphology, tectonic setting, and in subaqueous environments depth, temperature, chemistry, and current system of the water. Modern sedimentary environments are dominated by sedimentary processes like erosion, depositional equilibrium, and/or deposition. In general, subaerial environments are more exposed to erosion; subaqueous environments are largely areas of deposition.

The depositional environment is a particular type of the sedimentary environment, in which net sedimentation predominates. From the depositional environment derives the sedimentary facies. The relation between sedimentary environment, depositional environment, and sedimentary facies is shown in Table 23.

Facies is defined as the sum of lithological and faunal characteristics of a given stratigraphic rock unit. Facies analysis integrates all geological observations available taking into account sedimentary parameters such as: (1) inorganic sedimentary structures (cross-bedding, graded bedding, ripple marks, sole marks, etc.), (2) organic sedimentary structures (trace fossils), (3) fauna and flora, (4) grain-size distribution and sedimentary textures, (5) mineral composition, (6) geochemical composition (trace elements and stable isotopes), and (7) stratigraphic relations (depositional morphology, stratigraphic breaks and events, magnetic polarity). Facies analysis is aimed to design a facies model, practically hypothesizing on the depositional environment.

Table 23. Relationship between sedimentary environment and sedimentary facies (after Selley 1970, Table 1.1, p. 2).

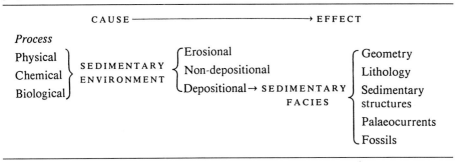

A continuous sedimentary facies development in a vertical direction is observed best in sedimentary sequences; it is subject to sequential analysis. In a lateral direction rocks of different facies are formed beside each other, though in vertical sections they lie on top of each other. This basic statement, the so-called Walther's rule (Walther 1894, p. 979) is of far-reaching significance: it says that only those facies could have been superimposed primarily which can be observed beside each other at present. This concept was successfully applied by alpine geologists to unravel the complex structures of the Alps, by discriminating different facies units, now juxtaposed, that must have been originally deposited far apart from each other.

It is recommended to use the term "facies" descriptively, such as bituminous-shale facies, red-bed facies, coral-limestone facies, with a subsequent interpretation of the environment. The terms lithofacies and biofacies refer to petrological or faunal and floral characteristics of a stratigraphic unit. Hallam (1981) gives an informative introduction into facies and environment interpretation within the stratigraphic record.

Analysis of the environment of deposition is an important goal in petroleum sedimentology. Such analysis requires a profound knowledge of sedimentary processes, acquaintance with literature on modern sedimentary environments, and experience in its geological application. The present approach to environmental analysis is based on the concept that modern depositional environments provide the "key to the past" when analyzing stratigraphic successions. Thus, when interpreting ancient depositional environments one applies the principles of actualism.

The concept of sedimentary models is important for environmental analysis. However, modern depositional environments can only be fully understood when the appropriate physical, chemical, and biological processes within a modern environment are known. Facies models, modern sedimentary environments, and applied aspects of sedimentology in hydrocarbon exploration are discussed in Brenchley and Williams (1985). Blatt, Berry and Brande (1991) deal with depositional environments and facies as well as lithostratigraphic and economic-stratigraphic aspects of certain sedimentary rock types. The model concept (Selley 1982, p. 265) can be applied to large-scale regional or small-scale punctual sedimentological settings, keeping in view well studied and well documented case histories or examples. Applying the model concept enhanced our thinking on one side, but also hampers the own mental agility and fancy.

Sophisticated environmental analysis of sedimentary rocks has reached such a refined level that deposition, for example, may be deduced to have taken place in the hollows and puddles present on the surface of the point bar of a meandering river in rocks more than 300 million years old!

Detailed examination of sedimentary rocks for all their primary sedimentary features is the first step. Various logging techniques (see Bouma 1962) are available to plot the data graphically. Sedimentary patterns lead to a purely descriptive lithofacies, e.g. coarsening-upward siltstone to sandstone facies or large-scale cross-stratified oolite facies. The lithofacies described has to be compared with the facies interpretation of recent environments of deposition. Possessing detailed knowledge of recent facies, realistic models can be assigned to ancient sedimentary rocks. However, we have to realize that one can never actually prove that an ancient facies was deposited in a certain environment (Selley 1982).

Reading (1978, 1986) is a valuable source of information on recent sedimentary environments and their ancient analogues. Further examples of facies models are given by Blatt (1982), Davis (1983), Reineck (1984), Selley (1988), and Walker (1984). Potter and Pettijohn (1977) refer to paleocurrents and basin analysis. Anderson (1986) is a modern text on marine geology.

Sediments are and sedimentary rocks have been deposited in a variety of environments. The broadest classification of environments are the environmental associations: continental, transitional, coastal-shelf, and deep-marine. However, further subdivisions are necessary. The classification of depositional environments shown in Table 24 follows Reading (1986) and Pettijohn, Potter and Siever (1987). It is comparable to most classifications of sedimentary environments but deliberately more restricted in order to better compare modern and ancient environments.

4.1.1. MODERN ENVIRONMENTS OF DEPOSITION

The main environments of deposition are shown in conceptual models (Figure 67). Their characteristic sedimentological features are compiled in the following. Mentioned are also some of the more important environmental variants.

4.1.1.1. Alluvial Environments (Figure 68)

4.1.1.1.1. Alluvial fans. Definition: coarse sediments deposited by a stream, leaving a narrow mountain valley, mostly upon a plain.

Geometry: mainly fan- or cone-shaped wedges, relatively flat to gently sloping (< 5°); area ranges from 1 to 900 km^2, radii of several hundred meters to a few kilometers, rarely up to 50 km; thickness normally < 1 km; mud fans generally larger and thicker than sand/gravel fans; coalescence of adjacent fans leads to a continuous wedge against a mountain front or fault scarp.

Table 24. Main environments of deposition (compiled and slightly modified after Leeder 1982 and Reading 1986).

Environmental Association		Environment	Examples of Environment Variants
Continental	1	Alluvial	Alluvial fans, alluvial plains, braided river, meandering rivers levee, crevasse splay, flood basin
	2	Lacustrine	(Saline, temperate stratified, tropical stratified, glacial, delta plain), lake terrace, shoreline, slope, basin, delta
	3	Desert	Wadi, dune, interdune, playa, duricrust
	4	Glacial	Supraglacial (lodgement till), intraglacial (melt out till), morainic complex, outwash fan, glacial lake, esker (also glaciomarine)
Transitional	5	Delta	Distributary channel, tidal channel, backswamp bay, mouth bar, prodelta
Coastal-shelf	6	Clastic shorelines	Beach, nearshore, offshore, barrier, lagoon, tidal flat, tidal inlet, coastal eolian dunes, estuarine channel, marginal flats, flood tidal delta
	7	Arid shorelines and evaporites	Sabkha, algal marsh, tidal flat, beach, lagoon, tidal delta, platform margin, marginal buildups
	8	Shelf siliciclastics	(Tide-dominated versus storm-dominated), various tidal bedforms, sand ribbons, linear tidal ridges, sand waves, shoal retreat massifs, buried channels, scarps (also glaciomarine)
	9	Shallow-water carbonates	Marine platform, reefs and atolls, subtropical carbonate shelves, temperate-water shelves
Deep-marine	10	Continental margin	Continental slope, continental rise, submarine fan, submarine channel
	11	Pelagic	Abyssal plain (hypersaline, euxinic, mid-ocean ridge, ridge-flank)

Lithology: mass of loose rock material; generally coarse clastic debris ranging from boulder to clay size; immature sediments with mineral composition close to that of the source area; usually distinct red color; sorting and roundness variable; grain size decreasing from fan apex to the periphery; often debris-flow deposits, i.e. coarse material in muddy matrix, mostly in a proximal position; armored mud balls common.

Sedimentary structures: internal structures commonly rather chaotic; indication of fining-upward; poorly stratified to unstratified, also cross-bedded, plane-bedded, or massive.

Current pattern: commonly radiating with low scatter of readings.

Organic content: generally rare or not preserved, normally abraded and reworked; plant fragments locally abundant; devoid of autochthonous marine fossils.

Examples: mainly in arid to semi-arid and subarctic regions, e.g. montainous semi-arid regions of the southwestern U.S.A. (Death Valley). Less common in humid climate, e.g. Kosi River draining area, Himalayas.

Literature: Rust and Koster (1984), Rachocki and Church (1990).

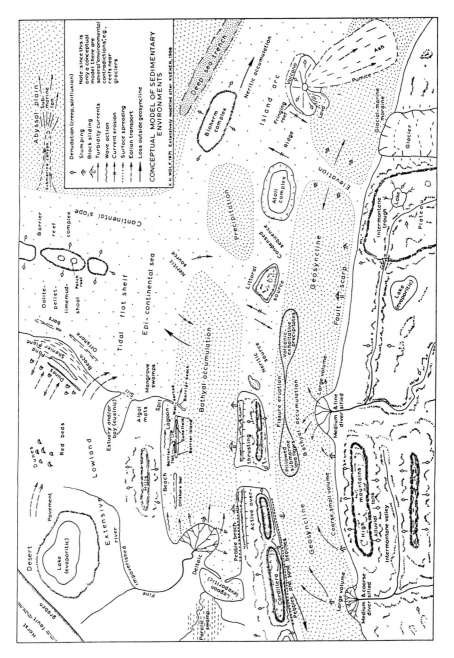

Fig. 67. Conceptual model of sedimentary environments (after Wolf 1973, Fig. 6, pp. 164-166).

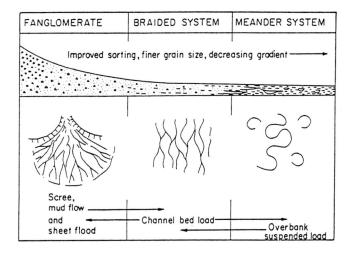

FANGLOMERATE	BRAIDED SYSTEM	MEANDER SYSTEM

Improved sorting, finer grain size, decreasing gradient ⟶

Scree,
mud flow
and ⟵ ── Channel bed load ──
sheet flood
Overbank
suspended load

Fig. 68. Recent alluvial environments: transitional relationship between fanglomerates, braided and meander systems (after Selley 1988, p. 171).

4.1.1.1.2. Alluvial plains

4.1.1.1.2.1. Braided rivers (low-sinuosity rivers). *Definition:* Braided-river deposits are characteristic of the upstream portion of a river. They have high slopes, a heavy load of coarse debris, and erodible banks. Braiding occurs also in semi-arid or arid climates.

Geometry: sedimentary facies as a whole prismatic, fan-shaped, or blanket-shaped; individual sand bodies form elongate bars; internally composed of interlacing network of small branching and reuniting shallow channels with sand of varying grain size; separated from each other by channel bars; anastomosing stream pattern (Figure 69).

Lithology: predominantly coarse-grained clastic sediments with gravel deposits and coarse sands; poorly sorted, but absence of debris-flow deposits; commonly red-colored; rarely finer sands and silts; sand/mud ratio very high.

Sedimentary structures: poorly developed fining-upward sequences; large cross-bedded or plane-bedded channels with occasional quicksand textures; rare gravel-floored abandoned channels filled by laminated silts with thin rippled sand layers and desiccation cracks.

Current pattern: very low scatter of unidirectional features at individual sample locations; may regionally describe fan-shaped arcs; gravels imbricated and oriented parallel to current direction; bars and channels parallel to regional slope.

Organic content: rarely plant debris and abraded disarticulated bones of terrestrial vertebrates; devoid of autochthonous marine fossils.

Examples: numerous streams in subarctic Canada and in the Alpine-Himalayan belt, e.g. Kosi River as tributary of the Ganges River.

Literature: Miall (1977), Allen (1982), Walker (1984).

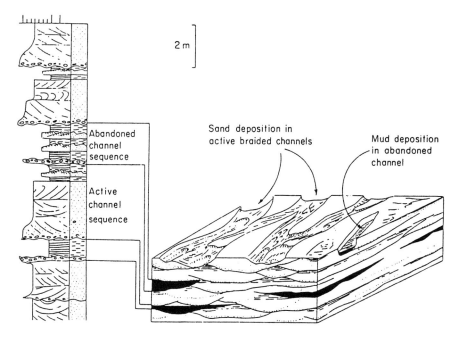

Fig. 69.   Recent braided alluvial channel system. Sedimentation takes place mostly in rapidly shifting channels (after Selley 1982, Fig. 132, p. 274).

4.1.1.1.2.2. Meandering rivers (high-sinuosity rivers). *Definition:* mature stream that swings side to side, i.e. meanders, as it flows across its flood plain.

Geometry: series of freely developing sinuous curves, bends, loops, turns, or windings; sedimentary facies as a whole often blanket-shaped; composed of coalescing channel complexes; forms also discrete shoestring sands enclosed in mud; crevasse splays, levees, and abandoned channel fills are discrete bodies of more limited extent; internally composed of alternating sand and mud layers in units about 1–3 m thick (Figure 70).

Lithology: fine- to coarse-grained sands and muds present in about equal proportions or sands even subordinate; some pebble and peat intercalations, without clast-supported gravels; occasionally red-colored; also with caliche-like concretions; sediments more reworked than in alluvial fans.

Sedimentary structures: fining-upward sequences common; above an erosional channel base heavily bedded gravel deposits, i.e. coarse "lag" deposits; overlain by trough-cross-bedded, coarse sands and plane-bedded or ripple-bedded fine sands that grade up into laminated silts with occasional mud cracks; crevasse splays and levees laminated and cross-bedded.

Current patterns: unimodal at a point with a wide scatter of dip directions; sand particles statistically oriented parallel to the current direction and, thus, to the regional slope.

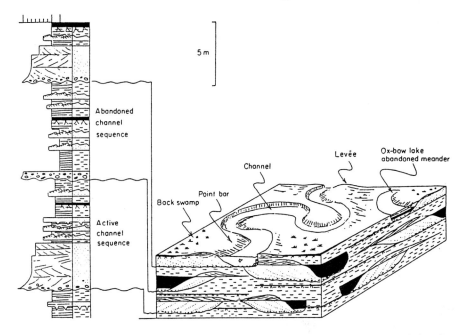

Fig. 70. Alluvial flood plain cut by meandering channels (after Visher 1965, reprinted with permission; from Selley 1982, Fig. 130, p. 270). Note lateral migration of channels and upward-fining of grain size.

Organic content: channel sands may contain plant debris, worn disarticulated bones of terrestrial animals, and freshwater fauna. Overbank flood-plain silts may carry plant debris, rootlet horizons, and peat beds. Devoid of autochthonous marine fossils.

Examples: world-wide, mainly in humid climate zones. The deposits of the Mississippi River, U.S.A., and Klarälven River, Sweden, are thoroughly studied examples. The main elements of a modern meandering river system, exemplified by the Mississippi River, consist of in-channel deposits (lateral accretion) with channel floor and point bars and overbank fines (vertical accretion).

Etymology: from Maiandros River in western Asia Minor.

Literature: Sundborg (1956), Allen (1970), Reineck and Singh (1980), Walker (1984), Collinson (1986).

4.1.1.2. Lacustrine Environments

Definition: lakes are inland bodies of standing fresh or saline water. Preservation of lake sediments is favored by their occurrence in depressions. Water composition and sedimentation are controlled by climate, water depth, and nature and amount of clastic and solute input.

Geometry: irregular sheets or elongate belts; mainly controlled by topography, e.g. lakes in rift valleys.

Lithology: depending on source area, water chemistry (freshwater versus saline), climate, and water depth; predominantly argillaceous, calcareous, or evaporitic pelites; coarse marginal clastics deposited during severe floods.

Sedimentary structures: fine lamination and varves in central lake areas; thermal stratification, cross-bedding, and channelling in marginal facies; with ripple marks and mud cracks; locally turbiditic features.

Current patterns: commonly complex current regimes by wind and water; mainly unidirectional and centripetal to the deepest part of the lake; bipolar flow patterns may indicate longshore currents; irregular wave-generated current patterns independent of bottom slope; in small lakes no tidal oscillations.

Organic content: diverse freshwater organisms (algae, fish, bivalves); washed-in land plant debris and terrestrial animals; marginal facies may contain peat; absence of autochthonous marine fauna.

Examples: sedimentological case histories known from Lake Geneva, Lake Constance, Lake Balaton, Great Lakes, East African Rift Valley Lakes, Lake Baikal, and from the Death Sea. Facies models range from deep, thermally stratified lakes to shallow, well mixed tropical saline lakes.

Literature: Picard and Lee (1972), Lerman (1978), Matter and Tucker (1978), Collinson (1986).

4.1.1.3. Desert Environments

Definition: areas with a mean annual precipitation of one centimeter or less and devoid of vegetation; four kinds of deserts distinguishable: (1) polar or high-latitude deserts, (2) middle-latitude deserts (e.g. Gobi), (3) trade-wind deserts (e.g. Sahara), and (4) coastal deserts (e.g. Peru). The trade-wind deserts (10°–30° N and S) are known best (Figure 71).

Geometry: rather irregular configurations of areal sediment distribution; eolian sand seas (ergs), alluvial fans, ephemeral streams, inland sabkhas, playa lakes, and bare rock areas; barchans, long-crested transverse, parabolic, longitudinal or seif dunes, and rhourds; mostly inland drainage; often fault-bounded basins or burying older topographies; occasionally below sea level.

Lithology: heterogenous sediment assemblage; clean, well-sorted, fine to medium eolian sands; poorly sorted, coarse ephemeral stream sands (wadis); fluvial processes near margin, eolian sedimentation in basinal parts; occasionally mudflow deposits ranging from boulder to clay size; evaporitic clays, silts, and sands with halite and gypsum (sabkhas or playas); locally carbonate-, sulfate-, and chloride-dominated facies; loess on desert fringes; light colors dominate, red colors are commonly post-depositional due to decomposition of detrital mafic constituents.

Sedimentary structures: common eolian bed forms are draas, dunes, and ripples with a wide variety of shapes; draas and dunes characterized by large-scale cross-bedding with set heights of up to 30 m, inclination up to 35°; plane bedding and penecontemporaneous deformation (e.g. sand avalanching) rare; wide-spaced, low-amplitude ripples (impact ripples), warty adhesion ripples, and fluid drag ripples; ventifacts, desert varnish, sand dikes, mud cracks, clay lags, paleosols, and duricrusts; windblown loess is structureless;

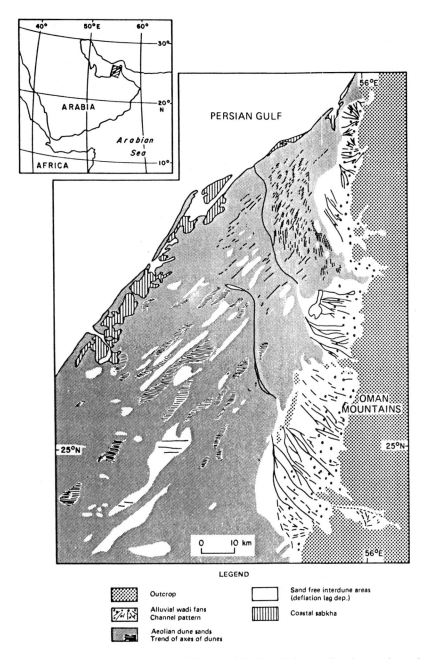

Fig. 71.   Recent desert environments in Trucial Oman: Alluvial wadi fans, aeolian dune sands, sand free interdune areas and coastal sabkhas (after Glennie 1970, redrawn by Seemann 1982, p. 59).

eolian sand grains display smooth surfaces and characeristic marks under scanning electron microscope.

Wind and current patterns: dip directions of cross-bedding may be variable at outcrops due to complex dune morphology; seif dunes with bimodal cross-bedding patterns; regional trends of current features may be present but are not necessarily slope-controlled. Wadi sands may show current directions opposite to associated eolian sands.

Organic content: generally absent or low; tetrapod footprints and plant debris; shell debris only in coastal dunes.

Examples: Sahara; Gobi; Iran; Danakil depression, Ethiopia; Central Australian Desert (Lake Eyre as inland sabkha); southwestern U.S.A. (e.g. Great Sand Dunes National Monument, Colorado, and Death Valley).

Literature: Bagnold (1954), Glennie (1970), Cooke and Warren (1973), McKee (1982), Brookfield and Ahlbrandt (1983).

4.1.1.4. Glacial Environments

Definition: the so-called glacial environments include deposits of many distinct environments such as rivers, lakes, continental shelves and margins, marine abyssal plains, but all characterized by the influence of grounded or floating ice and affected by rapid sea-level changes (Figure 72).

Geometry: as blanket deposits over wide areas or limited to glacial valleys; specific morphologies in form of eskers, drumlins, moraines, or till plains; glaciofluvial sands may develop as shoestrings.

Lithology: glacial: almost exclusively clastic fragments, till or boulder clay, commonly unsorted and unstratified, angular fragments; glaciofluvial: fine to coarse clastic sediments, well sorted, fairly rounded; glaciolacustrine: laminated muds, i.e. varves; periglacial loess (Eden and Furkert 1988); glaciomarine: muds with dropstones, sediments chemically and physically immature, rich in unstable minerals.

Sedimentary structures: in moraines lack of stratification; diagnostic striation on pebbles and boulders; glaciofluvial material cross-bedded or stratified; congeliturbation of periglacial soils and congelifluction.

Flow and current patterns: direction of ice flow reconstructable by means of striation (scratches) on glacial pavement and guide pebbles; glaciofluvial current patterns as in alluvial deposits.

Organic content: in glaciolacustrine varves pollen and spores; glaciomarine deposits contain restricted cold-water marine fauna; in tills allochthonous fossils from preglacial bedrocks.

Examples: all modern glacial, periglacial and glaciomarine areas of the northern and southern hemisphere, especially in Alaska, Canada, C.I.S., and Alpine Mountains.

Literature: Jopling and McDonald (1975), Eyles and Miall (1984), Dowdeswell and Scourse (1990), Brodzikowski and Van Loon (1991).

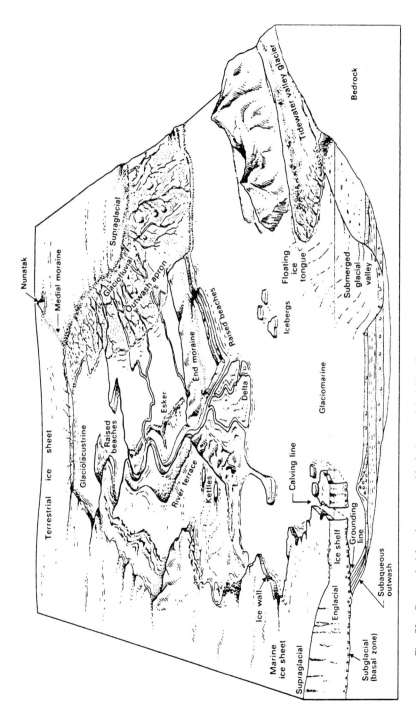

Fig. 72. Recent glacial environments with glaciers, periglacial landforms and glaciomarine facies (after Reading 1986, Fig. 13.2, p. 448).

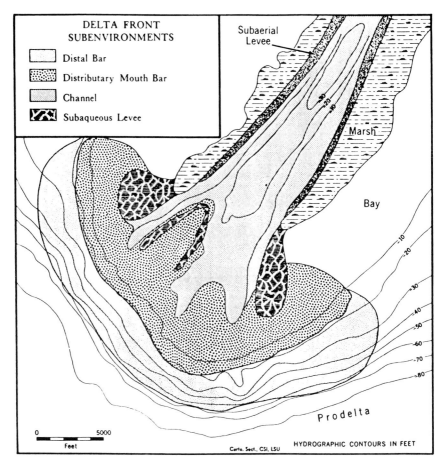

Fig. 73. Subenvironments in a delta front: Distributary mouth in river-dominated delta (after Coleman and Gagliano 1965, Fig. 9).

4.1.1.5. Delta Environments

Definition: low, nearly flat, alluvial tract of land at the mouth of a river. Most deltas are partly subaerial and partly subaqueous. Because delta environments are a chief goal of hydrocarbon prospection their variety and their variants have to be stressed.

In the delta environment numerous environmental variants such as distributary channels, tidal channels, backswamp, bay, mouth bar, prodelta occur. Such environmental variants of the present Mississippi delta are shown in Figure 73. Coleman and Wright (1975) undertook a multivariate analysis of 34 modern deltas, which they were able to group into six discrete delta models primarily on the basis of hydrographic conditions, topography, and sediment load (Figure 74). Galloway (1975) proposed a tripartition:

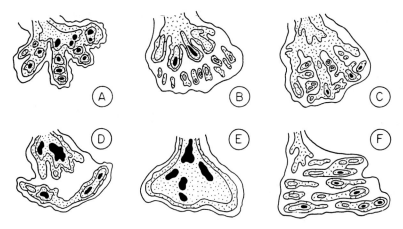

Fig. 74. Delta models of Coleman and Wright (1975): (A) River-dominated with low wave and tide energy, low littoral drift (e.g. Mississippi delta), (B) River-dominated with low wave energy, high tide range, low littoral drift (e.g. Ganges-Brahmaputra), (C) Intermediate wave energy tide, low littoral drift (e.g. Mekong delta), (D) Intermediate wave energy, low tide range (e.g. Brazos delta), (E) High wave energy, low littoral drift (e.g. São Francisco delta), (F) High wave energy, strong littoral drift (e.g. Senegal delta).

fluvial-dominated, tide-dominated, and wave-dominated. Another classification is based mainly on the sediment type carried by rivers (Sneider, Tinker and Meckel 1978): mud-type deltas, sand-type deltas, and coalescing deltas.

Geometry: wedge- or fan-shaped; cuspate, lobate, and elongate bodies define surface geometry; thickness varies from few meters to several tens of meters (e.g. Holocene Mississippi delta about 50 m thick); overall configuration dependent on climate, fluvial influx, tidal range, and wave activity.

Lithology: fluvial sands of the subaerial delta-plain and delta-front sands commonly clayey, micaceous, and carbonaceous; normally associated with overbank muds and silts; mouth-bar sands generally well washed, clean, and mineralogically mature.

Sedimentary structures: clays in lower part of the subaqueous sequence (i.e. prodelta) laminated, rarely rippled; middle portion of subaqueous sequence may comprise fans or channels filled with turbidite sands; slumps and slides may be present; overlying subaerial delta-platform or delta-top sequence consisting of a radiating complex of cross-bedded, massive distributary-channel sands with intervening interlaminated, finely laminated, rippled, and bioturbated clays, silts, and very fine sands; overall tendency of coarsening-upward clay/sand sequences, except for fining-upward in fluviatile subaerial delta-plain deposits.

Current patterns: rather complex; unimodal current pattern with little scatter in channel sands; may be regionally radiating both as slope turbidites and as distributary-channel sands; tidal currents crossing delta top produce bimodal current patterns with considerable scatter; longshore currents leave normally unimodal current patterns with little scatter but at right angles to tidal currents.

Organic content: marine fauna in lower fine-grained part of the subaqueous sequence;

brackish and freshwater faunas present in subaerial portion together with plant debris, rootlet horizons, and peat beds.

Examples: fluvial-dominated deltas: Mississippi, Po, Danube, and Ebro; wave-dominated deltas: Nile, Rhône, São Francisco, Senegal, Niger, and Orinoco; tide-dominated deltas: Copper, Ganges-Bramaputra, Mekong, and Gulf of Papua.

Literature: Classical accounts on modern delta sedimentation by Fisk *et al.* (1954), Morgan and Shaver (1970), Coleman and Wright (1975), Coleman (1981), and Elliot (1986a).

4.1.1.6. Clastic Shoreline Environments

Numerous attempts have been made to classify the wide range of present-day, non-deltaic clastic shorelines and, thus, those shoreline deposits. The following bipartition into lagoons and tidal flats (4.1.1.6.1) and beaches and barrier islands (4.1.1.6.2) is simple; it does not deal with further subenvironments such as salt marsh, etc. These environments are influenced by a wide variety of parameters and processes such as sediment influx, interaction of wind, waves and currents, and by the tide range. The tide range is expressed in terms of microtidal < 2 m, mesotidal 2–4 m, and macrotidal > 4 m. The terms supratidal, intertidal, and subtidal emphasize the role of the tides even more.

4.1.1.6.1. Lagoons and tidal flats. Geometry: tide-dominated sheets or shoestrings parallel with the regional strike; sand bodies reworked by tides show wider lateral extent than distributary fronts, tidal channels, and tidal-bar or point-bar deposits; small delta lobes can form at the mouths of small rivers or form the back of tidal channels.

Lithology: predominantly clays, silts, and fine sands; physically and mineralogically less mature than beach and barrier-island sands; detrital mica, carbonaceous fragments, solitary glauconite, and traces of pyrite may be present; tidal-delta sands cleaner and more mature than sands of river-mouth deltas.

Sedimentary structures: laminated, wavy bedded, rippled, and bioturbated sediments; with a great variety of laminations including flasers and wave- and current-formed ripple marks; rarely with desiccation cracks, armored mud balls, and conglomerate-lined channels infilled with obliquely inclined mud due to tidal gullies; tidal channels and bars cross-bedded; coarsening-upward and cross-bedded sand units may represent distributary fronts.

Current patterns: extremely variable, dependent on tidal currents and wave activity; difficult to measure; locally bipolar patterns in tidal-flat sands; i.e. herringbone cross-bedding.

Organic content: invertebrates and vertebrates ranging from marine, through brackish to freshwater; oyster reefs particularly characteristic; bioturbation intense and fecal pellets common; peat accumulation and halophytic plants in supratidal salt marshes.

Examples: thoroughly studied are lagoons behind barrier islands, e.g. Laguna Madre, Gulf of Mexico; the Dutch and German Wadden Sea in the North Sea (Figure 75), and macrotidal coasts of the English Channel as well as of the Bay of Fundy, eastern U.S.A.

Literature: Hubbard *et al.* (1979), Tietze (1979), Reineck and Singh (1980).

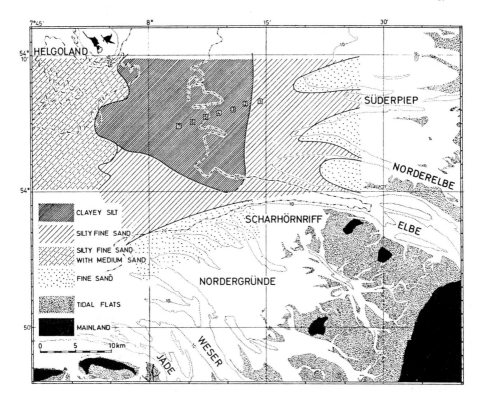

Fig. 75.   Tidal flats and subtidal deposits, eastern inner part of the German Bight (after Gadow and Reineck 1969).

4.1.1.6.2. Beaches and barrier islands. Definition: beach and barrier-island deposits are long, narrow sand accumulations aligned parallel with the shoreline. Beaches are attached to the land, and barrier islands are detached from the land by a shallow lagoon and dissected by tidal inlets.

Geometry: wave-dominated sheets or shoestrings parallel with the regional strike.

Lithology: very fine, fine, medium, and coarse sands, locally pebbly; pelitic sediments only in traces; either terrigenous (predominantly quartzose) or calcarenitic; predominantly matrix-free and well sorted; with dark heavy-mineral streaks; mineralogically and texturally rather mature to very mature; pyrite and detrital mica very rare.

Sedimentary structures: complete range of sedimentary structures; beaches with parallel stratification or low-angle, planar cross-stratification; the following coarsening-upward sequence may be present in regressive barriers: transitional base with laminated, argillaceous open-marine (subwave base) facies, grading up into interlaminated, rippled, and burrowed argillaceous silt and very fine sand. The higher upper shoreface is represented by regular parallel stratification, horizontal or gently seaward dipping, rare troughs and tabular planar cross-beds; local channels; upper contact with lagoonal facies sharp; coarse sands in backshore represent storm deposits; for transgressive barriers above sequence reversed, i.e. fining-upward (Figure 76).

Current patterns: commonly bipolar; behind the barrier high-angle cross-beds dip predominantly towards the land; shoreface dips more gentle and oriented towards the sea; in tidal channels bipolar current patterns; particle orientation complex, but commonly perpendicular to regional strike, e.g. swash and rill marks.

Organic content: bioturbation present, increasing towards the transitional zone between beach/barrier sand and open marine facies; sands contain, or may be largely composed of reworked and broken shallow-marine benthonic organisms; in front of barriers normal open marine fauna; carbonaceous matter rare.

Examples: German Bight, North Sea; barrier islands (Nehrungen), Baltic Sea; Gulf of Gaeta, Mediterranean Sea, Italy; barrier islands and chenier plains, Gulf of Mexico; Sapelo Island, Georgia; high-energy coasts, Oregon and California.

Literature: Komar (1976, 1983), Reineck and Singh (1980), Reinson (1984), Davis (1985), Elliot (1986b).

4.1.1.7. Arid Shorelines and Evaporites

Definition: arid shoreline and evaporite environments are characterized by barrier islands sheltering hypersaline lagoons with low tidal range (< 2.5 m); siliciclastic input low and evaporation high yielding essentially carbonates and evaporites.

Geometry: prismatic or tabular, progradational wedges; channels in lagoonal and supratidal areas with shoestring morphology and limited in extent.

Lithology: subtidal carbonate sediments (up to 15 m thick) comprise muds, lime muds, and bioclastic carbonate sands; pebble beds may be intercalated; intratidal lagoonal and supratidal sabkha carbonates with algal mats and evaporites; also fecal-pellet lime muds, oolite sands, and shell beds in tidal channels and deltas; interbedded with and passing shoreward into microcrystalline dolomite with laminae and nodules of anhydrite; locally replaced by calcite or gypsum and other evaporite minerals, also aragonite; halite only in surficial salt crusts; penecontemporanous reddening of evaporitic sediments common.

Sedimentary structures: subtidal portion with lamination, thin bedding, and hard-grounds; intratidal lagoon and supratidal sabkha with original lamination, commonly disturbed by mineral growth; disrupted algal mats and desiccation cracks; bands of anhydrite nodules and enterolithic anhydrite layers; displacive growth of anhydrite; gypsum and halite destroy original sediment fabric; characteristic hopper crystals indicate primary halite precipitation from brines; characteristic shallowing-upward sequences: (1) carbonate sediments and minor clastics, (2) similar sediments with anhydrite nodules, and (3) nodular-mosaic anhydrite.

Current patterns: wave- and wind-dominated sedimentation causing complex current patterns; barrier bar and shoal-carbonate sands with bipolar current patterns; in supratidal marsh areas wide scatter of azimuths caused by gullies and creeks.

Organic content: subtidal portion bioturbated with well-preserved fauna of pelagic and minor benthonic organisms; authochthonous algae absent; bioclastic grains of bivalves, foraminifera, gastropods, and coral/algae; locally small algal patch reefs; intertidal lagoons

WATER MOTION	OSCILLATORY WAVES	WAVE COLLAPSE	WAVES OF TRANSLATION (BORES); LONGSHORE CURRENTS; SEAWARD RETURN FLOW; RIP CURRENTS	COLLISION	SWASH; BACKWASH	WIND
DYNAMIC ZONE	OFFSHORE	BREAKER	SURF	TRANSITION	SWASH	BERM CREST
PROFILE						MLLW
SEDIMENT SIZE TRENDS	COARSER →	COARSEST GRAINS	COARSER →	BI-MODAL LAG DEPOSIT	COARSER	WIND-WINNOWED LAG DEPOSIT
PREDOMINANT ACTION	ACCRETION	EROSION	TRANSPORTATION	EROSION	ACCRETION AND EROSION	
SORTING	BETTER →	POOR	MIXED	POOR	BETTER →	
ENERGY	INCREASE →	HIGH	GRADIENT →	HIGH		

Fig. 76. Cross-section through a modern beach, displaying geometry, water motion, dynamic zones, and sedimentation characteristics (after Ingle 1966, Fig. 116). Zones of high-suspended-grain concentrations are hachured. MLLW = mean lower low water.

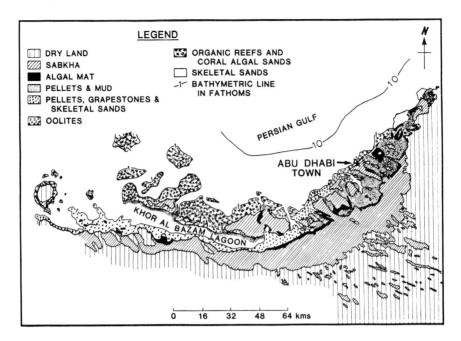

LEGEND

☐ DRY LAND	▨ ORGANIC REEFS AND
▨ SABKHA	CORAL ALGAL SANDS
■ ALGAL MAT	☐ SKELETAL SANDS
▨ PELLETS & MUD	⌐⌐ BATHYMETRIC LINE
▨ PELLETS, GRAPESTONES &	IN FATHOMS
SKELETAL SANDS	
▨ OOLITES	

PERSIAN GULF

ABU DHABI→
TOWN

KHOR AL BAZAM LAGOON

0   16   32   48   64 kms

Fig. 77.   Recent arid shoreline and evaporite deposits: coastal carbonate facies in Abu Dhabi, Trucial Coast (from Schreiber *et al.* 1982).

and supratidal sabkhas commonly burrowed, with sparse, well-preserved marine fauna; diagnostic algal mats; with extreme evaporation only banded algae survive.

Examples: modern sedimentation along the Trucial Coast (Figure 77) and Qatar Peninsula, Arabian Gulf; other examples are Ranns of Kutch, India; Baja California, Mexico; littoral portions of Shark Bay, western Australia; portions of the Bahama Banks.

Literature: Curtis *et al.* (1963), Holser and Kaplan (1966), Shearman (1970), Kinsman (1969), Purser (1973), Gavish (1974), Glennie and Evans (1976), Kendall (1984), James (1984).

4.1.1.8. Shelf Siliciclastics

On modern shelves, defined by 10–200 m water depth and gentle slope of 0.1°, a complex mixture of tidal, wind, wave, storm, oceanic, and density currents disperses the sediments. Because most modern shelves are relics of pre-Holocene sediments exposed and reworked by modern disposal systems, they reflect sedimentary disequilibrium. Relic morphology and shoreline retreat are evident on numerous modern shelves.

Water and sediment movements are variably tide- and storm-dominated. Thus, NW European tide-dominated shelves show tidal current transport paths with fluctuating grain-size distribution (Figure 78). Storm-dominated shelves, normally characterized by low tidal range (< 3 m) and weak tidal currents (< 0.3 m s^{-1}) show a general offshore decrease

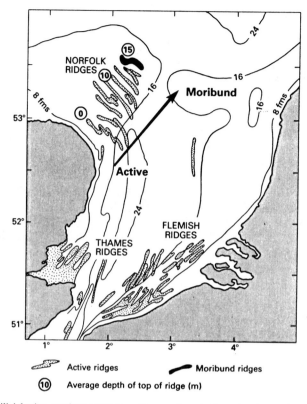

Fig. 78. Shelf siliciclastics, southern North Sea: Groups of sand ridges showing northward transition from nearshore active ridges to deeper, moribund ridges (from Stride *et al.* 1982, after Reading 1986, p. 243).

in grain size, e.g. on the Bering, Oregon, and southwestern Gulf of Mexico shelves.

The final distribution patterns of shelf siliciclastics depend on magnitude and direction of tidal and wave currents and on the availability of sediments. Grain size and bed forms change downcurrent along tidal transport paths. On storm-dominated shelves stirred-up bottom sediments can be transported by currents down to depths about 200 m. Only storm-generated density currents can transport significant amounts of sand from the shoreline out onto the shelf.

Geometry: in general oceanward-dipping prisms of sediments; modified through varying current and weather features. Tide-dominated sand bodies are second-order large-scale linear tidal sand ridges (up to 60 km long, 1–2 km wide, up to 40 m high, parallel spacing 5–12 km), second-order sand ribbons (up to 20 km long, 200 m wide, 0.1 m thick), and third-order sand waves (3–15 m high, 200–500 m wavelength). Storm-dominated sand bodies are first-order shoal-retreat masses (up to 72 km long, 21 km wide, 10–30 m thick); second-order linear shoal ridges (tens of km long, up to 10 m high); sand waves with lee face angles more than 25° (20 km long, 10 km wide, up to 8 m high) also formed by intruding ocean currents (e.g. southeastern tip of Africa).

Lithology: tide-dominated sediments: mud, silt, very fine, fine, medium to coarse sands; locally mud areas with thin graded sand and shell layers, interpreted as intertidal sands and faunas deposited by storm surge density currents; tidal ridges composed of shelly, well sorted, medium sands; sand ribbons commonly rest on gravels, blocks, pebbles, and shells. Storm-dominated sediments: mud, silt, very fine, fine, medium to coarse sands; generally sands on the inner shelf, muds on the middle and outer shelf; glauconite, chamosite, or phosphorite may be admixed.

Sedimentary structures: tide-dominated: common small- and medium-scale cross-bedding (sets up to 2 m); overall fining-upward sequences with a base of coarser lag; overlain by sand waves, also termed giant ripples; internal structures of sand ribbons and third-order sand waves little known; unimodal large-scale cross-bedding presumably predominating; tidal sand ridges with third-order megaripples and internal cross-bedding; sand patches and small sand waves with various ripple-bed forms and bioturbation at the distal ends of tidal-transport paths; tidal-transport paths end in mud areas (Figure 79). Storm-dominated: storm laminae composed of slightly bioturbated, graded sands (tempestites), up to 9 cm thick; wave-formed oscillation ripples; internal structure of shoal ridges little known, presumably with storm-erosion planes, medium-scale cross-bedding, and coarsening-upward gradient; difficult to discriminate between storm-dominated linear shoal ridges and similar tide-dominated ridges in terms of internal structures only; comet marks and sand shadows in the Baltic Sea; reworking by burrowing organisms (i.e. mottled structure) may destroy graded bedding and sedimentary structures.

Current patterns: in general shelf-current systems vary strongly in nature and intensity; persistent tidal-current systems and directions well known from many shelves; thus, grain-size and orientation trends of sand waves, sand ribbons, and tidal-sand ridges determinable. Wave- and storm-dominated current patterns are more irregular; shoal ridges may be obliquely oriented ($< 35°$) in respect to the shoreline; sparse data on particle orientation on deeper portions of the shelf.

Organic content: tide-dominated: animal population and bioturbation high; less bio-turbated than storm-dominated sediments; abundant fecal pellets; pelecypod shells with concave-up preferred orientation; even some allochthonous organisms (wood, freshwater diatoms, pollen, spores); sand ribbons occasionally with delicate epibenthonic fauna (bryozoans, lamellibranchs); storm-dominated: thoroughly bioturbated muds (shallow-water trace fossils) in relatively deep areas of low-wave activity; very abundant fecal pellets; diatoms, radiolarians, and foraminifera.

Examples: tide-dominated: German Bight and southern North Sea; off Great Britain; Strait of Gibraltar; Bay of Fundy, U.S.A. Storm-dominated: Texas shelf and Atlantic shelf off the eastern U.S.A.; Bering, Oregon-Washington, and southwestern Gulf of Mexico shelves. Dominated by intruding ocean currents: southeastern tip of Africa.

Literature: Moore and Scruton (1957), Stride (1963), Emery (1968), Houbolt (1968), Belderson *et al.* (1972), Swift *et al.* (1972), Stanley and Swift (1976), Banner *et al.* (1979), Reineck and Singh (1980), Johnson and Baldwin (1986), Walker (1984), Swift *et al.* (1991).

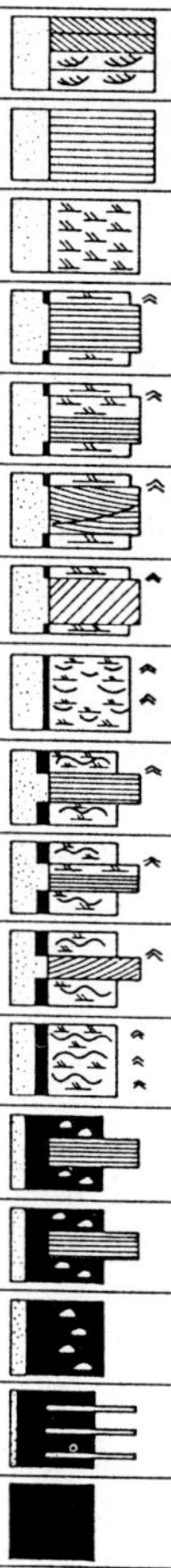

FACIES	SUBFACIES	TYPICAL LOG	INTERNAL STRUCTURE	SAND CONTENT	BED OR SET THICKNESS	INFERRED PROCESSES & NOTES
SANDSTONE FACIES	Sa Cross-Bedded		Tabular / Trough } Cross-bedding	90-100%	ca 10-200 cm	Cross-beds variable in type and set thickness. Represents dunes/megaripples (trough sets) and sand waves (tabular sets).
	Sb Flat Bedded		Parallel and low-angle lamination		variable	Wave- or current-formed lamination associated with high-energy conditions.
	Sc Cross-laminated		Cross-lamination		1-5 cm	Cross-lamination. Varies in relation to ripple type, notably current, combined-flow and wave ripples.
HETEROLITHIC FACIES	Ha Sand Dominated		Parallel lamination	75-90%	5-20 cm (max 200 cm)	Alternations of parallel and cross-laminated sheet sandstones. Thicker sheet sandstones may form 20-90% of this subfacies. Amalgamation may be common.
			Parallel to cross-lamination		5-20 cm (max 200 cm)	
			Low-angle and trough lamination		5-20 cm (max 50 cm)	Sand deposited from suspension & as bedload. Variable reworking by current and wave ripples.
			Isolated tabular cross-bedding		5-20 cm (max 50 cm)	Sheet sandstones commonly inferred to be the product of intense storm conditions. May contain transported shell debris.
			Sandy flaser bedding		1-5 cm	Bioturbation increases in the finer grained intercalations.
	Hb Mixed		Parallel lamination	50-75%	1-10 cm	Mainly ripple laminated sandstones & mudstones with subordinate parallel laminated sheet sandstones (10-50%).
			Parallel to cross-lamination		1-10 cm	Variable types of cross-lamination in response to current, combined-flow and wave ripples.
			Low-angle lamination		1-10 cm	Storm and fair weather increments may be recognized as above. Upper part of sheet sandstones bioturbated.
			Flaser-wavy bedding		1-3 cm	
	Hc Mud Dominated		Parallel lamination	10-50%	1-5 cm	Mainly linsen bedding with rare sheet sandstones (5-10%).
			Parallel to cross-lamination		1-5 cm	Sand lenses formed by current or wave processes. Sandstone interbeds deposited from suspension during storms. Suspension deposition of muds predominant fair weather process. Latter commonly intensively bioturbated.
			Linsen bedding		1-3 cm	
MUD FACIES	Ma		Graded sand &/or shell-rich layers	0-10%	0.1-2 cm	Mainly muds with thin sand interbeds and sand and silt streaks. Deposition entirely from suspension.
	Mb		Mud		< 0.5 cm	Wave and current activity only accompany rare storms. Intensive bioturbation, *in situ* or slightly transported benthic faunas.

Fig. 79. Facies scheme for siliciclastic sediments illustrating the main types of lithofacies in the sublittoral environment (after Johnson 1978).

4.1.1.9. Shallow-Water Carbonates

The shallow-water carbonates comprise subtropical carbonate shelves (protected shelf lagoons or rimmed shelves versus open shelves) and temperate-water shelves. Influx of terrigenous clastics is negligeable; waters are oxygenated. Banks are carbonate buildups of non-wave resistant material in contrast of reefs.

4.1.1.9.1. Platform carbonates. Geometry: linear, sigmoidal shoestrings; linear tidal ridges (e.g. 8 km long, 750 m wide, 5 m thick); spillover lobes (e.g. 1 km long, 500 m wide, 1.75 m thick); configuration dependent on current conditions and morphology.

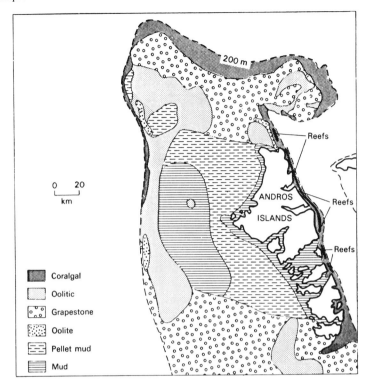

Fig. 80.   Recent shallow-water carbonates: sediment distribution on the Andros Platform, Great Bahama Bank (after Sellwood 1986, redrawn from Purdy 1963b).

Lithology: very fine, fine, medium to coarse carbonate sands, lime muds, and marls; locally with admixtures of wind-blown dust; also with thin, lithified subtidal hardgrounds (e.g. Persian Gulf); ooids, peloids, oncoids, and lithoclasts rather common in shallow areas; rounded, well sorted, and well washed under high-energy conditions (i.e. shallow coastal areas); angular, poorly sorted, and muddy under low- to moderate-energy conditions (i.e. deeper offshore areas); chiefly aragonite and calcite, locally dolomite crusts; only minor admixtures of siliciclastics and traces of glauconite.

Sedimentary structures: lamination and small-scale or large-scale cross-bedding; commonly bioturbated; low sedimentation rates.

Current patterns: cross-bedding and particle orientation of high-energy carbonate sand shoals and tide-dominated lagoonal areas controlled by current strength and direction; bimodal patterns under influence of tidal currents; strong tidal currents enhance oolite formation; current patterns comparable to siliciclastic environments.

Organic content: organisms are of prime importance, abundant fauna and flora; all organisms groups except for pelagic forms; sea-grass meadows; calcareous algae (even in hostile salt flats), coccoliths, foraminifera, bryozoans, molluscans, corals, and echinoids; mostly as skeletal sands composed of shell debris; fecal pellets common.

Examples: subtropical carbonate shelves: Bahama platform (Figure 80), South Florida Shelf (Figure 73); eastern Gulf of Mexico; Yucatan Shelf, off Mexico; Shelf of Guatemala and Honduras; Arabian Gulf; Shark Bay, off western and northern Australia. Temperate-water carbonate shelves: Manning Bay, off western Ireland and Norway.

Literature: Illing (1954), Houbolt (1957), Purdy (1963a, b), Pilkey and Noble (1966), Multer (1977), Purser (1973), Harris (1979).

4.1.1.9.2. Reefs and atolls. Major types of contemporary carbonate buildups are barrier reefs, fringing reefs, and atolls (oceanic reefs). Large reef complexes are composed of all forms. Karst-induced features can affect barrier reef and atoll morphology. Reefs occupy normally only a small proportion of a platform.

Geometry: in plan linear (e.g. Great Barrier Reef 2000 km long, 16–320 km wide), sub-circular, or isolated and atoll-shaped; three-dimensional form influenced by transgressions or regressions.

Lithology: three types of lithofacies generally recognizable: (1) back-reef lagoon with calcilutites, calcarenites, and pelletal lime muds; (2) reef core with biolithites, i.e. in-situ skeletons of calcium carbonte-secreting organisms that may be completely recrystallized, often dolomitized, obliterating original fabric; (3) reef talus with skeletal calcarenites and calcirudites with micritic matrix. The back-reef facies may grade into sabkha-type evaporites.

Sedimentary structures: back-reef facies: laminated, rarely bioturbated and with desiccation cracks. Reef core: no structure, massive. Reef talus: poor bedding dipping off reef front; slides and slumps; may grade basinward into carbonate turbidites.

Current pattern: no specific patterns; slumps, slides, and turbidites may indicate slope; mostly of local significance since reef-related.

Organic content: in reefs great faunal diversity; reef core characterized by rich and diverse fauna. Reef framework formed of calcareous algae, stromatoporoids, bryozoans, and corals; forereef areas with planktonic forms.

Examples: Great Barrier Reef, Australia (Figure 81); Pacific atolls; Bahamas, Florida, Guatemala, and Honduras carbonate platforms.

Literature: Maxwell (1968), Hill (1974), Purdy (1974), James and Ginsburg (1979), Chappell (1980).

4.1.1.10. Continental Margin Environments

4.1.1.10.1. Deep-sea clastics (turbidites). Modern turbidite fans are classified according to Reading (1986) into: abyssal cones (up to 10 km thick), deep-sea fans (up to 1 km thick), short-headed delta-front fans (e.g. in fjords, Swiss lakes, and North American lakes), continental rise fans, and mixed-type fans.

Geometry: distinct prisms or wedges of fans, sheets or cones covering large areas; individual sand layers may have considerable lateral extent (hundreds of meters); influenced by topographic features; channels with radiating shoestring morphology (Figure 82).

Lithology: proximal sediments: massive sands, graded pebbly sands, clast-supported gravel deposits (turbulent flow), and chaotic matrix-supported pebbly sands and gravel

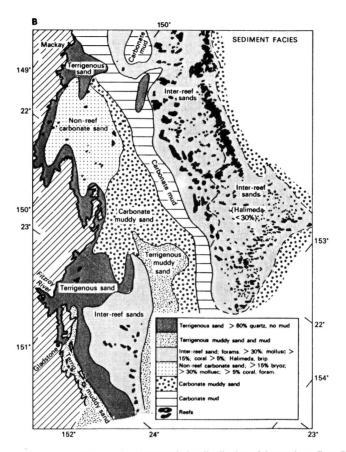

Fig. 81.   Recent shallow-water carbonates: Sedimentary facies distribution of the southern Great Barrier Reef, Australia (after Sellwood 1986, redrawn from Maxwell 1968).

deposits (subaqueous debris flow); more common interstratification of sand and mud; sands variable in composition and grain size ranging from skeletal shell sands, through quartz sands to lithic sands; sand-shale ratio high inside channels and low in interchannel areas; overbank muds rich in platy constituents, i.e. micas and organic matter.

Sedimentary structures: monotonous alternations of sands and muds are typical bedding features; graded-bedded often with a set sequence of structures from base to top (classic fining-upward Bouma cycle); sand layers seldom > 3 m, generally < 0.5 m thick; abrupt and sharp bases of sands with diverse erosional features (e.g. tool marks, scour marks), locally trails and burrows at the base of turbidite sands; dish structures in massive sands; slides, slumps, and channels may also be present; progradational nature of many fans cause large-scale coarsing-upward sequences with distal deposits at the base and proximal deposits at the top.

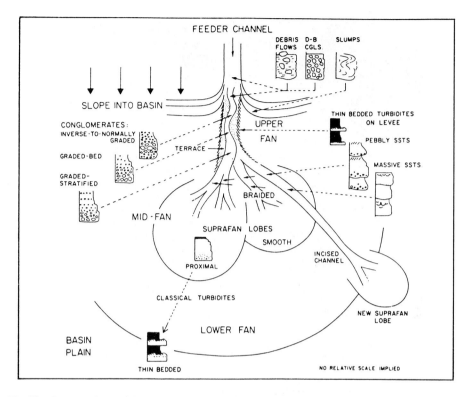

Fig. 82.   Deep-sea fan model proposed by Walker (1978, reprinted with permission). Downcutting channel, fan extension and development of new lobes like in the modern La Jolla Fan (California).

Current patterns: simple as well as complex; longitudinal and/or transverse basin filling; cross-bedding poorly preserved; tool and scour marks indicate local flow direction; long axis of pebbles normally parallel to flow; low azimuth scatter at any point of the fan; proximal and distal deposits contrast in current patterns.

Organic content: muds (nanno ooze) may contain pelagic planktonic and deep-water organisms; sands contain abraded debris of shallow-water benthonic, skeleton-forming animals and, locally, plant debris.

Examples: California Borderland deep-sea fans; Astoria deep-sea fan, off Oregon; Grand Banks; Bengal deep-sea fan; Mississippi, Amazon, Congo, Indus, Ganges, Rhône, Balearic, and Nile abyssal cones and plains.

Literature: Bouma (1962), Normark (1978), Middleton and Hampton (1976), Walker (1978, 1984), Stow and Piper (1984), Pickering *et al.* (1989).

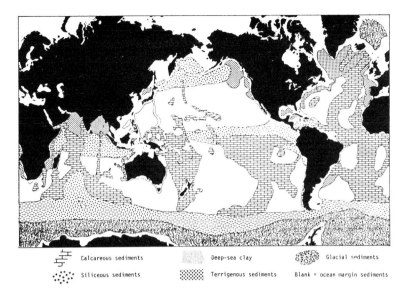

Fig. 83.   Global distribution of major types of pelagic and other sediments on the ocean floor (after Davies and Gorsline 1976).

4.1.1.11. Pelagic Environments

The term pelagic refers to ocean areas with limited input of clastic material; it does not necessarily mean deep water. The majority of pelagic deposits nevertheless occurs at bathyal and abyssal depth (Figure 83). The predominant type of modern pelagic sediments are oozes and clays (Figure 76). Their distribution and composition are controlled primarily by changing oceanic morphology, variations in surface-water productivity, and the calcite compensation depth (CCD). Remarkable is the pelagic sedimentation on present spreading ridges, normally rising to about 2 700 m depth. The sediments directly above the basalt contain normal smectite derived from the chemical breakdown of basalt and enriched in metals, especially iron and manganese, but with minor copper, lead, zinc, nickel, and cobalt (Jenkyns 1986). Dense colonies of suspension-feeding benthos, dominated by bivalves of unusually large size for a deep-sea fauna, occur at sites of hydrothermal discharge. These organisms probably feed on bacteria living near these warm submarine springs (black smokers).

Geometry: in general widespread veil-like distribution of lithologies.

Lithology: calcareous oozes (above CCD), red clays, siliceous oozes with opaline skeletons of diatoms, silicoflagellates, and radiolarians; volcanogenic muds, glacial muds, and turbidite muds; locally terrigenous with land- or shelf-derived material including wind-blown silt with "Wüstenquarz"; with phosphate and manganese nodules; interbeds of turbiditic silt and fine sand; locally influenced by submarine exhalations; sedimentary constituents are mostly pelitic of biogenic, terrigenous, authigenic, volcanogenic, and/or cosmogenic origin.

Sedimentary structures: laminated and thinly bedded like varves; locally with ripples or "hardgrounds"; deep-sea photographs reveal ripple marks, megaripples (abyssal dunes), current lineation, tool marks, scour marks, current marks, and small-scale cross-bedding; with burrows, partly pyritic, and surface trails; depth of bioturbation down to 15 cm.

Current pattern: bottom currents (4–40 cm/sec) observed.

Organic content: typical pelagic fauna with radiolarians; planktonic foraminifera, vagile thin-shelled benthic organisms (bivalves, gastropods, and sea urchins; sponges and corals), and marine vertebrates; high content of fecal pellets; shells may show corrosion, influence of carbonate compensation depth (CCD); shells of original calcitic composition better preserved than originally aragonitic shells.

Examples: deep-sea world-wide; small pelagic basins such as the Red Sea, the Mediterranean Sea, the Gulf of California, and the North Pacific back-arc basins (Bering Sea, Sea of Okhotsk, and Sea of Japan) may be distinct from large oceans in terms of salinity, productivity, water mixing, and other factors.

Literature: Shepard (1963), Lisitzin (1972), Berger (1974), Reineck and Singh (1980), Jenkyns (1986).

4.1.2. ANCIENT ENVIRONMENTS OF DEPOSITION

In analogy to Section 4.1.1, the ancient environments of deposition characterized by their typical sedimentological features are listed below.

The depositional systems of ancient environments are treated in detail in Davis (1983). Visher (1965, 1990) refers to modern hydrocarbon exploration tools and gives detailed instructions as to how to reconstruct ancient sedimentary environments.

4.1.2.1. Ancient Alluvial Deposits

The predominantly siliciclastic deposits of the alluvial realm range considerably in particle size from conglomerates in alluvial fans to the progressively more arenaceous and argillaceous sediments laid down in braided and meandering rivers. Patterns of sediment transport and deposition change as the base level is approached.

4.1.2.1.1. Ancient alluvial fans. Alluvial fans are cone-shaped deposits laid down in areas of high relief where sediments are supplied abundantly. Four types of alluvial fan deposits can be distinguished (Collinson 1986): debris-flow deposits, sheet-flood deposits, stream-channel deposits, and sieve deposits.

Ancient alluvial fan deposits in the stratigraphic record can be easily recognized, because they mainly consist of conglomerates. Stratified conglomerates with little matrix represent stream channels and sheet floods; unstratified conglomerates with dispersed clasts in an argillaceous matrix are indicative of mudflow. Fining-upward sequences may be caused by waning sediment supply as the source area was progressively eroded. Coarsening-upward sequences, which are less common, result either from uplift of the source area and/or from the progradation of fan lobes. The internal structure is chaotic.

Coalescing adjacent alluvial fans can form an elongate sedimentary wedge against a mountain front or a fault scarp. They may be recognized by seismic survey or defined on the dipmeter log by drape in overlaying sediments. The mineral composition closely matches that of the sediment source area. The uranium or potassium content may be slightly lower in the fan sequence than in the source area. Porosities are generally low to moderate due to poor sorting. The conglomeratic nature of alluvial fans may be recognized on the dipmeter log; non-correlatable log curves are characteristic.

Examples of ancient alluvial fan deposits are found in the Torridonian (late Precambrian) of NW Scotland, the Rotliegende of NW Europe, and the Triassic of central Europe. Hydrocarbons occur occasionally in ancient alluvial fan deposits, e.g. the Quiriquire Field, Venezuela, the Elk City Field, Oklahoma, and in Cretaceous fields (Zelten) of the Sirte Basin, Libya.

4.1.2.1.2. Ancient braided and meandering river deposits. Within the lower reaches of the alluvial realm low-sinuosity and high-sinuosity river deposits can be distinguished. Low-sinuosity rivers are braided. Their deposits contain more sand than those of high-sinuosity rivers. Sedimentation takes place commonly in rapidly shifting channels and silts are rarely deposited in abandoned channels. In the channels transverse sand waves, dunes, scour-and-fill structures, and low-angle erosion features, known as reactivation surfaces, are observed.

High-sinuosity or meandering rivers occur within gentle slopes. The river channel deposits contain more sand than the finer-grained flood plain deposits. Levees are normally composed of varying proportions of sand and silt. Fining-upward sequences are produced by lateral migration of channels.

Sedimentary sequences of meandering rivers are composed of interbedded sandstones and shales, generally in units up to 5 m thick, locally with thin coal seams. Intense red colors are less frequent than in other alluvial deposits because former vegetation kept the environment under reducing conditions. Nevertheless, many ancient alluvial deposits are distinctively red-colored due to the presence of hematite. This coloration seems to be post-depositional, either by diagenetic degradation of biotite and hornblende or by later ground-water circulation (Walker 1967). Sedimentary structures within levees and crevasse-splay deposits are not well documented, both may show lamination and current bedding. Except for occasional spores and plant debris in silty intervals, fossils are not preserved in this environment.

Apart from differences in grain size and sedimentary structures, paleocurrent analysis may assist to distinguish braided and meandering rivers. A high sand/shale ratio and a very narrow scatter of dips on the dipmeter log are diagnostic of the braided-river environment. Paleocurrent directions of meandering rivers are unimodal and parallel to the direction of flow and paleoslope, but with a wide scatter. This reflects the complex current regime at point bars with various bed forms and the sinuosity of constantly migrating river channels. Characteristic fining-upward sequences with a wide scatter of dips in the dipmeter log suggest point bars. In respect to external geometry, channels range from several hundred meters in width and 3 to 30 m in thickness. The general trend is perpendicular to the ancient coast line.

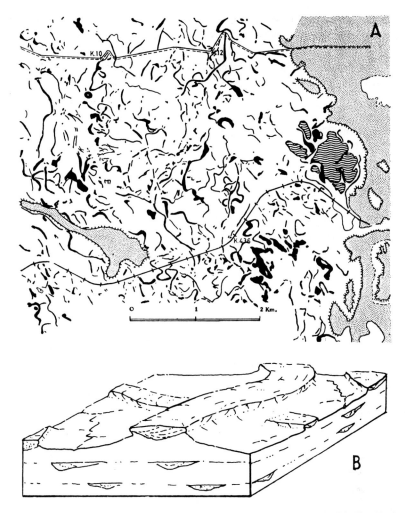

Fig. 84.   Aerial view (A) and cross section of Miocene meandering river channels (B), Ebro Basin around Caspe, N Spain (after Riba *et al.* 1967, Figs. 2 and 3).

Examples of ancient deposits of braided and meandering rivers are widely recognized in the stratigraphic record: the Torridonian of NW Scotland, the Old Red Sandstone in the Anglo-Welsh Basin, the Triassic Newark Group of the NE United States, the Upper Triassic Schilfsandstein, central Europe, and the Miocene Caspe sandstones, N Spain (Figure 84). The Tertiary paleochannels at Caspe, Spain, are one of the most spectacular examples of exhumed meandering-river sandstones (Riba *et al.* 1967).

Hydrocarbon reservoirs occurring in ancient river deposits are usually composed of coalescing multiple channels on top of each other. When channels are single and surrounded on all sides by shale, they form long, narrow oil fields called "shoestrings". Some of them are crooked like rivers, others are quite straight. Such reservoirs are known to occur in the Permo-Triassic intervals of the Prudhoe Bay Field, Alaska, in the Bush City Field, Kansas, U.S.A., in the Oligocene reservoirs of the Seeligson Field, Texas, U.S.A. (Figure 9), and in Miocene fields of the Maikop area, C.I.S.

4.1.2.2. Ancient Lacustrine Deposits

Ancient lacustrine deposits are difficult to recognize without paleontological and/or geochemical evidence. Otherwise lacustrine deposits do not markedly contrast from many marine deposits. Presently, lakes cover about 1% of the surface of the continents; in the past, however, lakes might have been more extensive.

Lakes with clastic deposition must be distinguished from those in which chemical and biological processes predominate. Climate, topography, and mineral composition of the bed rocks are the main environmental controls. Various types of annual-layered or varved deposits are more common than in marine deposits. Fauna and flora show low diversity compared with marine environments. Blue-green algae are normally abundant; they may form stromatolites or oncolites. Other plant fossils include diatoms and charophytes. Some ostracods, bivalves, and gastropods are characteristic. Bioturbation of sedimentary layers is minor due to the low diversity and poverty of burrowing faunas.

Mineralogically the presence of evaporitic minerals like trona, the bicarbonate evaporite mineral Na_2CO_3 $NaHCO_3 \cdot 2H_2O$, might be a direct hing at lacustrine deposition. Well-log techniques are no more diagnostic of lacustrine environments than the sedimentological approach using lithology and sedimentary structures.

Numerous ancient lake deposits such as the Eocene Green River Formation, Wyoming, U.S.A.; the oil shale of the Mae Sot district, Thailand; the Lower Cretaceous (Berriasian) paper shales of NW Germany; and the Miocene of the Nördlingen Ries, Germany, contain large amounts of organic matter; thus, they are actual or potential hydrocarbon-source rocks. The Lower Cretaceous lacustrine sequences of China are part of a rather prolific oil province with source rocks and sandstone reservoirs (see Section 4.2.7).

Classic example of an ancient lacustrine deposit is the Eocene Green River Formation. The following environmental variants are observed: deposits of alluvial fans, sand flats, dry mud flats, ephemeral lakes, saline mud flats, perennial lakes, and salt pans. The following association of sedimentary rock types occurs: sandstones, shales, algal and oolitic limestones, and dolomites. Of special interest is the bituminous, laminated oil-shale facies. Fossils in the oil shales, whose bituminous matter was derived from the decay of planktonic algae, include well preserved fish. Varve-like lamination and sedimentary structures indicate repeated subaerial exposure and current activity. The occurrence of trona deposits and the high amount of magnesium in the carbonates point to a model of a vast alkaline-earth playa fringing an alkaline lake.

Other well documented lacustrine deposits are the Triassic Lockatong Formation, Newark Group, eastern U.S.A., and the Devonian Caithness Flags of NE Scotland. Further examples of ancient lake deposits are given by Matter and Tucker (1978).

4.1.2.3. Ancient Desert Deposits

The identification of ancient desert deposits is well established due to the thorough work by Glennie (1970, 1987). Red coloration of ancient sediments can be taken as first hint at terrestrial deposits, mostly from a hot arid or seasonally humid climate. However, not all ancient desert sediments are red, nor did all red sediments form in deserts. Glennie (1987) compiled criteria for recognizing wind-deposited and water-deposited sediments of desert environments.

Characteristic sediments of the desert-environment are eolian, fluvial, desert-lake, and sabkha deposits.

Eolian deposits, important as reservoir rocks, accumulated as dunes. Eolian sandstone sequences may vary in thickness from a few centimeters to several hundred meters. A dune sequence starts commonly with horizontal or low-angle cross-bedded, coarse-grained, poorly to moderately sorted, locally pebble sandstones. Clean, well rounded, fine- to medium-grained dune sandstones follow, characterized by large-scale, high-angle cross-bedding (up to 30°). Horizontal stratification may occur locally. The laminae are normally planar with only sparse, poorly developed ripples. The individual laminae are well-sorted, especially in the finer grain size. Marked grain-size differences between adjacent laminae are common. The uncemented quartz grains show frosted surfaces. Sorting is very good; silt and clay content is low (below 5%). Clay drapes are very rare. Detrital mica is mostly absent. Grain-size distribution is commonly unimodal; if bimodal, two well-sorted grain-size maxima are observed. Grain sizes usually range from silt (60 μm) to coarse sand (2 mm) with an average of about 125–300 μm. Maximum size for grains transported under wind action is in the order of 1 cm; but grains over 5 mm are rare. Large sand grains (0.5 to 1.0 mm) tend to be well rounded.

Such eolian-sandstone units display normally characteristic dipmeter patterns with consistent high-angle sedimentary dips. Individual units may become slightly argillaceous towards the base with a typical blue dipmeter motif. Dipmeter patterns may even help to identify the dune type. Paleowind directions are not necessarily related to the paleoslope. Wadi and eolian sandstones may even show opposite paleocurrent directions. Thorium : uranium ratios, shown by the natural gamma-ray spectrometer log (NGT), are high, generally exceeding 6.

Near the basin margins and at the base of desert sequences, hamada and serir deposits may occur, generally followed by a sequence of fluvial conglomerates and sandstones (wadis). The contact between conglomerates and sandstones is rather sharp in desert environments.

Fluvial deposits in desert environments are characterized by sedimentary structures similar to those of fluvial sediments of non-desert continental environments. These structures may be modified by the following features: locally mudflow conglomerates and flash-flood deposits. Horizontally laminated and foreset bedding of flood-plain deposits with deformed bedding features. Common channeling in streamflow sediments with widespread clay laminae. Some sandstones are not well sorted, being locally argillaceous or pebbly. Sandstones are commonly cemented by calcite, locally by gypsum or anhydrite. Fluvial sediments, deposited in a desert environment, may also display features inherited from eolian sediments. Subaerial exposure is indicated by the presence of curled clay flakes, clay pebbles, mud cracks with sandy infill, sand dikes, and interbedded eolian sandstones.

Sandstones, intensely reworked by the wind, but finally deposited in ephemeral streams, shallow lagoons, desert lakes, or sabkha plains are normally more argillaceous and generally cemented by calcite or anhydrite. They may contain argillaceous or micaceous laminae as well as curled mud chips or pebbles.

Examples of ancient desert deposits, especially of eolian sandstones, are known from different stratigraphic intervals world-wide: Torridonian Sandstone (Precambrian), N Great Britain; Lyons Sandstone (Permian), Colorado, U.S.A.; Coconino Sandstone (Permian), Colorado Plateau, U.S.A.; parts of Buntsandstein, central Europe; Navajo Sandstone (Jurassic), Colorado Plateau, U.S.A.; and Botucatu Formation (Upper Jurassic/Lower Cretaceous), Parana Basin, Brazil. For more details see Bigarella (1972) and Glennie (1987).

Some of the best known ancient desert deposits are those of the Permian Rotliegende in NW Europe because they form prolific natural gas reservoirs (see also Section 4.2.1).

4.1.2.4. Ancient Glacial Deposits

Basin-wide studies of ancient glacial deposits are few, though five major periods of glaciation are currently known in geologic history: late Cenozoic (Pleistocene), late Paleozoic, late Ordovician, late Precambrian, and early Proterozoic. During Pleistocene glaciations the ice covered a maximum of about 30% of the Earth's surface.

The most characteristic glacial deposit is the boulder clay or till, an unsorted and unstratified, chemically and physically immature product of glacial abrasion. The criteria for the glacial origin of tillites, the ancient equivalent of tills, comprise exotic, far-travelled clasts, striations on boulders and underlying pavement, and wide lateral extent. The underlying strata may also have been deformed by the ice movement. These proglacial deposits cover substantial areas of marginal to glaciated regions.

Glaciofluvial deposits, laid down by meltwaters, consist dominantly of braided-stream deposits with interstratified sand and gravels. In addition, the following sedimentary features occur: laminites, drumlins, outwash fans, kames, eskers, kettleholes, and freeze-and-thaw-structures.

In periglacial areas beyond the limit of glaciofluvial deposits, finer-grained particles of silt and clay may be transported by wind and deposited as loess. Eolian dust and eolian dust deposits such as loesss are a much overlooked subject. The monograph of Pye (1987) deals with all aspects of dust formation, transportation, and deposition. In glacial lakes, seasonal meltwater influx causes the deposition of thin silt layers alternating with clay laid down through the remaining year, the so-called varves.

Confident identification of ancient glacial deposits depends on the examination of the whole facies association. Subglacially deposited massive or banded tillites normally exhibit a wide variety of clast types. These deposits, normally with an erosive base, are traceable laterally for at least several kilometers. The long axes of clasts commonly show a preferred orientation parallel to the ice flow. Laminated mudstones with isolated dropstones in a random orientation, released by melting icebergs, are characteristic of glaciomarine sedimentation.

Chemical immaturity and complex lithology of the bedrock might be reflected by the electric-log responses. Glacial origin can also be indicated by characteristic surface marks of sand grains under the scanning electron microscope, especially in young sedimentary rocks. Such features may be even diagnostic of various glacial subenvironments. The common conglomeratic structure of glacial deposits and the erratic bedding-features may also be detected by the dipmeter log.

Besides the Pleistocene deposits ancient glacial deposits are well known from the late Precambrian of Norway and Australia and from the late Ordovician and the late Paleozoic of the southern hemisphere and India (Gondwanaland).

Hydrocarbon reservoirs are found in Cambro-Ordovician glacial deposits of the Marsul Field, Oman; in the Permian of the Cooper Basin, S Australia; and outwash sands in N Argentina. Recently, the interest in glacial deposits increased considerably, recognizing their hydrocarbon potential (Eyles and Miall 1984).

4.1.2.5. Ancient Delta Deposits

Deltas are prograding sedimentary wedges which build out into the ocean or into lakes from the mouth of rivers. As rivers reach the coast, currents slow down, drop their load of sediment, and form a delta. Normally the delta plain subsides slowly as a result of compaction of the soft, recently deposited muds. This association is also called paralic.

Delta sediments are known to contain large quantities of hydrocarbons because deltaic processes inject porous sands far out into marine basins where simultaneously rich source beds are found. The potential sand-reservoirs are deposited by fluvial, tidal, and wave processes. Within the spectrum of various delta types the relative importance of each type of sand body (e.g. channels or beaches) and the distribution of the prime reservoir facies within the delta are very variable. Successful subsurface interpretation of a delta depends on recognizing the dominant processes constructing the delta and the genetic type of individual sand bodies. Moreover, deltas contain different types of traps, not only conventional structural traps, but also growth faults, roll-overs, and diverse stratigraphic traps. Rapid sedimentation of organic-rich sediments in a dominantly reducing environment favors quick coalification of peat, and thus, generation of hydrocarbons. Recognition of appropriate delta models is also important for efficient exploitation of hydrocarbons from ancient deltas.

The general shape and the sedimentary features of deltas depend on various factors, such as the nature of the hinterland, sediment input, climate, rate of subsidence, topography, and relative importance of river, wave, and tidal influence (see section on modern delta environments). Consequently, no two deltas are identical, neither at present nor in the past.

The paleocurrent pattern in ancient deltas is rather complex (e.g. Wurster 1964). Distributary channels normally radiate across the delta top; individual channels show unimodal current patterns with little scatter. Tidal currents across the delta top display bimodal current patterns with fair scatter. Longshore currents sweeping across the delta

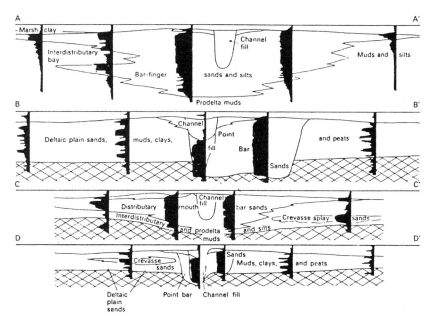

Fig. 85.   Major sedimentary environments in delta facies registered by electric well logs.   Tertiary flu-vial-dominated Holly Springs delta system, Gulf Coast, U.S.A. (after Galloway 1968).

episodically can be unimodal over wide intervals, or occasionally bimodal. They move with less scatter at right angles to tidal currents.

The major sedimentary patterns in a delta facies can be well recognized by means of electric well logs (Figure 85). These patterns can be further refined by appropriate evaluation of dipmeter logs, provided the direction of the shoreline is known. In addition, glauconite present characterizes marine sediments; abundant carbonaceous matter marks winnowed sequences.

Distributary channel sands can be normally recognized on electric well logs (e.g. GR, SP, and short spacing resistivity curves). The base is abrupt. Usually both SP and resistivity are a maximum in the lower portion, because the sands contain less clay and are more porous than the overlying beds. The laminated, cross-bedded and graded-bedded (both coarsening- and fining-upward) sequences of distributary channel sands can be normally identified by the dipmeter log with red patterns corresponding to channel fill.

At their mouth distributary channels deposit their sand load forming distributary-mouth bars which are prograding lobes of sediment with well developed foreset beds. These crescent-shaped river-mouth bars have a lower contact grading into the underlying clay and a sharp upper contact. The electric log shows weak SP and resistivity at the base with shale interbeds, increasing and becoming more constant upwards, with a sharp drop at the top of the sand. They are recognized as distinctive blue patterns on the dipmeter log. When deposited by fast-flowing currents the foresets are fairly flat and the distributary

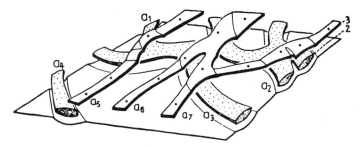

Fig. 86. Ancient delta-sandstone strings: Upper Triassic Schilfsandstein sedimentation in S Germany (modified from Wurster 1964, Fig. 24). The sandstone strings represent distributary channels of a Mississippi-like delta. Note two sand-string levels 2 and 3. a_1–a_7 stand for local names of sand strings. Width of diagram about 11 km.

front is fan- or crescent-shaped. When currents are slow the dips of the foresets may become much steeper and the front tends to be elongate.

Scour channels are generally oriented normal to the distributary channels because they are side-filled by turbulences across the top of the distributary front.

Sand bars built by tidal and longshore currents or wave action are cross-bedded, internally rarely structureless. Tidal bars are normally parallel to distributary channels, wave-built bars perpendicular to them.

Sands reworked by tides or longshore drift currents are characterized by a rapidly changing bimodal distribution of dips. Persistent dip sequences reflecting currents parallel to the shoreline may indicate periods of marine transgressions.

Between active distributaries are bays of shallow, quiet water where mostly mud is deposited. Occasionally when the river is in flood it breaks through its natural levee and spreads sandy material out over the interdistributary bays. These deposits are termed crevasse splays.

Log motifs are widely used to define fining-upward or coarsening-upward sequences. Such a discrimination assists to distinguish between channel sands and distributary-front deposits. Gamma-ray or porosity logs indicating the clay content of different sands are a measure to determine the degree of winnowing and reworking.

Even in ancient deposits it is possible to recognize both fluvial-dominated and wave-dominated delta sediments.

Examples of ancient deltaic deposits are common in the geological record (Akramkhodzhaev *et al.* 1989); e.g. the Ordovician of the Appalachians, U.S.A.; the coal-bearing Pennsylvanian of the Illinois Basin, U.S.A.; the Pennsylvanian Yoredale series of northern England; the Upper Triassic Schilfsandstein, central Europe (Figure 86); the Upper Cretaceous Castlegate Sandstone of Utah, U.S.A.; and the Eocene Coaledo Formation of Oregon, U.S.A.

Ancient delta deposits are preferred environments for hydrocarbon source and reservoir rocks such as black shales and coals and reservoir sandstones (Whateley and Pickering 1989); e.g. the Pennsylvanian sandstones and conglomerates of the Elk City Field, Oklahoma, U.S.A.; the wave-dominated deltaic sandstones of the Middle Jurassic Brent Group,

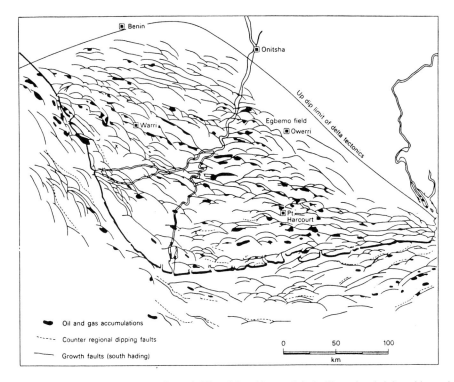

Fig. 87. Ancient delta environment: Cenozoic Niger delta with growth faults illustrating their lateral impersistence, slightly curved concave-to-basin boundary, and the general parallelism with the delta front (after Weber and Daukoru 1975).

North Sea (Scotchman and Johnes 1990); the Tertiary proto-Mississippi delta, U.S.A.; and the Cenozoic Niger delta, Africa (Figure 87). The Niger delta is large and complex, with many depositional subenvironments and characterized by numerous growth faults (Weber 1971).

4.1.2.6. Ancient Clastic Shoreline Deposits

4.1.2.6.1. Ancient beach and barrier-island deposits. With a general idea of the ancient shoreline trend it is possible to define paleogeography and evaluate character and extent of potential sandstone reservoirs using dipmeter and porosity logs. The dipmeter log assists to define the major subenvironments in clastic shoreline deposits, especially in well developed fining-upward and coarsening-upward cycles. Beach deposits normally show coarsening-upward successions. Beach and barrier-island deposits commonly show straight trends, i.e. elongate, bi-convex shoestring sandstones. Grain size and cleanness of the sand decrease in an onshore direction or toward the lagoon. When beaches are buried by layers of mud, they become completely enclosed in shale, forming stratigraphic hydrocarbon traps.

Main criteria used in recognizing ancient beach and barrier-island deposits are shape, orientation, and stratigraphic framework of the sandstone bodies. The majority of beach and barrier-island deposits are composed of fine- to medium-grained sandstones. Progradational sequences produced by beach faces of differing wave energy can be distinguished.

Examples of ancient beach and barrier-island deposits are Carboniferous orthoquartzites of the Pocahontas Basin, Virginia, U.S.A.; Carboniferous sandstones of northern England; Lower Jurassic sandstones of England; Lower Jurassic sandstones in Luxemburg (Berners 1983); the Lower Cretaceous Muddy Sandstone, Montana, U.S.A. (Figure 88); and Upper Cretaceous sandstones in Utah and New Mexico, U.S.A.

Only a few transgressive barrier deposits have been identified in the geological record so far, presumably because they are thin (e.g. Pre-Cambrian Ekkeroy Formation, northern Norway and Lower Silurian of southwest Wales).

Ancient beach and barrier-island deposits are highly potential hydrocarbon reservoirs because the sandstones are fine to medium-grained, well sorted, low in primary clay matrix, and, thus, porous and permeable. Examples are the Devonian Third Venango Sand, Pennsylvanian, U.S.A.; the sandstones in the Bell Creek Field, Montana, U.S.A.; the Jurassic Piper Field in the northern North Sea; and the Cretaceous Bisti Field (Sabins 1962) in the San Juan Basin, New Mexico, U.S.A. (Figure 89).

4.1.2.6.2. Ancient tidal-flat and estuarine deposits. Criteria to recognize tidal processes in ancient sedimentary rocks include the presence of paleocurrent patterns indicating bidirectional flow, the abundance of reactivation surfaces and clay drapes, and the presence of sedimentary features indicating repeated, small-scale alternations in sediment transport. In the bimodal paleocurrent patterns the inferred ebb stream component is dominant and produces low-angle and sigmoidal foresets, whereas the inferred flood stream produced mainly planar foresets. The intertidal realm, moreover, is characterized by repeated emergency and the presence of surface run-off features formed by the ebb stream.

Thus, recognition of ancient tidal-flat and estuarine deposits is mainly based on the association of sedimentary structures and sediments considered to indicate tidal processes and on the presence of fining-upward sequences reflecting tidal-flat progradation. Ancient tidal channels comprise an erosive-based conglomeratic lag overlain by coarse-grained sandstones with planar, trough, sigmoidal, and complex low-angle sets of cross-bedding. The overlying tidal-flat deposits include fine- to medium-grained, ripple-laminated sandstones, and thin, tidal-current and storm-generated sandstone units interlaminated with silty claystone interbeds. Trace fossils showing escape structures and the presence of stromatolites, birdseyes, and desiccation cracks as well as symmetrical and asymmetrical ripple marks draped by clays, and causing flaser, wavy, and lenticular bedding may be additional evidence. The presence of sand waves might also be representative of tidal activity (Teyssen 1984).

Estuarine sequences may comprise a basal erosion surface overlain by a thin, intraformational conglomerate and sands with bimodal trough cross-bedding.

Examples of tidal-flat and estuarine deposits have been recognized from Pre-Cambrian to Recent: e.g. the Precambrian-Ekkeroy Formation, northern Norway; the Cambrian

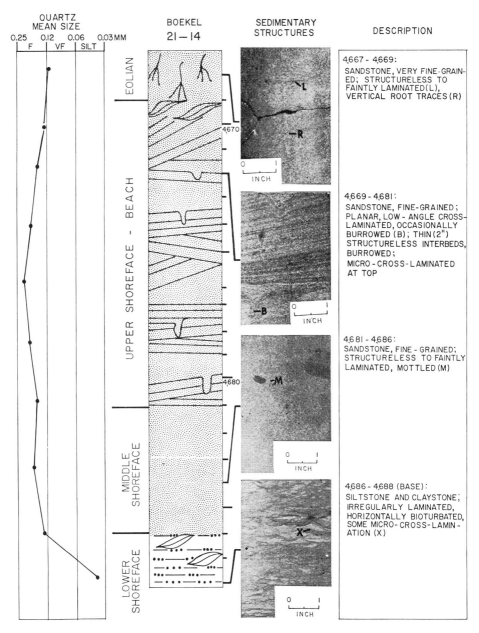

Fig. 88.   Ancient beach and barrier-island environment, Lower Cretaceous Muddy Sandstone, Montana, U.S.A.: Coarsening-upward sequence caused by barrier-island progradation (after Davies *et al.* 1971, p. 555, reprinted with permission; simplified by Reading 1978, Fig. 7.35, p. 165).

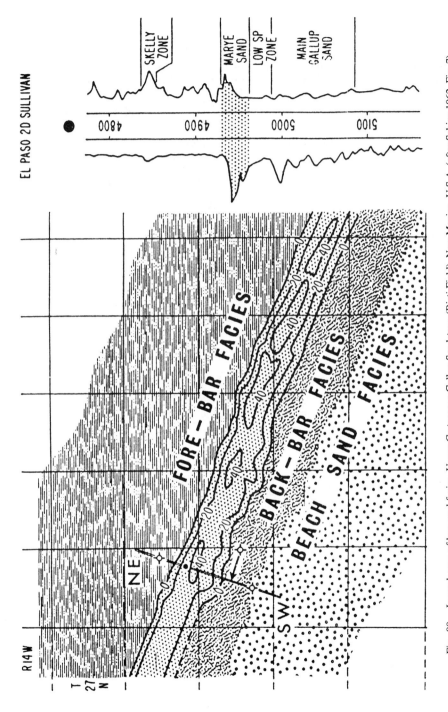

Fig. 89. Isopach map of bar-sand facies, Upper Cretaceous Gallup Sandstone (Bisti Field), New Mexico, U.S.A. (after Sabins 1962, Fig. 7).

Eriboll Sandstone, northwestern Scotland; Lower Jurassic tidal sandstones, Bornholm; the Mid Jurassic Minette sandstones, Lorraine, France; and Eocene sandstones, Pyrenees.

In contrast, only few examples of ancient estuarine deposits are known. A portion of the Tertiary Lower Bagshot Beds of southern England as well as several Pleistocene sequences in the Netherlands are interpreted as estuarine deposits.

In contrast to the massive and porous sandstones in ancient beach and barrier-island deposits, ancient tidal-flat and estuarine sandstones are locally oil-impregnated, but do not easily release the oil. Especially enhanced oil recovery is much hampered by the predominance of thin sedimentation units with complex sedimentary structures.

4.1.2.7. Ancient Arid Shoreline and Evaporite Deposits

Discussion of models of ancient arid shoreline and evaporite deposits is somewhat problematic. Especially the deep versus shallow-water origin of ancient evaporite deposits is still debated (Hsü 1972). Moreover, many large gypsum/halite deposits of the geological past are without modern analogue. Examples of such ancient halite deposits (saline giants) are the Upper Elk Point Basin (Middle Devonian), Canada, and the Zechstein Basin (Upper Permian) of northwestern Europe. The Upper Elk Point Basin is of economic interest because the carbonate rocks of the basin contain large oil fields.

The gross lithology of ancient arid shoreline and evaporite deposits is mainly controlled by the climate, by a high net evaporation and the extent of influx of terrestrial clastic debris. Consequently, a complete spectrum of various lithologies occurs from carbonates, evaporites, mixed carbonate-clastics to clastics.

In carbonate-dominated arid shorelines mudstones and packstones develop in the deeper-water marine areas, although argillaceous and sandy terrigenous material may be swept in by longshore currents. Barrier bars are made up by coarsening-upward packstone to grainstone biosparites and oosparites. Intertidal lagoons are composed of bio-pelmicrite mudstones and wackestones, locally with oolite banks and shell beds. In supratidal marsh areas behind the lagoons, algal mats and hypersaline ponds with interbedded dolomitic mudstones and gypsum are deposited. Such a sequence represents a simple model. Variations are conceivable if the terrigenous influx increases or evaporitic conditions become more pronounced. The presence of pyrite is indicative of reducing conditions, especially in lagoonal areas. Glauconite reflects open-marine deposits. Terrigenous clays may still be swept into offshore areas by longshore currents, or blown into lagoonal areas by wind, even along shorelines without influx from rivers.

Lagoonal and supratidal salt-marsh deposits form sedimentary wedges by lateral accretion. However, channels and associated features in the lagoonal and supratidal areas have limited extent.

Physical processes and environmental diversity along a carbonate-dominated shoreline are similar to shoreline deposits dominated by terrigenous influx. Thus, sedimentary structures and paleocurrent patterns are the same in carbonate-dominated as in terrestrial clastic-dominated arid shoreline deposits. However, mineral composition may vary significantly. Without much of terrigenous input, calcium carbonate deposition predominates.

The resolution of the dipmeter log is normally sufficient to detect small variations in formation resistivity. However, the motifs of other well logs are difficult to use for recognizing textural and grain-size variations. For instance, a clean lime mud shows the same response as a cemented grainstone on both porosity and gamma-ray logs.

The hydrocarbon potential of ancient arid shoreline and evaporite deposits is not as large as in shallow-water carbonates. And seismic prospection for this type of deposits is still difficult.

The grade of diagenesis depends on the original depositional environment, on the movement of ground waters, and much on tectonic interference. Nevertheless, the presence of minor clay admixtures in the deeper-water marine and lagoonal areas, the presence of evaporites in the supratidal zones, and better porosities in barrier-bar grainstones may allow environmental modelling in such carbonate sequences mainly by log evaluation. Sedimentological analysis of cuttings or sidewall cores may ultimatively lead to a better environmental interpretation than normally achieved in terrigenous sequences.

Well-documented examples of ancient arid shoreline and evaporite deposits are sparse. Thoroughly studied cases of carbonate-sulfate evaporite-dominated facies are: the late Pre-Cambrian of Kitwe, Zambia; the Lower Devonian (Middle Siegenian), Normandy, France; the Givetian of northern France and southern Belgium; Frasnian deposits of northeastern Belgium and of the North Sea (Argyll field); the Upper Devonian of western Canada; the Lower Carboniferous (Visean) of central England and Ireland; the Mid-Carboniferous Windsor Group of the Maritime Provinces, Canada; the Upper Permian of Cumberland, England; the Germanic Muschelkalk in central Europe; the Upper Jurassic of the Arabian Gulf; the Lower Purbeck (uppermost Jurassic) succession of the Warlingham borehole, south of London; and the Lower Eocene of Jamaica.

Actual oil and gas reservoirs of this type are found in central Europe. The Upper Jurassic (Malm) carbonate reservoirs (Schmidt 1965, Schönfeld 1979) of the Weser–Ems area, northwestern Germany, are small. Fossiliferous beds were deposited in a supersaline epeiric sea at the base of a major saline cycle. Bioclastic and oolitic calcarenites and micrites form most of the shallow-water deposits (Figure 90). Shale and sulfate predominate in the basin.

The Zechstein salt deposits in northwestern Europe are up to 1 200 m thick. Four main evaporite cycles (Z1 to Z4) are recognized in Germany. Each cycle, when complete, consists of:

> Top: Anhydrite (retrogressive)
>
>   Potash salt
>
>   Halite
>
>   Anhydrite (progressive)
>
>   Dolomite
>
>   Limestone
>
> Base: Siltstone, clay, etc.

The halite and potash beds are rhythmically banded on a centimeter scale with halite-rich and clay-dolomite-anhydrite-rich laminae, interpreted as varves or annual layers. Both shallow and deep environments occur in the Zechstein. The edge of the Zechstein

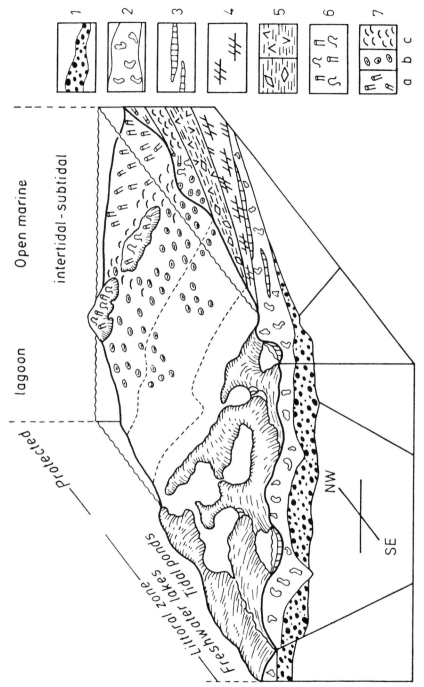

Fig. 90.   Transgressive arid shoreline and evaporite sequence, Tithonian/Berriasian, NW Germany (courtesy M. Schönfeld). Legend: (1) basal conglomerate, (2) lacustrine marls with nodular calcrete, (3) blue-green and green-algal limestones (lacustrine), (4) oolitic limestones and dolostones (supratidal-intertidal), (5) marls with celestite and gypsum (intertidal-subtidal), (6) serpulid-stromatolite patch reefs, (7) reef debris, (7a) serpulid tubes, (7b) ooids with reef debris, (7c) bivalve coquinas with reef debris.

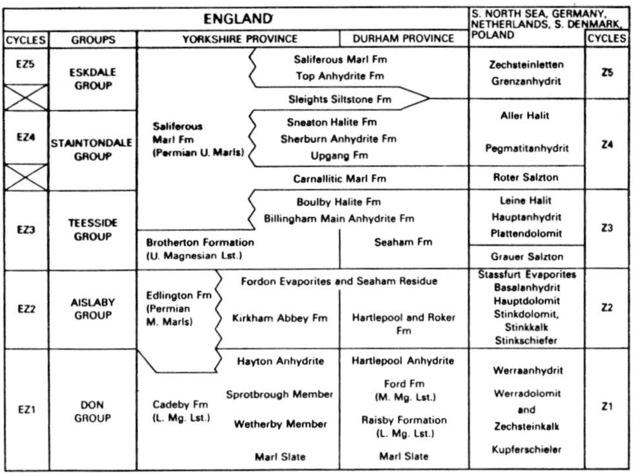

CYCLES	GROUPS	YORKSHIRE PROVINCE	DURHAM PROVINCE	S. NORTH SEA, GERMANY, NETHERLANDS, S. DENMARK, POLAND	CYCLES
EZ5	ESKDALE GROUP	Saliferous Marl Fm / Top Anhydrite Fm / Sleights Siltstone Fm		Zechsteinletten / Grenzanhydrit	Z5
EZ4	STAINTONDALE GROUP	Saliferous Marl Fm (Permian U. Marls)	Sneaton Halite Fm / Sherburn Anhydrite Fm / Upgang Fm / Carnallitic Marl Fm	Aller Halit / Pegmatitanhydrit / Roter Salzton	Z4
EZ3	TEESSIDE GROUP	Brotherton Formation (U. Magnesian Lst.)	Boulby Halite Fm / Billingham Main Anhydrite Fm / Seaham Fm	Leine Halit / Hauptanhydrit / Plattendolomit / Grauer Salzton	Z3
EZ2	AISLABY GROUP	Edlington Fm (Permian M. Marls) / Kirkham Abbey Fm	Fordon Evaporites and Seaham Residue / Hartlepool and Roker Fm	Stassfurt Evaporites / Basalanhydrit / Hauptdolomit / Stinkdolomit, / Stinkkalk / Stinkschiefer	Z2
EZ1	DON GROUP	Cadeby Fm (L. Mg. Lst.) / Hayton Anhydrite / Sprotbrough Member / Wetherby Member / Marl Slate	Hartlepool Anhydrite / Ford Fm (M. Mg. Lst.) / Raisby Formation (L. Mg. Lst.) / Marl Slate	Werraanhydrit / Werradolomit and Zechsteinkalk / Kupferschiefer	Z1

Fig. 91. Stratigraphic nomenclature and correlation of the Zechstein across the NW European Basin (England–Germany–Poland) (after Reading 1986, Fig. 8.45, p. 222, simplified from Smith 1980 and Harwood *et al.* 1982).

basin in the North Sea area reaches the shores of northeastern England. Exploration for oil and gas over the past twenty years has enabled correlations across the basin (Figure 91).

The following carbonate rocks are gas reservoirs in the area between Weser and Ems of northwestern Germany; Hauptdolomit (Main Dolomite) in the Zechstein 2 and Plattendolomit (Banked Dolomite) in the Zechstein 3. The gas is found in biostromes, 180 km long, 30 km wide, and up to 90 m thick. Main reservoir rocks are sucrose dolomites (dolomitic oncolites) with moldic porosity from oncoids. Dolomitization is early diagenetic and subaquatic. The overlying anhydrites act as a seal. Source rocks for the gas are Carboniferous coals. Also in the U.K. North Sea minor reservoirs occur in Zechstein carbonates (e.g. Argyll field). Minor amounts of oil occur in Zechstein carbonate reservoirs only at Volkenroda, Thuringia, F.R.G.

4.1.2.8. Ancient Shelf Siliciclastic Deposits

Sedimentation of modern shelf siliciclastics was mostly studied on continental shelves which still preserve numerous features of Pleistocene sedimentation. Work on ancient shallow-marine sandstones, on the other hand, is governed by conceptual models dominated by epeiric sea settings. Thus, it is somewhat difficult to interpret ancient deposits in terms of what is known about modern shelf siliciclastics. No single sound shallow-marine facies model is in use.

Johnson and Baldwin (1986) and Swift *et al.* (1991) have attempted to compare ancient and recent settings. It proved useful to distinguish between tide-dominated and storm-dominated shelf sedimentation.

Modern tide-dominated shelf sedimentation in partly enclosed seas and gulfs is characterized by a tidal range exceeding 3–4 m. On the other hand, wind-driven waves control the more extensive modern storm-dominated shelves with a tidal range below 2–3 m and a low tidal current velocity. When sand passes into quieter, usually deeper mud environments discrete bodies of sand may form in a muddy matrix. Frequently, however, sand and mud are mixed by bioturbation. This facies of mixed sediments is ideal for the preservation of trace fossils.

Various types of inorganic sedimentary structures are likely to be preserved in the stratigraphic record. Wave-ripple cross-lamination is a distinctive type of sedimentary structure characteristic of such deposits.

The benthic biomass reaches a maximum on the continental shelf, 150–500 g m^{-2} in contrast to a mere 1 g m^{-2} on the abyssal plains. Consequently, benthic organisms are important in reworking the sediment. While there is little benthic life in and on mobile substrates where active transport of bed load takes place, finer-grained deposits exhibit greater stability and contain a rich fauna. Deposit feeders are abundant and continuous reworking and flocculation can destabilize the top few centimeters of a sediment to a thixotropic condition. On the other hand, sea grass, algal baffles, and organic surface films can have a stabilizing influence.

Modern studies in widely separated shelf regions suggest that vertical burrows are almost the only forms in the shore zone between high-tide level and wave base, a phenomenon evidently related to the intensity of physical disturbance. Below the wave base a much greater diversity of forms occurs including horizontal burrows. In the stratigraphic record bioturbated mudstones and clays with rich benthic faunas, indicating a quiet but fairly shallow-marine depositional environment, are common.

Because the flood tide may be much stronger than the ebb tide, or vice versa, unidirectional current patterns may predominate. Moreover, sets with opposite dipping may represent longer-term fluctuations than diurnal.

Reliable criteria for recognizing ancient shallow-marine deposits are sedimentary features that are controlled by salinity and water depth. Such features are marine body fossils, trace fossils, characteristic minerals, and geochemical parameters: invertebrate fossils (corals, cephalopods, articulate brachiopods, echinoderms, bryozoa, and certain calcareous foraminifera), trace fossils (e.g. *Cruziana* facies), authigenic minerals (glauconite, chamosite, and phosphates), and trace elements (e.g. boron-gallium and boron-lithium ratios in illite and mudstones as well as carbon-oxygen isotope ratios). Sedimentological data such as texture, sedimentary structures, lithofacies associations, sand-body geometry, and paleocurrent patterns are usually not in themselves diagnostic. For instance, shallow-marine sandstones of Cambrian and Ordovician age lack body fossils but contain trace fossil assemblages which prove a shallow-marine environment. Tillman *et al.* (1985) review the main characteristics of shelf sands and sandstone reservoirs and figure many instructive outcrops and cores.

High textural maturity is typical of shallow-marine sandstones because of long transport and winnowing of fine constituents. Sedimentary structures alone are rarely diagnostic as to the environment. But bidirectional current structures and patterns are considered to be indicative of tidal currents. Similarly, wave-ripple cross-lamination is characteristic of wave action. Submarine bar sandstones sensu lato are recognized in well logs as well sorted, mostly mature sandstones. They may be developed as coarsening-upward sequences or as massive units with a scoured base. Size, type, and direction of cross-bedding can be identified in the dipmeter log.

Although not diagnostic per se certain lithofacies assemblages and vertical sediment sequences are typical of shallow-marine deposits (Figure 79). Three major facies have been distinguished by Johnson (1978): Sandstone facies (S), heterolithic facies (H), and mud facies (M). The heterolithic facies is subdivided into (1) sand-dominated (Ha, 75–90% sand), (2) mixed (Hb, 50–75% sand), and (3) mud-dominated (Hc, 10–50% sand). Classifications of shallow-marine lithofacies, however, in terms of the dominant physical process have to be taken with caution.

Tide-dominated shallow-marine sandstones can be recognized based upon: (1) open marine macro- and/or micro-fauna usually occurring either as shell beds, or as dispersed and comminuted shell fragments, (2) assemblages and/or communities of characteristic trace fossils, and specific bioturbation (see Section 4.1.1.8), (3) sandstones interbedded with open-marine shales, (4) presence of accessory glauconite and/or phosphate, (5) high textural and mineralogical maturity of the sandstones, and (6) mud drapes, low-relief erosion surfaces of wide lateral extent and absence of deep channels. Tidal current action can be inferred from the following criteria: bidirectional current structures and patterns (herringbone cross-bedding) with two modes approximately 180° apart, the presence of reactivation surfaces, multimodal paleocurrent patterns reflecting either temporal fluctuations in direction or rotation of tidal currents, possibly superimposed by storm activity, and abundant cross-bedding reflecting submarine dunes and sand waves. Unidirection paleocurrent patterns, however, are inconclusive.

Johnson (1978) distinguished the following three facies models for tide-dominated, shallow-marine sandstones based on paleocurrent patterns as well as on other sedimentological and paleontological criteria: (1) blanket sandstones, (2) sand-wave deposits, and (3) linear sand-bar deposits.

An example of a blanket sandstone is the unfossiliferous late Pre-Cambrian Jura Quartzite, well exposed in islands off western Scotland (Anderton 1976). It is thought to have been laid down in a shallow-marine environment as an extensively cross-bedded, texturally and mineralogically mature blanket sandstone. The following lithofacies are discriminated based on grain-size and sedimentary structures: (1) coarse facies (sandstone facies Sa), (2) fine facies (heterolithic facies Hb and Hc, mudstone facies M), and (3) fine/coarse facies alternations. These facies reflect the interaction of fair weather and storm conditions superimposed on tide-dominated deposits.

The Eocene Roda Sandstone Formation of the southern Pyrenees is an example of a shallow-marine sandstone deposited in form of sand waves (Nio 1976). In the open-marine portion of this sandstone deposit, five major sandstone bodies, called 'sand-wave facies', each 10–30 m thick, are separated by 5–30 m thick intervals of finer-grained, fossiliferous marls, bioturbated silty sandstones, and bioclastic limestones, called 'inter-sand-wave facies'. The sandstones, enclosed in marls with a rich, open-marine fauna, contain abundant marine shell debris. Traced in a downcurrent direction each sandstone body displays a recurrent sequence of sedimentary structures consisting of the following five subfacies: (1) initial sand-wave subfacies, (2) sand-wave (sensu stricto) subfacies, (3) proximal slope subfacies, (4) distal slope subfacies, and (5) abandonment subfacies. This sand-wave complex seems to have developed under the influence of abundant sand supply and strong tidal currents.

Ancient linear sand-bar deposits are composed either by parallel sandstone bodies deposited by a unidirectional paleocurrent, or by sandstone bodies with large-scale inclined surfaces normal to the paleocurrent direction representing bar flanks. Well documented examples of linear sand-bar deposits occur as isolated, elongate sand bodies surrounded by marine muds in the Mesozoic of the western U.S.A. (Figure 92). The dimensions of these ancient sand-bar deposits, their internal coarsening-upward sequence, and their paleocurrent patterns resemble those of both tidal current and storm-generated modern linear sand ridges.

A well studied example of ancient linear sand-bar deposits are Upper Jurassic sandstones from Wyoming and Montana, western U.S.A. (Brenner and Davies 1973, 1974). A shallow-marine origin of the cross-bedded, fine-grained quartzose sandstones is inferred by the textural and mineralogical maturity, glauconite content, local bioturbation, and concentrations of broken bivalves. The sheet sandstone is composed of several coalescent sandstone bodies. The individual sandstone bodies are at least 1 to 5 km long, 200 m to 2 km wide, and up to 21 m thick. The coarsening-upward is explained by the migration of linear sand bars and their finer-grained interbar troughs over a low-energy mud shelf. Within each sand bar, interbar trough, and shelf-mud layer, coquina sandstones occur, having formed during storms and consisting of channel lags, storm lags, and swell lags.

Another example of linear sand bodies is known from the Cretaceous of the western U.S.A. North-south trending seaways generated strong tidal, oceanic, or storm-generated currents which built up coast- and current-parallel, elongate shelf-sand bodies. These sandstone bodies are generally characterized by: (1) well-sorted, glauconitic quartzose sandstones, (2) cross-bedding, (3) coarsening-upward, (4) paleocurrent modes parallelling sand-body elongation, (5) sand-body dimension: 10–30 m thick, 2–60 km wide, and up to 160 km long, (6) planar bases and convex-upward tops, and (7) surrounded by or interfingering with marine muds. The sandstone bodies frequently form oil reservoirs because they may be both surrounded and capped by potentially hydrocarbon-bearing, impermeable shales, e.g. Sussex Sandstone (Berg 1975).

Wave- and storm-dominated shallow-marine sandstones are characterized by symmetrical and asymmetrical wave ripples. The latter are distinguished from current ripples by their morphology. Because bedding planes are normally not exposed it is essential to distinguish wave- from current-formed cross-lamination.

Storm-dominated sandstones of the sublittoral are represented by well-sorted, fine- to medium-grained sheet sandstones, normally interbedded with fossiliferous or bioturbated shales and siltstones (sublittoral sheet sands or storm lag deposits).

The storm-sandstone layers are commonly 5 to 30 cm thick (maximum circa 2 m), and persist laterally for tens of meters to kilometers with sharp, planar to gently undulating erosive bases and non-erosive tops. The dominant sedimentary structure is parallel lamination, which may display primary current lineation and/or hummocky cross-stratification. These sandstone beds resemble structure-wise turbidites and are interpreted as deposits

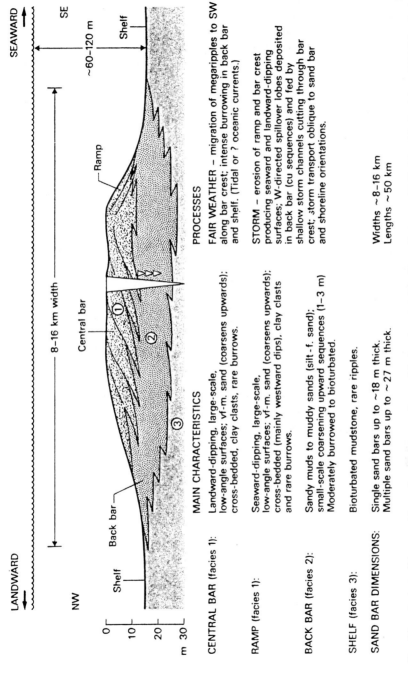

LANDWARD

SEAWARD

NW

SE

Shelf

Back bar

Central bar

~60–120 m

Shelf

Ramp

8–16 km width

m 30

① ② ③

MAIN CHARACTERISTICS

CENTRAL BAR (facies 1): Landward-dipping, large-scale, low-angle surfaces; vf-m. sand (coarsens upwards); cross-bedded, clay clasts, rare burrows.

RAMP (facies 1): Seaward-dipping, large-scale, low-angle surfaces; vf-m. sand (coarsens upwards); cross-bedded (mainly westward dips), clay clasts and rare burrows.

BACK BAR (facies 2): Sandy muds to muddy sands (silt-f. sand); small-scale coarsening upward sequences (1–3 m); Moderately burrowed to bioturbated.

SHELF (facies 3): Bioturbated mudstone, rare ripples.

SAND BAR DIMENSIONS: Single sand bars up to ~18 m thick. Multiple sand bars up to ~27 m thick.

PROCESSES

FAIR WEATHER – migration of megaripples to SW along bar crest; intense burrowing in back bar and shelf. (Tidal or ? oceanic currents.)

STORM – erosion of ramp and bar crest producing seaward and landward-dipping surfaces; W-directed spillover lobes deposited in back bar (cu sequences) and fed by shallow storm channels cutting through bar crest; storm transport oblique to sand bar and shoreline orientations.

Widths ~8–16 km
Lengths ~50 km

Fig. 92. Ancient offshore-sand bar, schematic cross-section, and facies characteristics, Upper Cretaceous Duffy Mountain Sandstone, NW Colorado (after Boyles and Scott 1982, p. 500, reprinted with permission).

from waning sediment-laden currents. Each sandstone layer is considered to record a single storm event.

Sublittoral sheet-sandstone facies occur in three associations (Reading 1978): (1) shoreface-shoreline association, (2) tide-dominated shelf association, and (3) nontidal (storm- and wave-dominated) open-shelf association.

A heterolithic shallow-marine clastic sequence in the Lower Carboniferous of southern Ireland exemplifies the wave-dominated facies model. It has been divided into four facies based on grain size and sand-mud ratio (De Raaf *et al.* 1977): (1) streaked muds, (2) lenticular beds, (3) parallel- and cross-laminated sandstones, and (4) large-scale structured sandstones. The absence of emergent features indicates sedimentation in a sublittoral environment. Main criteria for interpreting these rock associations as wave-dominated are types of lamination and unidirectional paleocurrent patterns, directed onshore.

Thick, mud-dominated shelf deposits probably accumulated rapidly in areas of low wave and current activity if modern environmental analogues are valid. Thus, during Jurassic times a vast portion of northwestern Europe was covered by a shelf sea in which marine shales were the dominant sediment. This epeiric basin contains examples of various mud facies which may serve as models for other basins. For instance, during the Lower Jurassic three major shale facies may be recognized in northwestern Europe: normal, restricted, and bituminous (Morris 1979). The same facies associations can also be observed in the Lower Cretaceous of northwestern Europe.

Ancient mud-dominated shelves occur in many areas and at various times, especially during the Lower Paleozoic and Cretaceous in both Europe and North America.

Tide and storm-dominated sand ridges on a muddy shelf are exemplified by the Upper Pensylvanian Cottage Grove Sandstone, northwestern Oklahoma (Fruit and Elmore 1988).

No example of an oil field in an ancient sand ridge is known yet (Coneybeare 1976); but such sands should make excellent reservoirs. However, such sandstones are likely to be cemented by calcite if detrital calcite grains or calcareous fossils are abundant. Some ancient productive transgressive shelf sandstones show good reservoir properties. They vary in thickness from 6 to 90 m, have porosities of 15–28% and permeabilities of 80–300 md. Because they commonly fine-upwards and grade into overlying marine shales the reservoir properties are normally poorest at the top. For this reason commercial oil and gas fields are less common in transgressive than in regressive sandstones.

4.1.2.9. Ancient Shallow-Water Carbonates

Most facies interpretation of ancient shallow-water carbonates relies strongly on the studies of modern environments such as the Great Bahama Bank and the Persian Gulf.

The main facies of an idealized carbonate shallow-water platform have been summarized by Wilson (1975). Those belonging to the shallow-water carbonates proper are the talus slope, the buildup, and platform-sand belt.

In the talus slope the carbonate debris is swept off the shallow-water platform into deeper water during storms. Chaotic, unsorted conglomerates and breccias set in a matrix of terrigenous argillaceous sediments result normally rich in pyrite and organic matter. Planktonic fauna with the shallow-water forms is found in this generally hostile environment. The sedimentary rocks are composed of clay minerals (dominantly illite), mica, pyrite, and variable amounts of calcite decreasing away from the platform. The more argillaceous sequences are finely laminated.

The buildup and platform-sand belt, a high-energy zone, consist of wave-resistant reefs and reworked carbonate banks or shoals. The structural framework of reefs is mainly represented by boundstones composed of calcareous skeletons or bindstones, lime mud held together by algal crusts. Reworked carbonates are oolites, shell fragments or coral-bearing grainstones. They are mostly cross-bedded and commonly cut by channels. These bank sediments, locally built up above sea level, may also contain littoral and eolian components. The limestones consist of almost pure calcite, except for partial or complete dolomites. Glauconite and quartzose sandstones are rare.

Ancient carbonate-platform associations seem to have been more extensive than modern platforms, especially during the Paleozoic (Wilson 1975). Well-documented examples are the Cambrian to Devonian sequences of the central Appalachians and the Upper Triassic and Lower Jurassic sequences of the Mediterranean area. Shallow subtidal subenvironments are inferred from oolites and shelly micrites. Lagoonal environments are characterized by a variety of micritic, pelletal, and oncolitic limestones with a low faunal diversity. Intertidal to supratidal environments are represented by stromatolitic limestones with bird's-eyes, vadose pisolites, and karstic erosion features. A classical setting of platform carbonates can be observed in the early Jurassic of southern Europe and southern England (Figure 93).

Large ridge- or mound-like bodies of massive or poorly bedded limestones are termed reefs or buildups. Reefs are wave-resistant buildups produced by sedentary calcareous organisms, especially corals grown in turbulent water. Whereas some buildups have a shape and facies association suggesting growth on the edge of a bank, like large modern reefs, others have a symmetrical mound- or dome-like shape suggesting isolated growth on a flat surface. Fossils vary widely in type. The lithology ranges from micrites to bioclastic limestones. Characteristic is the sedimentary structure called stromatactis, a post-depositional sparry-calcite cavity fill produced by burrowing activity or decay of a soft-bodied organism.

In consequence of evolution, the faunal spectrum and composition changed through time. Extensive stromatolites are confined to the Precambrian. Archaeocyathid limestones occur only in the Cambrian. Silurian buildups are dominated by tabulate corals and stromatoporoids. Rugose and tabulate corals are characteristic of the even more extensive Upper Devonian reefs. The Lower Carboniferous limestones contain mainly micrites with stromatactis and subordinate bryozoans. The famous Triassic reefs of the Alps are composed of scleractinian corals, calcisponges, and hydrozoans. Buildups with siliceous sponges in the Upper Jurassic of southern Germany contrast with the coeval coral buildups of the Paris Basin and the Jura Mountains. Rudistic bivalves are the characteristic component of many Cretaceous reefs. In numerous buildups dolomitization and dissolution of fossils produced additional porosity with cavernous structures for good hydrocarbon reservoirs.

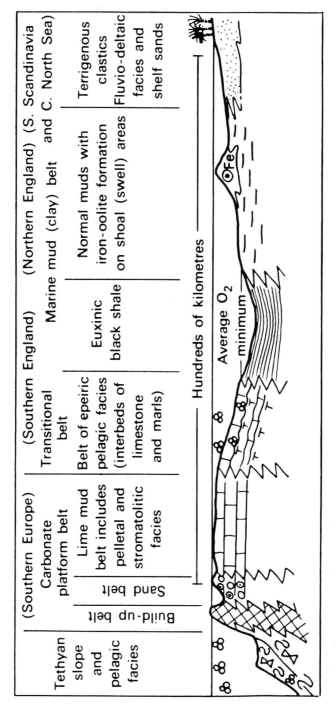

Fig. 93.   Ancient carbonate-platform sediment type: an epeiric shelf based on the early Jurassic of Europe (after Sellwood 1986, Fig. 10.19, p. 303).

Figure 94 shows examples of large ancient barrier reefs with a characteristic sequence of environments: deep basin, fore-reef talus slope, reef front, and back-reef lagoon. The most classical is the Capitan Reef Complex, Texas and New Mexico, U.S.A., for which several environmental models have been proposed.

The problem of using logs for subsurface environmental analysis in carbonate-rock sequences is that characteristic log motifs known from terrigenous clastic sequences and produced by changes in grain size and texture, are not applicable to the essentially monomineralic carbonate-rock environments. Variations in clay content in the forereef area or variations of anhydrite or gypsum content in the supratidal zones may be used as indications of transgressive or regressive tendencies. However, variations in the mineralogy within the high-energy platforms or low-energy lagoons may rather reflect diagenetic changes not necessarily related to the original sedimentary environment. Petrographic and paleontological analyses of cuttings and sidewall cores will generally supplement the environmental picture. Dipmeter logs and well-to-well correlations are of great value in determining reefal trends in the subsurface.

About 38% of the world's oil reserves is contained in shallow-water carbonate reservoirs, either in shelf calcarenites or in reefs. The widespread Ordovician and Silurian carbonate reservoirs of the Williston Basin, U.S.A., were deposited in a regressive shallow-shelf environment. During the Silurian, tall pinnacle reefs rising from the slope formed reservoirs in Michigan, U.S.A. Devonian barrier reefs contain prolific oil reservoirs in Alberta, Canada. The reefs are scattered randomly over the shelves. Devonian pinnacle reefs form the Zama and Rainbow reservoirs. During the Paleozoic, mounds formed by calcareous algae at the shelf edge, are important oil reservoirs in Oklahoma and Utah, U.S.A. And the giant Jurassic oil fields of Saudi Arabia like Ghawar (see Section 4.2.2) are composed of regressive shoaling-upward carbonates, each sequence terminating with an anhydrite unit. Reefs at the edge of a major carbonate platform are known from the Mid-Cretaceous Golden Lane trend, Mexico (see Section 4.2.5), and from the Oligo-Miocene Kirkuk field, Iraq. The Tertiary *Lithothamnion* Limestone, a shelf carbonate interfingering with shelf siliciclastics, is a minor gas reservoir in southern Germany and Austria.

Exploratory efforts are still aimed at Paleozoic carbonate buildups in northwestern Europe.

4.1.2.10. Ancient Continental-Margin Deposits

4.1.2.10.1. Deep-sea sandstones. Deep-sea sandstones, most distal signs of terrestrial sand influx, appear on the surface of modern fans as distributary channels with levees, inter-levee areas, and depositional lobes below the mouths of the channels (Walker 1984, Pickering *et al.* 1989).

On the upper fan, bulk-sand transport and deposition are restricted to elongate channels, often with persistent levees which may be recognized on seismic cross-sections through ancient fans. On the middle part of the fan numerous meandering channels occur. These channels lack persistent levees and pass downcurrent into a braided and rapidly migrating system that disappears downfan. These supra-fans with low convex-upward segments on an otherwise concave-upward profile, may be recognized on seismic cross-sections. The

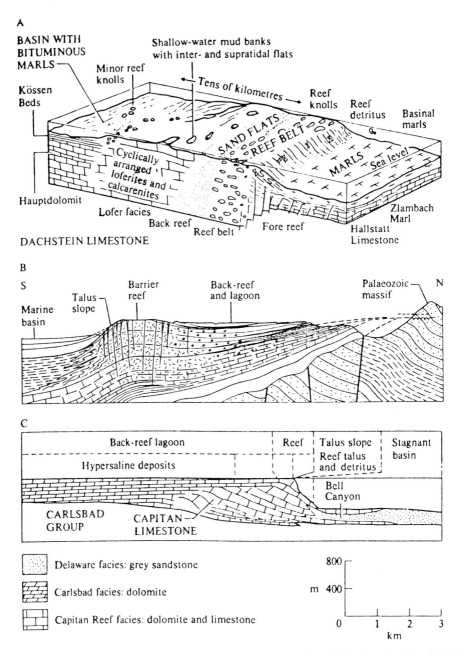

Fig. 94.   Ancient carbonate buildups interpreted as barrier reefs (after Hallam 1981, p. 53): A = Upper Triassic Dachstein Limestone, Austria (after Zankl 1971), B = Lower Jurassic Bou Dahar reef, Atlas Mts., Morocco (after Du Dresnay 1977), and C = Permian Capitan Reef Complex of Texas and New Mexico, U.S.A. (after Newell *et al.* 1953).

lower part of the fan, characterized by sheet flows and the lack of channels, grades into the basin plain with pelagic sediments alternating with thin-bedded, fine-grained turbidity deposits.

Proximal turbidites, i.e. turbidites nearer to the sediment source, are characterized by thick, coarse-grained, poorly graded or ungraded sandstone beds, commonly with scoured bases. Distal turbidites are thinner and finer grained, and show more even bedding surfaces. Scours and channels are rare, tool marks more frequent, and grading good (Table 25).

A depositional model in which sand is supplied through canyons and channels on an essentially non-depositional slope, like in modern deep-water fans, to the basin floor, where it accumulates as a submarine fan, has been developed by Mutti and Ricci Lucchi (1978). The slopes can be tectonically controlled. This model applies for both ancient and modern fans, together with criteria for recognizing and interpreting the main turbidite-facies associations. The majority of deep-water fan sediments is deposited from turbidity currents. Slides, slumps, and mass flows are mostly restricted to the relatively steep slopes. Thick-bedded, medium- to coarse-grained, graded or ungraded sandstones occur on the floors of depositional channels on the inner and middle fan. Commonly graded, medium-grained sandstones are found in depositional lobes on the outer part of the middle fan. Argillaceous sediments usually predominate in the inter-channel areas of the inner and middle fan and on the outer fan. Thinning/fining-upward sandstone sequences are interpreted as channel fills on the inner and middle parts of the fan, and the thickening/coarsening-upward sequences as sediment lobes accumulated on the middle fan beyond the downstream end of fan channels.

However, not all sediments of deep-sea fans and their surroundings must be attributed to deposition from turbidity currents. Rupke (1978) distinguished three major ways of sediment transport: mass-gravity transport, high-density turbidity currents, and low-density turbidity currents, producing the following characteristic types of deposits: the mass-gravity transport, triggered by earthquakes, tsunamis or storm waves, carries mainly siliciclastic sediments to the deep-sea at present. The term debris flow is used for sediments containing matrix-supported clasts. Thick-bedded, coarse-grained sediments may be the result of grain flow. Slumping is common. It is largely restricted to areas of rapid deposition such as delta fronts and canyon heads. It can take place on slopes as gentle as $1°$.

High-density turbidity currents are suspensions of sand and mud with a specific gravity between about 1.5 and 2 g cm^{-3}. When the flow decelerates, deposition of graded beds may result. The characteristic Bouma cycle occurs in this realm. A complete Bouma cycle (Figure 11) comprises from bottom to top:

A   A massive, coarse-grained, pebbly sandstone with shallow-water shells on a scoured surface; deposited by rapidly moving traction currents; fining-upward to:

B   A medium-grained, plane-bedded sandstone deposited by upper flow regime currents; fining-upward to:

C   A silty, fine-grained sandstone with cross-bedding or rippled and wavy laminae deposited by lower flow regime currents; fining-upward to:

Table 25.   Characteristic features of proximal, medial, and distal turbidites (from Reading 1986).

	Proximal (coarse-grained)	Medial (medium-grained)	Distal (fine-grained)
Bed thickness	Thick	Medium and thin	Thin beds and laminae
Bed shape	Irregular; lensing, channels and washouts common	Parallel-sided; regularly bedded	Parallel-sided beds and laminae, also discontinuous laminae.
Sand/mud ratio	SS/MD high, amalgamation of sandstones, thin mudstones partings and layers	SS/MD medium, rare amalgamation, well-developed mudstone layers	SS/MD low, mudstone dominant
Grading	Beds often ungraded or poorly graded, some negative grading	Grading commonly well-developed	Grading often subtle and on very small scale
Facies models	Bouma T_{ae} sequences and Lowe sequences common	Classical Bouma sequences common (T_{abcde}, T_{bcde}, etc.)	Stow/Piper sequences common, Bouma ($T_{(c)de}$ and T_e divisions only)
Stratification	Large-scale parallel and cross-stratification common	Lamination, ripples and convolute lamination common	Interlaminated siltstone and mudstone common, micro-cross-lamination, etc.
Top and bottom structures	Base sharp, commonly scoured; top often sharp	Base sharp, minor scours; top usually graded	Base sharp, more rarely gradational, micro-scours; top sharp or gradational
Bioturbation	Mostly absent	Can be well developed in mudstone layers	Can be well developed; micro-bioturbation also common
Deformation structures	Slump and dewatering structures common	Minor slump and dewatering structures	Siltstone loading and balling in mudstone layers can occur
Grain size	Gravel and coarse-sand size dominant	Medium-fine sand size and interbedded mud-grade	Very fine sand and silt-size with mud-grade dominant
Sorting	Often poor	Moderate	Moderate to well-sorted
Composition	Immature and mixed components	Moderate maturity, compositional grading common	Mature, compositionally well-sorted
Associated facies	Slumps and debrites	Fine-grained turbidites, some hemipelagites	Medium-grained turbidites, contourites, hemipelagites and pelagites

D   Parallel, argillaceous and silty laminae laid down by very slow currents; overlain by:

E   Very fine pelagic sediments with a deep-water fauna.

In the proximal parts of a fan, near the mouth of the canyon, the coarser lower parts of the Bouma cycle will be commonly represented. In the distal parts on the ocean floor, only the finer upper layers are represented and erosive basal surfaces may be difficult to recognize. While each individual Bouma cycle fines-up, the stratigraphic section as a whole normally exhibits a distinct coarsening-upward character. Thick sequences of such interbedded sandstones and shales are referred to as 'Flysch'.

Low-density turbidity currents lead to the deposition of finely laminated and graded claystone-siltstone layers between turbiditic sandstones rather than pelagic background sedimentation.

Two tectonic models have been developed for deposition of deep-water sandstones (Parker 1977): a canyon model, primarily in active tectonic settings and influenced by extra-basinal control, and a slope-wedge model, primarily in non-tectonic settings, resulting from the interaction of intrabasinal depositional processes. These models are considered to represent end members with all transitions possible. Such conceptual models help to assess the hydrocarbon potential and reservoir distribution of a basin. In the canyon model deep-water sand accumulations are elongated perpendicular to the slope. In the slope-wedge model sediments are mainly deposited on the slope and close to the base of the slope; sand accumulations are parallel to the slope.

A deep-water environment of deposition can be inferred evaluating core, paleontological, and log data from wells. In cores, sedimentary structures such as oscillation ripples diagnostic of shallow-water deposition are absent. On the other hand, graded bedding, load casts, contorted bedding, and groove marks are common in thinner beds. Even when graded bedding is common, the classic Bouma cycle is rarely fully developed. Dish structure may occur in thicker beds. Shales within and around a fan may contain deep-water faunas and typical trace fossils (Section 2.2). Gamma-log shapes are diagnostic of the various parts of a deepwater fan. Interbedded thin sandstones and shales of the outer fan show a serrated pattern. Depositional lobes of the middle fan display thickening/coarsening-upward sections with sharp tops. Channel deposits of middle and inner fans and slopes are characterized by thinning/fining-upward profiles with sharp bases. Logs of deep-water sands from the North Sea Forties Field were published by Walmsley (1975). Selley (1976) presented a case history of a North Sea well, integrating exemplarily both core and log data. Dipmeter logs commonly reveal micro-fining-upward sequences in macro-coarsening-upward sections.

However, interpretation of deep-water environments exclusively based on log shapes should be made with caution, because similar patterns are found in shallow-water sequences as well. Sedimentary structures in cores and outcrops are probably reliable single features to recognize ancient deep-water deposits. The various modes of deep-sea sedimentation and their transport mechanisms are shown in Figure 95.

Examples of ancient turbidite-fan deposits are known from different stratigraphic horizons all over the world: Upper Devonian/Lower Carboniferous of the Rhenish Schiefer-

PROCESSES   CHARACTERISTICS   DEPOSITS

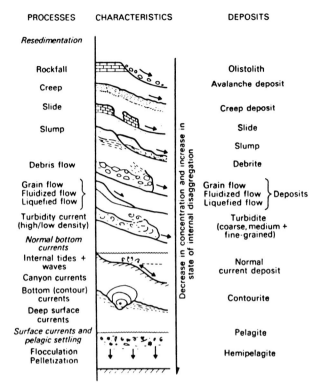

Fig. 95.   Transport mechanisms in deep-sea deposits (after Stow 1986).

gebirge and the Montagne Noire; Namurian basin fill of the Pennines, northern England; Upper Pennsylvanian Cisco Group of the Midland Basin, Texas; Upper Carboniferous, Pesaguero Fan, Cantabrian Mountains, Spain; late Jurassic/early Cretaceous, eastern Greenland; Cretaceous flysch belt of the eastern Alps, Bavaria; Cretaceous and Tertiary of the northern Apennines; Eocene flysch of northern Spain; and in the Eocene/Oligocene of the Santa Ynez Mountains, California, U.S.A.

Turbidite deposits contain only a small portion of the world's oil reserves. However, in certain tectonic settings prolific reservoir sandstones occur such as in the Permian Spraberry Sandstone of Texas, U.S.A.; in the Paleocene of the North Sea (Figure 96), especially in the Forties Field (Kulpecz and Van Geuns 1990); in the Oligocene of Austria; in the Tertiary (Miocene-Pliocene) of the Los Angeles and Ventura Basins, U.S.A.

Deep-sea fan deposits are the sandstones and conglomerates of the Upper Oligocene Puchkirchen Series in numerous gas fields of Oberösterreich, Austria (Polesny 1983). The reservoir facies is restricted to a relatively narrow zone paralleling the trend of the Calcareous Alps. Towards N the coarse clastics interfinger with shales of the basin. Sedimentological features and microfaunal evidence show that the alpine debris, brought in by rivers, was carried into deeper marine environments through canyons and deposited

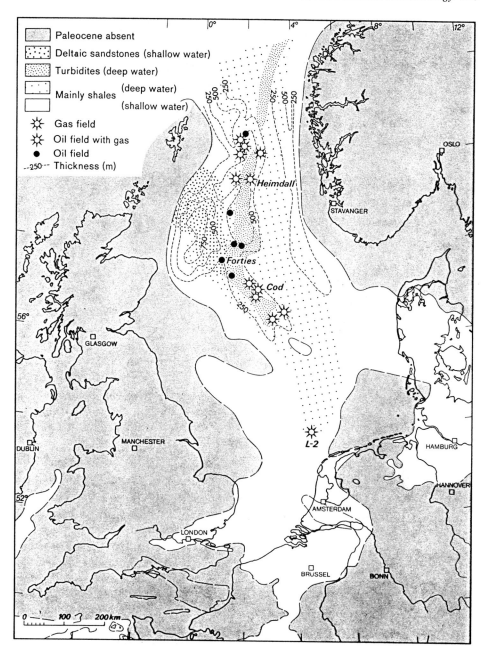

Fig. 96. Turbidite deposits (distribution, facies, and hydrocarbon occurrences) in the Paleocene of the northern North Sea, especially in the Forties Field (compiled by the Bundesanstalt für Geowissenschaften und Rohstoffe, Hannover).

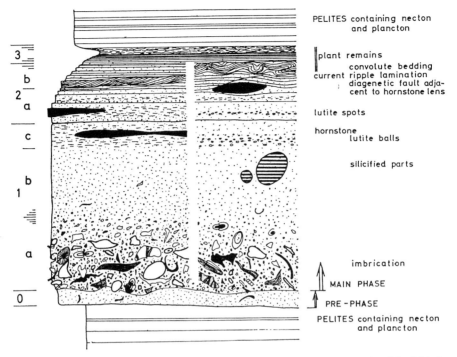

PELITES containing necton and plancton

plant remains
convolute bedding
current ripple lamination
diagenetic fault adjacent to hornstone lens

lutite spots

hornstone
lutite balls

silicified parts

imbrication

MAIN PHASE

PRE-PHASE

PELITES containing necton and plancton

Fig. 97.   Facies model of allodapic limestones. Thickness of the limestone bed about 1 m (after Meischner 1964).

as submarine fans. Such fan deposits can amalgate to a maximum thickness of 1 500 m. Dipmeter logs are useful to establish the direction of transport.

4.1.2.10.2. Allodapic limestones.   Allodapic limestones are limestones deposited by turbidity currents (Meischner 1964, p. 156). Limestone turbidites differ from siliciclastic turbidites by the rare occurrence of Bouma cycles, specific vertical zonation, absence of scour marks, and indistinct lamination. The vertical zonation of an individual allodapic-limestone bed is shown in Figure 97.

Source areas of allodapic limestones are shallow-water environments (platforms, reefs, banks) and upper-slope environments. The thickness of allodapic-limestone beds range from 1 cm to several meters.

Examples of ancient limestone turbidites are known from the Cambrian, New York and Vermont, U.S.A.; Devonian, Cornwall, England; Oxfordian, Poland; and Cretaceous, Bavarian Alps.

Major hydrocarbon reservoirs do not occur in allodapic limestones.

4.1.2.11. Ancient Pelagic Deposits

Pelagic deposits on the Recent ocean floors are the key to the past; only red-brown clay, radiolarian and diatom oozes occur. Their distribution is chiefly related to depth, temperature, fertility of surface waters, and distance from land. In this context, the role of the carbonate compensation depth (CCD), i.e. the depth below which calcite does not accumulate on the deep-sea floor, is important. The CCD fluctuated since Jurassic.

The identification of pelagic sediments in outcrops as truly oceanic might be assisted by the interpretation of associated basic and ultrabasic igneous rocks as oceanic crustal remnants. Ancient pelagic oceanic facies usually occur as parts of obducted or subducted ophiolites.

Well-studied examples of ancient pelagic deposits, now partly exposed on land, are the bedded cherts of the Ordovician Ballantrae ophiolite in southern Scotland; the Tethyan Triassic-Jurassic; the NW European chalk; the Troodos Massif of Cyprus; the Cretaceous red clays of Timor; and Paleogene clays of Barbados.

Impressive examples of ancient pelagic sedimentary rocks within condensed pelagic carbonates, cherts, and marls overlying shallow-water Bahamian-type carbonate platforms in the late Triassic-Jurassic of the Tethyan ocean, are exposed in the Alps (Bernoulli and Jenkyns 1974).

Condensed facies comprise red biomicritic limestones rich in ammonites and Fe-Mn crusts. Thicker sequences include pelagic limestones and radiolarites. Slump and turbidite deposits are found at the basin margins with much evidence for fissuring. Modern equivalents of oceanic hardgrounds of erosional origin with Fe-Mn crusts and fissures are observed on the Carnegie Ridge, eastern Pacific. Facies almost identical to the Mesozoic seamount-and-basin examples of the Tethys are thought to occur in the Late Paleozoic Rheno-Hercynian zone of southwestern England and Germany (Tucker 1973, 1974).

Another pelagic facies deposited over continental shelves are the Cretaceous chalks of North America, northwestern Europe, and the Middle East. Hiatuses marked by penecontemporaneously lithified hardgrounds with glauconite and phosphate nodules are known in almost all chalk sequences. And chalks are potential reservoirs in the U.S.A. and western Europe.

Dipmeter logs have been used to recognize turbidity sequences. Fracture identification logs are preferentially applied in defining productive zones. Paleocurrent directions are impossible to determine by logs, except in areas with turbidite deposition.

The Troodos Massif of Cyprus is the site of a pelagic sedimentary sequence, up to 800 m thick, overlying an ophiolite sequence. The basal sediments, brown shales of Campanian age, are rich in iron, manganese, and other metals (umbers). These shales, deposited in ponds on a spreading ridge, grade upwards into marls, radiolarian chert, and chalk (Figure 98). The ophiolites contain pockets of metallic sulphide ores. The umbers are derived from pillow lavas and basalts. Modern analogues are observed in the East Pacific Rise, Carlsberg Ridge, and Mid-Atlantic Ridge. These pelagic facies represent deposition above the CCD because they contain redeposited chalk.

Evaporite deposition in a major ocean basin occurred in the western Mediterranean during Miocene (Hsü *et al.* 1973). Deep Sea Drilling Project (DSDP) cores revealed deep

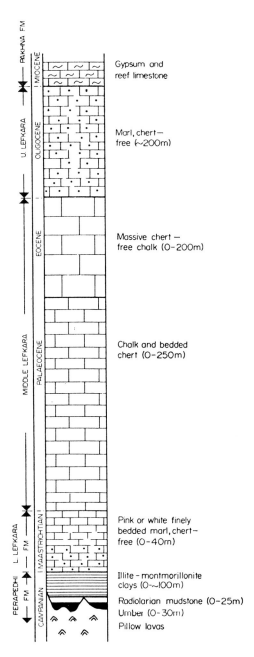

Fig. 98.   Ancient pelagic sequence of Upper Cretaceous and Tertiary sediments overlying the Troodos ophiolite complex, Cyprus (after Robertson and Hudson 1974).

pelagic and hemipelagic oceanic sediments overlain by anhydrite and halite evaporites up to 1 500 m thick with clear evidence of shallow-water playa to sub-aerial sabkha precipitation. The Mediterranean had completely desiccated during Miocene time following tectonic blockage at the western outlet to the Atlantic (Straits of Gibraltar). A spectacular deep basin more than thousand meters below sea level was formed, with streams and rivers deeply incising their valleys into the basin margins. The evaporites are overlain by early Pliocene deep-water pelagic and hemipelagic sediments.

Deep-water carbonate environments range in age from Cambrian to the present; examples are known from Canada, U.S.A., Mexico, and Africa (Cook and Enos 1977). Their geologic settings include ancient deep open oceans, continental slopes, and continental-interior basins.

The deep-water Mississippian Ranchera Formation of New Mexico and Texas, U.S.A., covering an area of 60 by 120 km, consists of coarse skeletal grainstone deposited by sediment gravity flows at the northern edge of the basin. The grainstone is replaced southward by deeper water hemipelagic sediments and by low-energy turbidity-current deposits further basinward (Yurewicz 1977).

The Pennsylvanian and Permian of the Sverdrup Basin in the Arctic Archipelago consist of pelagic carbonates, turbidites, and debris sheets deposited in trough and slope environments seaward of a shallow-water shelf, at water depths ranging from 300 to 1 200 m.

Processes leading to ancient deep-water carbonate rock deposition, however, are not well known yet. Research on deep-water carbonates is less advanced than research on shallow-water carbonates, because most hydrocarbon reservoirs form in shallow-water carbonates.

4.2. Sequence Stratigraphy, Basin Analysis, and Sediment Provenance

Stratigraphy is a basic discipline for any work in sedimentology. Pertinent texts recommended are Geyer (1973, 1977), Conybeare (1979), Schoch (1989), and Cotillon (1992).

4.2.1. SEQUENCE STRATIGRAPHY

Sequence stratigraphy evolved from seismic stratigraphy, the principles of which were published by Vail *et al.* (1977). Numerous advances comprising computer-simulation studies, outcrop documentation, and subsurface studies based on well-log and seismic control have been made, especially by Exxon. Sequence stratigraphy has become a key topic during the past decade.

Fundamentals of sequence stratigraphy and key definitions in sequence stratigraphy are summarized by Van Wagoner *et al.* (1988) and Van Wagoner *et al.* (1990). Sea-level changes as responsible for sedimentological settings of exploratory significance are stressed by Wilgus *et al.* (1988). The fundamental aspects of stratigraphy and sequences are further discussed in James and Leckie (1988), Fraser (1989), Schoch (1989), Schlager (1992), Posamentier *et al.* (1993), Hailwood and Kidd (1993), Williams and Dobb (1993), and Katz and Pratt (1993). Sarg (1988) and Loucks and Sarg (1993) refer to carbonate sequence stratigraphy in particular.

Sequence stratigraphy is the study of rock relationships within a chronostratigraphic framework of repetitive, genetically related strata bounded by erosion or non-deposition surfaces, or their correlative conformities. The basic unit is the sequence (Mitchum 1977).

The boundaries of sequences, parasequence sets, and parasequences provide the chronostratigraphic framework. However, absolute thickness, amount of time during which they form, and interpretation of regional or global origin are not used to define sequence-stratigraphic units.

Sequences and their stratal components are interpreted to form in response to changes in relative sea level and to reflect the interaction between rates of eustasy, subsidence, and sediment supply which depend on climate as well as on sediment and water discharge. The terms used in sequence stratigraphy are fully defined and discussed by Bally (1987).

Sequence boundaries can be identified in seismic-reflection profiles, well logs and cores, and outcrops from evidence for offlap within underlying strata and a downward shift in coastal onlap in overlying strata. Within sequences, stratigraphic position, stratal geometry and facies arrangements can be used to define three system tracts: lowstand (and shelf-margin), transgressive, and highstand. Unconformity-bounded depositional sequences are also composed of higher-order sequences or parasequences, which normally shoal upwards and are bounded above and below by marine flooding surfaces.

Eustasy is thought to play a major role in the formation of unconformity-bounded depositional sequences (Posamentier *et al.* 1988). The advantage of the sequence-stratigraphic approach is, that unlike transgressions and regressions of the shoreline or changes in paleobathymetry, the formation of prominent sequence boundaries seems to be relatively insensitive to sediment supply. Major boundaries of eustatic origin may correlate or nearly correlate globally. Vail *et al.* (1977) postulate that – in the global cycle curve they constructed for the Phanerozoic time – three major orders of cycles are superimposed on the sea-level curve: two first-order global cycles (200–300 ma duration) and 14 second-order global cycles (10–80 ma duration) of relative change in sea level (Figure 99). A global cycle of relative change of sea level is a geologic time interval during which a relative rise and fall of the mean sea level takes place on a global scale.

The stratigraphic relationship of the various members of the Upper Cretaceous Cardium Formation (about 100 m thick) hosting oil and gas fields in the subsurface is also seen in the light of sequence stratigraphy (Swift *et al.* 1991).

In the evaporite-to-red bed sequence of the Triassic Gipskeuper, SW Germany, Aigner and Bachmann (1989) demonstrated step by step the dynamics of a evaporite accumulation in an exemplary manner. An 'ideal' shallowing-upward cycle is shown in Figure 100. In the light of sequence stratigraphy the overall regressive cycle probably represents part of a 'third-order sequence', while the small-scale cycles might be considered as 'paracycles'.

Criticism directed at sequence stratigraphy refer to the assumed pre-eminence of eustasy, the lateral extent of sequence boundaries (Cartwright *et al.* 1993), and to the uncritical application of the concepts of sequence stratigraphy. The usage of essentially physiographic terms like basin fans and slope fans in sequence stratigraphy needs to be clarified.

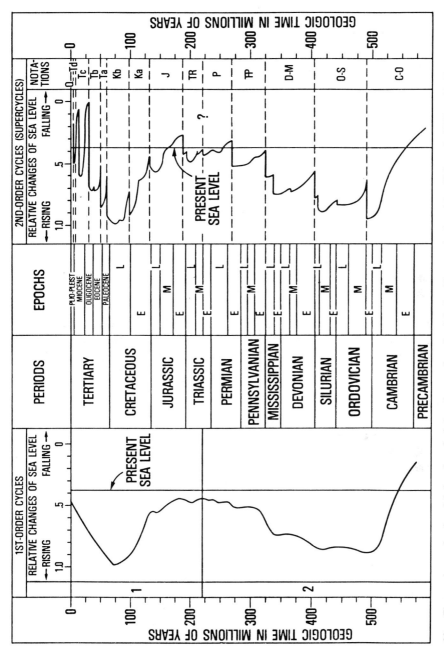

Fig. 99. First- and second-order global cycles of relative change of sea level during Phanerozoic time (after Vail *et al.* 1977, p. 84, Fig. 1, reprinted with permission).

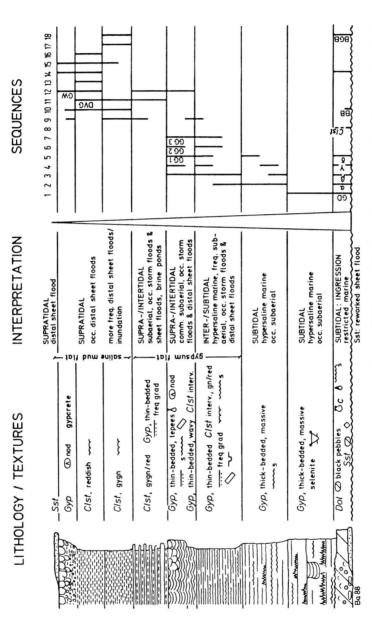

Fig. 100. An 'ideal' shallowing-upward cycle of the Gipskeuper, SW Germany. Black bars show sequences of actually observed cycles from bottom to top (after Aigner and Bachmann 1989).

4.2.2. BASIN ANALYSIS

Basin analysis is one of the ultimate goals of petroleum sedimentology. It integrates geological, paleontological, geophysical, geochemical, structural, and sedimentological analyses and interpretations and is aimed at improving discovery rates. The principles and application of basin analysis are reviewed by Kuenen (1966), Davis (1983), Beaumont and Tankard (1987), Allen and Allen (1990), Miall (1990), Visher (1990), and Einsele (1992). Shannon and Naylor (1989) compiled case histories of petroleum-basin studies. The thermal history of sedimentary basins is likewise of great importance to hydrocarbon exploration. Methods to reconstruct the thermal history as well as some case histories are summarized by Burrus (1988) and Naeser and McCulloh (1989). Microstructures as applied to basin analysis are discussed by Boersma and Terwindt (1983). Physical fundamentals and computer programs for simulating clastic deposition in sedimentary basins were discussed by Slingerland *et al.* (1994).

Basins are rather localized depocenters that persist for periods of tens to hundreds of millions of years. The bulk of sedimentary deposits near the surface of the Earth are preserved in basins.

Size and configuration of sediment bodies are fundamental parameters in basin analysis. The overall dimensions of the entire basin and the size and shape of each of the lithostratigraphic units are important to know. Such three-dimensional data is obtained through field mapping and measurement of stratigraphic sections, evaluation of wells, examination of outcrop samples, cores or cuttings, and from geophysical data. Most stratigraphic sequences in a given basin are incomplete; unconformities may be present, indicating the removal of certain parts of the rock record. Large-scale sedimentary sequences are generally wedge-shaped regardless of composition or location. Several major depositional environments may fall within the general lithology of quartz arenite with moderate sorting. A blanket sand body could indicate the shoreface or inland dunes. Linear sand bodies may represent barriers if they are straight or gently curved. Anastomosing patterns normally suggest fluvial systems. Branching or bifurcating linear sand bodies could develop as tributary or distributary systems.

4.2.3. SEDIMENT PROVENANCE

Basin analysis includes a comprehensive understanding of where the sediments came from, i.e. their source-rock type and location (provenance of the basin sediments). Provenance characteristics can be studied in the field with the unaided eye or the hand lens and in the laboratory by detailed petrographic analysis. In some cases even more detailed mineralogy or trace-element composition is required in order to trace the derivation of some sediments. Pebbles and cobbles form the best basis for provenance analysis in the field or by hand lens. One must look for readily identifiable and traceable detrital particles: rock-fragment types, chert varieties, particles containing recognizable fossils, and any metallic ores in large clasts. Only a few of these particle types are traceable to a definite source rock.

Provenance analysis of sediments and sedimentary rocks is of particular importance for paleogeographic and paleotectonic reconstruction and basin analysis. It provides evidence of: (1) the climate, relief, and composition of source areas and (2) type of transport and depositional processes. Provenance analysis becomes more difficult and ambiguous with increasing age of the strata due to diagenetic alteration and decomposition of mineralogically unstable mineral and rock fragments. Reconstruction of the compositional suites of terrigenous deep-sea sands of present continental margins is simple (Valloni and Mezzadri 1984) but the evaluation of sand debris in fossil sandstones is more difficult than shown by Mack *et al.* (1983) and Mack (1984). Bhatia (1983) demonstrated how the geochemical composition of sandstones is related to plate tectonics. Recent advances in sedimentary provenance studies were summarized by Morton *et al.* (1991).

Provenance analysis can be carried out by analyzing conglomerates, sandstones, siltstones, argillaceous sediments, and even carbonates. As a general rule it can be said that provenance analysis becomes the more difficult the finer-grained a sediment gets. Conglomerates with a high amount of rock-fragment pebbles give most information. Provenance analysis using sandstones, particularly heavy minerals, is common as shown by the compilation of Zuffa (1984). Leggewie, Füchtbauer and El Najjar (1977) present a good example of provenance analysis in the Germanic Trias Basin by means of detrital rock fragments. Le Ribault (1977) examined the internal and surface features of quartz from various sources. Zimmerle (1994) summarized the criteria known for determining the provenance of quartz (see Figure 18). Diagnostic features in both light and heavy minerals which can be used as provenance indicators have not yet received the attention they deserve. Most workers in this discipline tend to overinterpret the mere mineralogical composition. Provenance analysis of argillaceous sediments is still in an early stage. A promising example is that of the Eocene Messel Oil Shale, NE of Darmstadt, F.R.G. (Kubanek *et al.* 1988) which was interpreted to have been derived from altered basalts in the immediate neighborhood.

4.3. Case Histories in Petroleum Sedimentology

The following case histories will be discussed with emphasis on their sedimentological aspects:

      NW Europe – Rotliegende sandstones

      Saudi Arabia – Ghawar Oil Field (Jurassic)

      U.S.A. – Devonian Gas Shales

      U.S.A. – Mineral cementation, Permian Lyons Sandstone

      Mexico – Golden Lane, Tamabra Limestone (Cretaceous)

      C.I.S. – West Siberian gas fields (Jurassic/Cretaceous)

      China – Cretaceous lake deltas

These case histories were selected world-wide (Figure 101) for their importance as major oil and/or gas producing areas or for some striking peculiarities. There is a single giant field like Ghawar in Saudi Arabia or areas with several giant fields (NW Europe, Rotliegende sandstones; Mexico, Golden Lane; West Siberian gas fields; NE China, Cretaceous lake deltas). Under striking peculiarities the following features are meant: eolian reservoir sandstones 'comme il faut' as in the Rotliegende of NW Europe; a vast reservoir extent of more than 100 km like in the Ghawar Oil Field, source and reservoir

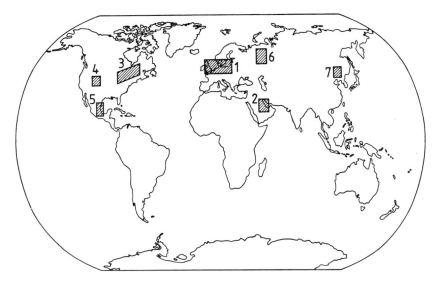

Fig. 101. Global location map of the case histories: 1 = NW Europe, Rotliegende sandstones, 2 = Saudi Arabia, Ghawar Oil Field, 3 = U.S.A., Devonian Gas Shales, 4 = U.S.A., Lyons Sandstone, 5 = Mexico, Tamabra Limestone, 6 = C.I.S., W Siberian Hydrocarbon Province, and 7 = China, Cretaceous Lake Deltas.

rocks together as in the Devonian Gas Shales or in the West Siberian gas fields; the detailed mineral cementation story as in the Lyons Sandstone; deep-water limestones as prolific oil reservoirs like the Tamabra Limestone in Mexico; the presence of bazhenite as a peculiar type of source and reservoir rock in the West Siberian gas fields; fine-grained sandstones in a rift-graben setting and oil with an exceptional high wax content in the Cretaceous lake deltas in NE China.

The case histories are summarized in the following order: (1) General data; (2) source rock, migration, trap type, and cap rock; (3) diagenetic history; (4) production data, reserves, and summary.

4.3.1. NW EUROPE – ROTLIEGENDE SANDSTONES

4.3.1.1. General Data

The Rotliegende (Lower Permian) of NW Europe was deposited in a depression, the Northwest-European Permian Basin, extending from Poland to Scotland for a distance of about 2 000 km. The basin can be subdivided into the Southern Permian Basin (with halite deposition in a large desert lake), the Northern Permian Basin, and the Moray Firth Basin, the latter two without halite deposition. In 1959 the great gas potential of the Rotliegende became evident by the first gas strike at Slochteren, N Holland. Ever since the Rotliegende remained the prime exploration target in Poland, F.R.G., N Holland, and in the southern North Sea. Major Rotliegende gas fields in NW Europe are Rough, West Sole, Amethyst, Leman, Hewett, Viking, Indefatigable, Groningen, Emsmündung, Groothusen, Halmern,

Söhlingen, Munster, Wustrow, and Salzwedel-Peckensen.

Geology, sedimentology, diagenesis, and reservoir characterization have been reviewed in numerous publications (Stäuble and Milius 1970; Katzung 1972, 1975; Lutz *et al.* 1975; Plein 1978; Van Wijhe *et al.* 1980; Drong *et al.* 1982; Glennie 1982, 1983, 1990; Smith 1982; Schröder and Schönaich 1986; Gast 1988; Gralla 1988; Eiserbeck *et al.* 1990).

In western Europe the Upper Rotliegende is presumably of Saxonian age. The Saxonian, however, is poorly defined by paleontology. Only tetrapod footprints are diagnostic fossils.

The Rotliegende of NW Europe consists of a continental sequence of clastics and evaporites over 600 m thick, overlain by up to 1 000 m of marine evaporites and carbonates of the Zechstein. The Permian strata rest unconformably on Upper Carboniferous (Westphalian to Stephanian) and older formations. The coal-bearing Upper Carboniferous is 2 500 m thick. Late Carboniferous deposition was interrupted by uplift in most areas towards the end of the Westphalian. Subsequent Late Hercynian tensional movements, associated by wrench faulting, led to the subsidence of trough- or graben-like elements, which became major depocenters in the Stephanian and early Permian, e.g. the Ems Trough. A basin-and-range topography with the corresponding sedimentation pattern evolved.

Rotliegende deposition took place under semiarid to arid climatic conditions. In the basin, clays and evaporites, essentially rock salt, were deposited in a desert-lake environment. These desert-lake sediments are fringed in the south, and to a minor extent in the north, by sandstones and conglomerates which form the major reservoirs for Rotliegende gas. Within this coarse clastic belt in the south three main facies belts are distinguished (Figure 102): (1) Alluvial sandstones and conglomerates, (2) eolian sands, and (3) sabkha deposits.

The alluvial sands and conglomerates were introduced into the basin by wadis. In the south a transport direction from south to north is inferred from the decrease in both pebble size and thickness of conglomeratic beds. In N–S trending zones of marked subsidence, alluvial sandstones are particularly abundant.

The eolian sands are common along the southern basin edge due to the reworking of alluvial sands by wind. These sands were deposited by the interaction of the N directed drainage system and the W to SW directed Permian trade winds on this NW–SE trending edge. Extensive dune deposits developed between the alluvial fans.

The sabkha deposits of fine-grained sands and silts characterize the basinward edge of the sand belt. The amount of sand decreases rapidly towards the basin. The sabkha sands were often tightly calcite-cemented during early diagenesis; they are rimmed by beach sands.

Marine ingressions are restricted to the uppermost Rotliegende; they are thought to be related to the Zechstein marine transgression. Gast (1988) summarized the interaction between sedimentology and tectonics, i.e. rifting, in the Rotliegende basin of NW Germany, mainly based on seismic survey (Figure 103).

In Great Britain desert conditions prevailed from late Carboniferous (Stephanian) to the end of Early Permian (Smith 1982).

The Yellow Sands, a special facies within the Rotliegende, are patchily distributed, largely unconsolidated eolian sands in NW England, resting unconformably on Upper

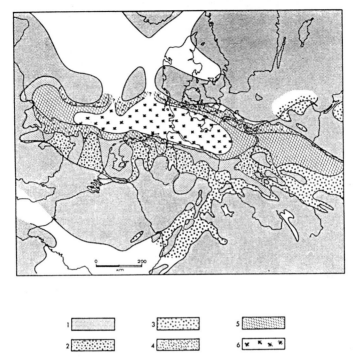

Fig. 102.   Facies distribution in the NW European Rotliegend Basin: Dunes, alluvial fans, sabkha, and evaporites (after Plein 1978, Fig. 8). 1 = mainly eolian (dunes), 2 = eolian/fluviatile, 3 = mainly fluviatile, 4 = fluviatile/sabkha, 5 = mainly sabkha, 6 = sabkha/halite.

Carboniferous and overlain by Zechstein carbonates (Steele 1982). The sands are aligned in a series of 8–10 parallel, NE–SW trending ridges, normally 20 m high and 2 km wide, separated by 1 km wide, sand-free corridors. The ridges are considered to be eolian sand bodies, i.e. longitudinal draas, slightly modified by a subsequent marine transgression. Modern analogues of the Yellow Sands bed configurations are the 'whale backs' from the Great Sand Sea of western Egypt. The yellow and grey colors of the Yellow Sands are the result of bleaching of an original, probably red color turning to grey by strongly reducing pore water coming from underlaying coal measures during burial.

The Rotliegende of the Netherlands consists mainly of sandstones; volcanic rocks occur only at the base and evaporites in the basin center. The sedimentary Upper Rotliegende was subdivided into the sandy Slochteren and the argillaceous Ten Boer Members (Figure 104).

In NW Germany the Rotliegende sediments (thickness 100–500 m) are grouped into three distinct lithological units: (1) the basal member of Schneverdingen Sandstone, occurring mainly as eolian filling of grabens, (2) the Slochteren Sandstone (Slochteren

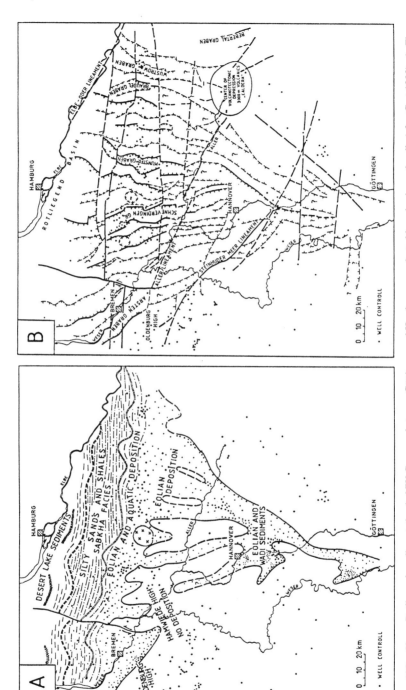

Fig. 103.   Upper Rotliegende sedimentation in NW Germany between Göttingen and Hamburg as evaluated by seismics and wells (after Gast 1988). A: Time of Slochteren-Hauptsandstein deposition, sabkha and desert lake sediments. B: Graben systems with major faults and wrench faults.

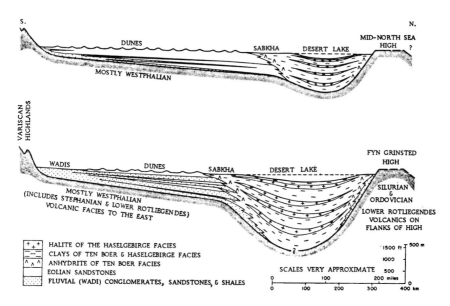

Fig. 104.   Conceptual N–S cross-sections through the Rotliegende Basin in the eastern Netherlands and NW Germany (after Glennie 1972, reprinted with permission, and Plein 1978).

Hauptsandstein), and (3) an alternating sequence termed the Hannover Beds (Hannover Wechselfolge). Cyclic deposition, lateral thickness variations, and facies changes indicate an onlap basin fill, controlled by basin morphology and climatic factors. The sandstones within the alternating sequence are subdivided from bottom to top into the Ebstorf Sandstone, Wustrow Sandstone, Bahnsen Sandstone, Niendorf Sandstone, and Munster Sandstone (Figure 105). The most important reservoir is the Wustrow Sandstone.

Within the limit of the Northwest-European Permian Basin, Rotliegende sedimentation began with the extrusion of volcanic rocks and the local deposition of volcanoclastics. Volcanism was active in parts of SW Scotland and SW England. In the Netherlands volcanic rocks are known from the eastern part of the country only. The conglomerates of the Lower Slochteren Sandstone from the Groningen area, however, contain numerous basic volcanic-rock pebbles and lithoclasts. In the F.R.G., the volcanic rocks (Eckhardt 1979), the so-called 'Lower Rotliegende', reached its greatest thickness in trough-like depocenters. Abrupt thickness variations at the edges of these troughs suggest depositional control by synsedimentary faulting. These faults may have facilitated the rise of igneous material which intruded into pre-Permian rocks or extruded onto eroded Carboniferous surfaces.

The environment of deposition of Rotliegende reservoir sandstones early influenced diagenesis and reservoir behaviour.

In this context, earlier environmental ascriptions of Rotliegende deposits were corroborated by the exemplary studies on modern desert environments by Glennie (1972). Based on Glennie's work the eolian origin of many Rotliegende reservoir sandstones was

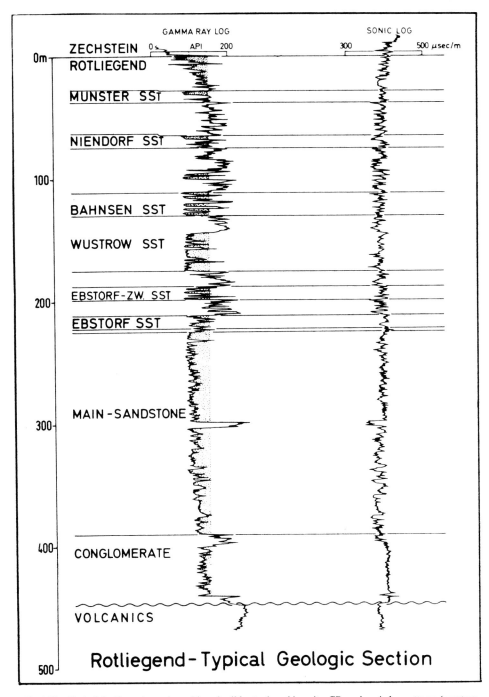

Fig. 105.   Typical Rotliegende section with major lithostratigraphic units, GR, and sonic log patterns (courtesy J. Zimdars and Zimdars 1982).

established and generally accepted for the reservoir sandstones in the southern North Sea, N Holland, and NW Germany. Pryor (1971), on the other hand, has questioned the interpretation of the Permian Yellow Sands at the top of the Rotliegende of NE England and the southern North Sea as being eolian. Maturity, good sorting, and high-angle foresets were interpreted by Pryor to be diagnostic of subaqueous dunes in a shelf sea. However, the prime portion of the Rotliegende sandstones remains unquestioned in their environmental ascription.

Log interpretation and evaluation are of great assistance to Rotliegende exploration everywhere in NW Europe (Bradel and Draxler 1982, Oehmig 1988). Sandstones are easily differentiated from shales by means of the gamma-ray log; this makes the gamma-ray log a reliable tool for regional correlations. The eolian sandstones, which are more porous than the fluviatile sandstones due to better sorting, show somewhat lower sonic velocities (higher interval transit time) than the fluviatile sandstones. Continuous dipmeter logs register high depositional dips (10°–30°) within eolian sands, angles which had been observed also in cores (Zimdars 1982). Even the dip azimuth which cannot be measured on conventional cores can be determined. It indicates the predominating wind direction in the Rotliegende dune areas that is in the U.K. North Sea, the Netherlands and NW Germany predominantly towards the west (Figure 106). Rotliegende winds blew from the NE in the Southern Basin and from the NW in the Northern Basin, thus outlining a barometric high over the intermediate Mid North Sea High. Strong Rotliegende wind activity coincided with lacustrine halite deposition (Glennie 1982).

4.3.1.2. Source Rocks, Migration, Trap Types, and Cap Rocks

Based on geological considerations and geochemical evidence the Late Carboniferous, essentially Westphalian coal beds are considered to be the main source of the natural gas encountered in the Rotliegende (Patijn 1964a, b, Colombo *et al.* 1968, M. and R. Teichmüller 1968, Karweil 1969, Hedemann and Teichmüller 1971, Boigk *et al.* 1976). This conclusion has been corroborated by C-isotope analyses.

Coal beds form approximately 3% of the total Westphalian section. The geological factors controlling coal rank and gas potential were studied, temperature being of prime importance.

During early coalification mainly water and CO_2 are expelled. Above a vitrinite reflectance equivalent (VR/E of 1.04) methane is released increasingly. Above VR/E of 3.20 the amount of volatile matter expelled becomes very small for any further depth increment. The coal beds within the coalification range of VR/E 1.20–3.20 are considered to be the main source of methane. It corresponds to the gas-window interval below 2 000 m.

Characteristic elements and trace elements in Rotliegende gas are nitrogen, helium, and mercury. In contrast to the methane generation nitrogen seems to be released in rather constant, but small amounts through a wide coalification range. Thus, the amount of nitrogen in respect to the total amount of gas generated is relatively low during main methane formation, but increases when methane generation is markedly reduced. This is also evidenced by high percentages of nitrogen adsorbed on high-rank coals. Besides

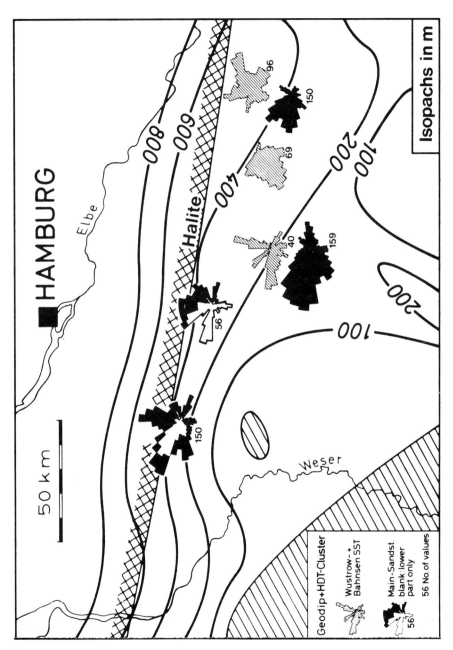

Fig. 106.   Paleowind directions, Rotliegende sandstones, NW Germany (courtesy J. Zimdars and Zimdars 1982).

from coal and dispersed carbonaceous matter, nitrogen seems to be generated from nitrogen compounds and 'inclusions' in argillaceous rocks during late diagenesis and early metamorphism, too. Smaller volumes of nitrogen are derived from coalification, larger volumes come from deeper rocks. This interpretation is corroborated by similar trends of helium. Traces of mercury were trapped in the Late Carboniferous coals from which they were again mobilized during the coalification process. Because no geophysical indications for deep-seated intrusive massives exist, a sound explanation for the high mercury content cannot be given yet.

The burial of coal beds through geologic time is mainly controlled by the structural history of a given area; it can be estimated reasonably evaluating borehole and seismic data. Transient heat flow increases are locally associated with intrusions or extrusions of igneous rocks. In various parts of the Northwest-European Permian Basin intrusions of igneous rocks raised the geothermal gradient, especially during Permian and Late Jurassic-Cretaceous times, e.g. the Zuidwal subvolcano at the southeastern end of the Central Graben. Such intrusions are frequently associated with magnetic anomalies, too.

A magnetic anomaly NE of the Groningen Field coincides with an area of exceptionally high coalification of the Upper Carboniferous (Van Wijhe *et al.* 1974, Kettel 1983).

Lateral variations in heat flow decrease fast with depth; nevertheless, they are significant for understanding the temperature-dependent diagenetic processes in the Rotliegende reservoirs proper, such as the authigenesis of illite.

Relating gas generation to the burial history of the source rocks, Patijn (1964a, b) attributed only local influence to heat anomalies caused by intrusions; he stressed the process of re-coalification. Beginning and duration of the the main gas-generation phases can be consequently derived from subsidence-time plots for any location (Figure 107). Gas migration postdates an earlier liquid hydrocarbon migration whose traces can be recognized in the sediments locally.

Both, length of the migration path and its permeability strongly influence the gas composition. In particular, during migration the relative amounts of nitrogen tend to increase. In structures with superimposed accumulations, gases in Buntsandstein reservoirs show commonly a higher nitrogen content than gases in deeper Zechstein, Rotliegende and/or Carboniferous reservoirs (Boigk *et al.* 1976).

Gas migrated presumably early, almost contemporaneously with its generation. Thus, the timing of migration is related to the structural development of any given part of the Northwest-European Permian Basin. First gas generation is inferred for Early and Middle Mesozoic. The gas migrated towards the basin flanks present then, being trapped for the first time (first migration phase). Late Cretaceous inversion-tectonics in most subbasins forced the gas to remigrate into newly formed structures in the center of former Early to Middle Mesozoic basins (second migration phase). In this phase the Rotliegende reservoir properties, especially the permeability, deteriorated mainly due to the authigenic growth of fibrous illite and chlorite during deep burial. Economic production is hampered or made impossible as observed in numerous wells. In areas characterized by weak diagenesis and moderate depth of burial, economic gas accumulations formed in and around inverted basins, e.g. Leman Bank and Sole Pit Fields. During Tertiary and Quaternary, subbasins, e.g. the southern North Sea basin, underwent pronounced subsidence after Late Cretaceous

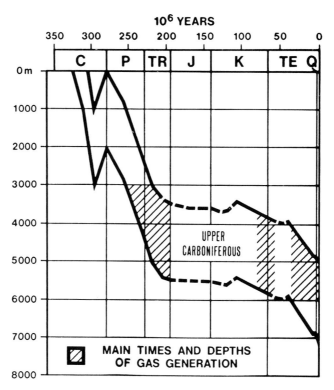

Fig. 107. Subsidence-time plot of gas-generating Upper Carboniferous strata in NW Germany (after Grotewold *et al.* 1979).

– Early Tertiary uplifting. In numerous reservoirs a new phase of gas generation and migration occurred, the so-called re-coalification (Patijn 1964b).

There is a variety of traps in the Rotliegende: both tectonic and sedimentary in nature: wide doming, stratigraphic traps, and trapping in tectonic grabens. Moreover, the gas-filled structures in the Northwest-European Permian Basin differ considerably in size. Age and geological development of regionally structured blocks can generally be reconstructed based on the thickness distribution of post-Zechstein sediments. For example, the Texel-IJsselmeer High with the giant Groningen Field apparently was a tectonic high relative to its surrounding tectonic units rather continuously from Late Carboniferous to Early Cretaceous. The uplifts in the center of the Late Jurassic-Early Cretaceous basins, on the contrary, are latest Cretaceous in age.

The evaporites of the Late Permian Zechstein seal most of the Rotliegende gas accumulations. The distribution of these evaporites is controlled by the primary depositional

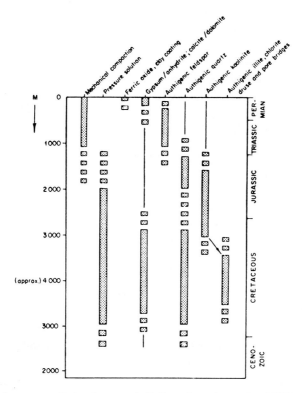

Fig. 108. Diagenetic sequence of mineral cements in Rotliegende sandstones from NW Europe (redrawn after Nagtegaal 1979).

patterns, later halokinetic movements, and/or later erosion and subsurface solution.

4.3.1.3. Diagenetic History

The areal distribution of Rotliegende sandstone reservoirs and the reservoir quality are controlled (1) by the environment of deposition and (2) by the subsequent diagenetic history.

Diagenetic processes of Rotliegende reservoirs were studied in much detail by numerous authors (Stalder 1973, Glennie, Mudd and Nagtegaal 1978, Hancock 1978, Drong 1979, Nagtegaal 1979, Seemann 1979, Almon 1981, Drong *et al.* 1982, Steele 1982). One of the numerous cementation schemes for Rotliegende sandstones is shown in Figure 108.

The primary porosity and permeability of the red Rotliegende sandstones decreased during initial subsidence by mechanical compaction and by penecontemporaneous to early diagenetic neoformation of hematitic clay crusts around the detrital sand grains. The red color is due to these hematitic clay crusts as well as to pore-lining red hematitic clays. Some of these pigments accumulated before final deposition. The pore-lining pigment was derived from infiltration and intrastratal breakdown of unstable grains. The clay coatings have inhibited subsequent mineral cementation. As diagenesis advanced, quartz – and to a minor extent – feldspar dissolution and reprecipitation further diminished the reservoir quality. However, the temperature-controlled diagenetic growth of fibrous illite (Photoplate 4) reduced pore throats to such an extent that sandstones may become practically impermeable, even for gas migration. Thus, gas-bearing sandstone intervals cannot produce gas at economic rates.

Depth and temperature required for authigenic illite growth put a threshold for economic Rotliegende reservoirs. Present data indicates that unrestrained illite grew below the base of economic reservoirs, i.e. below the 'reservoir floor' that is located at approximately 3.5 km.

Illite neoformation is inhibited by hydrocarbons in the reservoir. If hydrocarbon emplacement occurs at an early stage of diagenesis, porosity may be preserved deeper than the normal 'reservoir floor'. This possibility is significant to exploration, likewise the 'reservoir floor' may be deeper-seated than normal when the compaction-fluid flow is retarded or inhibited by tight vertical or lateral seals. Early, the sealing prevents compaction, later, it impedes authigenic mineral growth such as illite. Inhibited compaction-fluid flow also causes overpressures in porous reservoirs. Thus, overpressure may be responsible for reservoir preservation in greater depths than under normal hydrostatic pressure conditions. All these specific conditions can lower the 'reservoir floor' to 4.5–5.0 km. On the other hand, the 'reservoir floor' may be shallower than 3.5 km if the Rotliegende was once in geological history buried appreciably deeper than at present.

Hence, the occurrence of exploitable Rotliegende gas accumulations is commonly restricted to areas in which the Rotliegende sandstones were never buried deeper than 3.5–5.0 km (Figure 109).

In addition, salt plugs overlying some of the sandstone reservoirs, e.g. in the East Hannover gas province, seem to influence petrophysical properties of the underlying reservoirs favorably (Färber 1984). Generally better porosities and permeabilities are observed underneath the salt plugs than outside because heat flow is accelerated in rock salt and, thus, diagenesis is retarded.

Grain size and sorting, reflecting length and type of transport (fluviatile vs. eolian), define primary porosity and permeability. Decrease of sand thickness and grain size as well as early carbonate cementation, e.g. in sabkha sands at the basinward side of a sandstone belt, lead to a marked deterioration of primary reservoir qualities towards the basin.

Regional diagenetic trends preserving porosity and permeability to varying degrees, e.g. in sandstones from the southern North Sea, are supposed to depend on (1) the depositional environments, i.e. early diagenetic cements differ between wadi and eolian deposits, and (2) on the facies distribution in the overlying Zechstein, influencing intermediate and late diagenesis (Almon 1981). Thus, the maximum depth of burial, the interrelation of depth and temperature to diagenesis, as well as presence and magnitude of factors influencing these relationships have to be taken into consideration for prediction of the expected potential of a Rotliegende reservoir at any location.

The maximum thickness of overburden which affected Rotliegende reservoirs can be

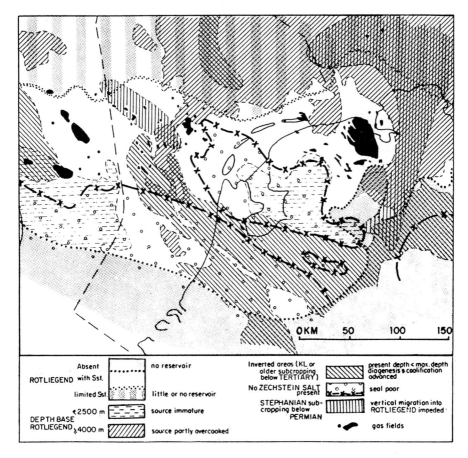

Fig. 109. Prospects of Rotliegende gas accumulations in NW Europe (after van Wijhe *et al.* 1980: Fig. 18).

approximated by adding the thickness of the underlying Zechstein sequence and, where present, Rotliegende claystones and evaporites to the maximum burial depth of the Lower Buntsandstein, obtained by cross-plotting sonic log-derived velocities and depths of the Lower Buntsandstein.

4.3.1.4. Production Data, Reserves, and Summary

Exploration for gas in the Rotliegende of the F.R.G., the Netherlands, and the southern North Sea is mainly concentrated on the southern eolian-fluviatile belt and less to the sabkha belt. In the Northern Permian Basin no reservoirs occur. In the Central Graben area between the Northern and Southern Permian Basin no gas is found in Rotliegende sandstones which are of poor reservoir quality. Moreover, the distribution of coal-bearing

Carboniferous is uncertain. About 95% of the Dutch production comes from the Groningen Field, the third largest gas field (1988) in the world. Total Rotliegende gas production (1988) of the F.R.G. was approximately 400×10^9 m^3. Total reserves of Rotliegende gas in NW Europe (1988) are estimated as 3.6×10^{12} m^3.

Because some of the natural gas of NW Germany contains up to 80% nitrogen and significant amounts of helium (0.1–0.5%), mercury (max. 5 mg/m^3), and carbon dioxide (< 1%), the nitrogen has to be eliminated. Nitrogen is removed by a cryogenic process in the presence of carbon dioxide and mercury in order to increase the calorific value.

In summary, the favorable conditions leading to the prolific Rotliegende gas accumulation in NW Europe are:

– Thick, good-quality gas source rocks (underlain by up to 2 500 m of Late Carboniferous coal-bearing strata);

– Discontinuous deep burial of these source rocks down to depths of 4 000-6 000 m allows methane generation over wide areas;

– Close spatial relationships between source and reservoir rocks facilitated short-distance migration;

– Excellent reservoir conditions in alluvial and eolian Rotliegende sandstones, only locally affected by diagenesis;

– Thick, first-rate seals by Zechstein evaporites;

– Presence of abundant, locally large structural traps, mostly formed prior to main gas generation.

4.3.2. SAUDI ARABIA – GHAWAR OIL FIELD

4.3.2.1. General Data

The Ghawar Oil Field, one of the world's largest oil fields, is located about 90 km inland from the southwestern shore of the Persian Gulf, Saudi Arabia (Figure 110). The field was discovered by a wildcat at Ain Dar in 1948. Surface mapping prior to World War II had furnished the first clue to the presence of the very large anticlinal structure. Gravity mapping gave additional hints. Structure drilling finally led to the discovery.

The Persian Gulf Basin is the largest petroliferous region of the world; it contains about 1/3 of the world's known extractable petroleum reserves. The basin is located at the junction of two large tectonic plates.

The oil field is a structural accumulation in a huge anticline, extending north-south for approximately 224 km and covering approximately 2 275 km^2. The vertical oil column reaches a maximum of 433 m (ARAMCO 1959). The oil-saturated interval extends about 83 m below the anhydrite that separates the Arab D reservoir from the overlying Arab C carbonate beds.

The field is normally blanketed by Miocene-Pliocene continental deposits, chiefly sandy 'marls' and sandy limestones with minor shales and sandstones. The stratigraphic rock sequence is shown from top to bottom (Figure 111): (1) indurated Miocene-Pliocene sandstones, sandy marls, and sandy limestones (about 165 m thick), (2) Eocene-Paleocene

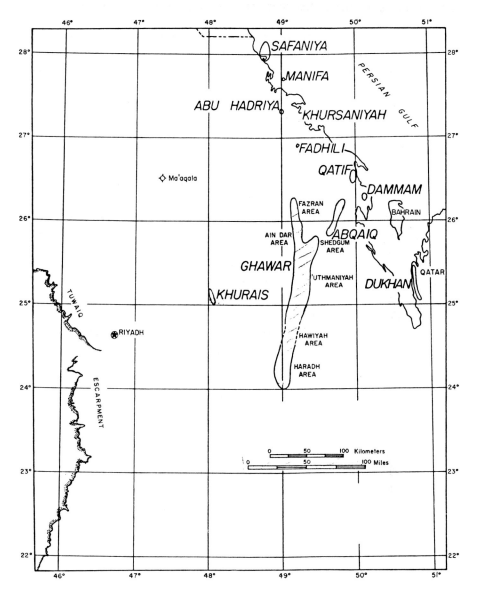

Fig. 110. Location of the Ghawar Oil Field, NE Saudi Arabia (after Powers 1962, p. 127, Fig. 2, reprinted with permission).

marls, limestones, clays, and dolomites, locally including anhydrite and gypsum (about 300 m thick), (3) Cretaceous limestones, dolomites, shales, and sandstones (about 1070 m thick).

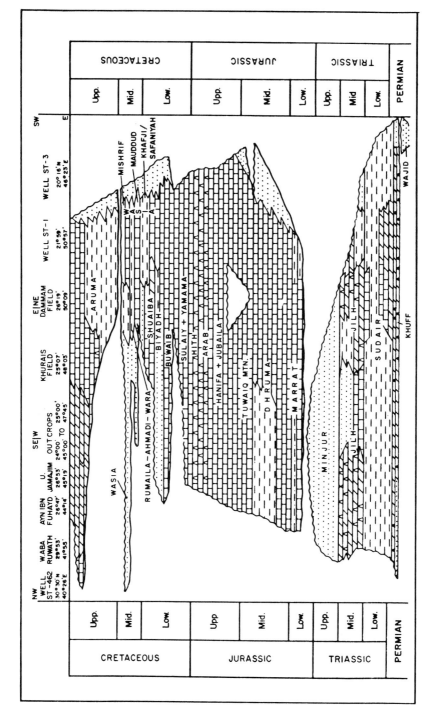

Fig. 111.   Stratigraphic rock sequences of Mesozoic formations, NE Saudi Arabia (after Powers 1968, from Beydoun 1988, p. 55).

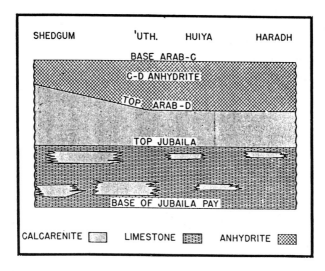

Fig. 112. Generalized cross-section of the upper Jubaila-Arab-D member, Ghawar Oil Field, NE Saudi Arabia (from ARAMCO 1959, reprinted with permission).

Platform carbonates, pelletoid-ooidal pack-grainstones cyclically alternating with argillaceous pelletoidal-bioclastic mud-packstones (Murris 1980), were deposited during Middle Jurassic (Bajocian to Callovian). During late Oxfordian to Tithonian, carbonate rocks and evaporites of the Arab formation were deposited in shallow marine to sabkha environments. Jurassic sedimentation terminated with four main depositional cycles. Each cycle is composed of a shallowing-upward normally marine carbonate sequence which is overlain by an anhydrite horizon (Powers 1962). Each carbonate cycle is composed of ooidal-pelletoidal grainstones. These grainstones contain most of the oil accumulated in the Jurassic Arab A, B, C, D reservoirs. The Hith anhydrite, averaging 167 m in thickness, forms the extensive seal.

The carbonate rock members of the Arab formation have been denominated from top to bottom by the letters A to D. The largest single oil accumulation occurs in the lowermost grainstone cycle, the Arab D. Textural changes in the Arab D reservoir are gradual; they are not controlled by early structural growth. Production is mainly derived from the grainstones which retained the original pore space. Other reservoir rocks are fine-grained limestones, calcarenitic limestones, dolomitic limestones, and dolomites. Partial dolomitization, usually megablastic, is extensive.

Figure 112 shows a generalized cross-section of the carbonate sediment distribution of the upper Jubaila-Arab-D member of Ghawar.

The productive beds have been partly or completely cored in many wells. A working carbonate-rock classification was used for both geological and reservoir studies, mainly of the upper Jubaila and the Arab-D member rocks (Bramkamp and Powers 1958, Powers 1962) (Table 26). The primary subdivision is based on the structure and compositon of the original carbonate sediment. The rocks are further subdivided based on type and degree of

Table 26.   Carbonate-rock classification after Powers (1962, reprinted with permission).

	Original Texture Not Visibly Altered (except by cementation)						Original Texture Altered					Original Texture Obliterated
	Original Particle Type						Moderately		Strongly			
	More than 25% Skeletal[2] Remains	More than 25% Aggregate[3] Grains	More than 25% Ooliths	More than 25% Detritus[4] from Older Ls.	10–50% Non-Carbonate Sand	10–50% Non-Carbonate Mud	Weakly Developed Calcite Mosaic (<10% Dolomite)	More than 25% Discrete Dolomite Rhombs	Strongly Developed Calcite Mosaics (<10% Dolomite)	25–75% Interlocking Dolomite	More than Dolomite Relic Structure	More than 75% Dolomite / 75% Dolomite
					Sandy	*impure* (marl)	*Partially recrystallized*	*Partially dolomitized*	*Strongly recrystallized*	*Strongly dolomitized*		
APHANITIC LIMESTONE (Lime mud with less than 10% sand- or gravel-size clastic carbonate grains.)	Origin of mud-size particles generally indeterminate (chalk)											
CALCARENITIC LIMESTONE (More than 10% sand- or gravel-size clastic carbonate grains set in more than 10% original mud-size matrix.)	*Skeletal²* calcarenitic limestone	*Aggregate³* calcarenitic limestone	*Oolite* calcarenitic limestone	*Detrital⁴* calcarenitic limestone								
CALCARENITE (Sand-size clastic carbonate grains dominant; contains less than 10% original mud-size matrix.)	*Skeletal* calcarenite	*Aggregate* calcarenite	*Oolite* calcarenite	*Detrital* calcarenite								
COARSE CARBONATE (Gravel-size clastic carbonate grains dominant; contain less than 10% original mud size matrix.)	Coarse *skeletal* carbonate (coquina)	Coarse *aggregate* carbonate	Coarse *oolite* carbonate (pisolite)	Coarse *detrital* carbonate								
RESIDUAL ORGANIC (Rocks composed dominantly of attached reef-building organisms still in growth position.)	*Residual* algae, *residual* coral, etc.											

[1] Chart shows main rock groups in UNDERLINED CAPITAL LETTERS; modifiers of main rock groups in *italics*. For example: sandy, oolite calcarenite or impure, partially dolomitized, foraminiferal calcarenitic ls.

[2] Skeletal is used here as a general modifier; specific modifiers includes foraminiferal, crinoidal, algal, coral, etc.

[3] Aggregate grains is a general term used for all discrete, penecontemporaneous, sand- and gravel-size grains formed on the sea floor by (1) the tearing-up, movement, and redeposition of fragments of semi-consolidated bottom sediment or (2) the aggregation of finer particles by cementation (Illing, 1954). Specific types are: angular aggregate, pellet (rounded aggregate), 'foecal' pellet, and 'algal' module.

[4] The detrital carbonates contain more than 25% grains which have been (a) formed by the mechanical disintegration of older, well-consolidated limestones, (b) transported and (c) redeposited as part of a younger sediment. Detrital grains are commonly distinguished by their 'inappropriate' fossil content, color, lithology, etc.

recrystallization and dolomitization, and on visually estimated porosity. This classification is normally applied to cuttings and conventional cores by using a hand lens and low-power binocular microscope and also on thin sections. Such a 'taylor-made' classification takes into account the local and regional geological setting and the practical requirements of exploitation and reservoir engineering. At the same time it clearly demonstrates the severe limitation of all classification systems, especially of the carbonate-rock classification (see Section 2.1.4).

The Arab-D interval (Figure 113) is characterized by the presence of the following components (Powers 1962): calcareous algae, stromatoporoids, and foraminifera as main skeletal grains, as well as aggregate pellets, algal nodules, and fecal pellets as most important non-skeletal particles. These carbonate rocks are products of mechanical deposition within certain environments characterized by specific particle sizes and types. According to the working carbonate classification the carbonate rocks are subdivided based on original particle size and sorting or on obliteration of original texture into five groups: (1) aphanitic or fine-grained limestones, (2) calcarenitic limestones, (3) calcarenites, (4) coarse carbonate clastics, and (5) crystalline dolomites (Table 26). This classification permits distinction of stratigraphic units for correlation and reservoir zonation, delineation of the original depositional patterns and environmental conditions, as well as for relating reservoir properties to original textures and secondary alterations. Lithofacies, diagenesis, and depositional sequences were demonstrated in a core workshop by Mitchell *et al.* (1988).

Shallow-water conditions during upper Arab-D time are indicated by the pronounced increase in skeletal sands and clean-washed calcarenites derived largely from dasyclad algae and stromatoporoids. Calcarenites, i.e. sand-size carbonate sediments and identical with grainstones (Wilson 1975), are the dominant carbonate rock type of the Upper Jubaila and Arab-D. Outside of Ghawar true oolites occur locally or even dominate. Particles resulting from the attrition of organic remains and pre-existing carbonate rocks usually are subordinate. These calcarenites are characterized by the lack of original fine-grained matrix, indicating current and wave agitation.

Conglomeratic sediments are of minor importance in the Arab members. Portions of the Arab are considered to be biostromal. The only reef-building organisms are stromatoporoids. Actual reef structures, however, were not observed. The environment of deposition must have included rocky bottom areas with scattered fragments of various types of reef-builders, like those of the present Persian Gulf.

The Arab C-D reservoirs are composed of poorly cemented bioclastic-oolitic lime sands with porosities of 20–25% and very good permeabilities. Initial oil-production potential from wells is in the order of 950 to 2 860 m^3/d. This type of reservoir falls under the category 'grainstones with preserved primary porosity' (Wilson 1975). It furnishes half of the world's carbonate reserves only from this giant oil field of the Middle East. Ghawar alone holds 25% of the world's petroleum reserves in carbonate rocks.

According to the typification of petroliferous basins (Klemme 1981) the large Middle East basin is a type A 4 Extracontinental-Borderland downwarp closed basin as the result of the collision of the African and Indian plates with the Eurasian plate. The Type 4

254   *Chapter 4*

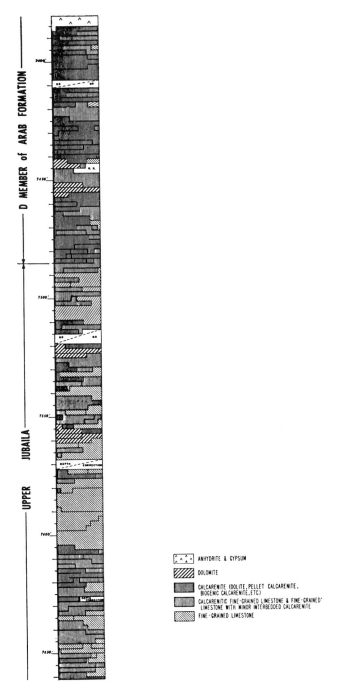

Fig. 113.   Lithologic sequence of the Upper Jubaila-Arab-D member, Ghawar Oil Field, NE Saudi Arabia (from ARAMCO 1959, p. 224, reprinted with permission).

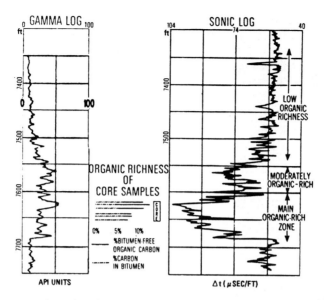

Fig. 114. Log patterns and organic carbon content of Jurassic source rocks (Callovian and Oxfordian), N Saudi Arabia (after Ayres *et al.* 1982, reprinted with permission).

downwarp closed basins represent 18% of the world's basin area, but contain 48% of the world's reserves.

4.3.2.2. *Source Rock, Migration, Trap Type, and Cap Rock*

Magara (1986) reviewed petroleum generation and migration in the entire Arabian Gulf region. Geochemical studies of sediments and oils (Ayres *et al.* 1982) suggest that the late Middle to younger Jurassic reservoired oils from the large fields in southern Saudi Arabia were derived from thermally mature, thinly laminated, organic-rich peloidal carbonate rocks of Callovian-Oxfordian age (Figure 114). In general, the organic matter is concentrated in dark laminae (0.5 to 3 mm thick) with alternating light-colored organic-poor laminae. The source rocks show features of both deep-water and shallow-water origin.

Thus, the depositional setting of this source-rock facies is interpreted to have been a restricted intrashelf basin separated by carbonate-grainstone and dolomitized facies from the open-marine environments of the Tethyan sea to the east. The total organic-carbon content averages 3 to 5 wt. %. The bitumen content is high and the hydrocarbon content of the rocks is frequently several thousand parts per million. The kerogen is predominantly an amorphous alginite. Kerogen maturation follows type II path on the van Krevelen diagram. These rocks show distinct log characteristics (high resistivity and low sonic velocity). Because the log response to the source facies is proportional to the amount of organic matter, well logs can be used to estimate organic richness.

Migration started immediately after structuring. Since there were no marked barriers, such as faults or diagenetic tightening, we face the rather unusual case of short-distance migration and very long-distance exchange of hydrocarbons.

The Ghawar structure started to grow as a huge anticline in the Early Cretaceous but the present structure first became apparent being truncated at an early/late Upper Cretaceous unconformity roughly post-Cenomanian and pre-Maastrichtian. Growth continued into the middle Eocene. No fundamental structural disturbance, except for post-Miocene regional tilting, affected the structure later.

The terminal Jurassic Hith Anhydrite and the Albian Nahr Umr Shale are the two principal regional seals (Murris 1981).

4.3.2.3. Diagenetic Alterations

Diagenetic alterations of the carbonate reservoirs are in general very weak. Impermeable capping anhydrites prevented subsequent extensive freshwater influx and, thus, noticeable late diagenetic cementation.

The presence or absence of the carbonate-mud matrix, one of the original textural elements, exerts a dominant control on the reservoir-rock behavior. The initial porosity of the high-energy bioclastic and oolitic sands was between 40 and 75% depending on grain size and sorting. As the mud content increases, porosity and permeability uniformly decrease. The original sedimentary textures of most carbonate sediments underwent only minor alterations, either by recrystallization or by dolomitization or both. An early event in lithification was the precipitation of fine-crystalline calcite cement, usually as coatings on the carbonate particles. Such an early fringing cement protects the fabric from solution-compaction; coating and micritisation of the particles may also inhibit cementation by crystal overgrowth. Locally, recrystallization led to holocrystalline textures by solution and redeposition. Only in such cases the original texture may be nearly or entirely obliterated.

Commonly, dolomitization is associated with recrystallization. However, dolomitization is generally not common in grainstones. The dominant type of dolomitization is the growth of large dolomite rhombs, increasing in number as dolomitization progressed. End members of holocrystalline recrystallization and dolomitization are sucrose dolomites. Dolomite growth has a strong affinity for mud-size particles; sand-size grains remain essentially unaltered. Replacement by dolomite affects markedly the reservoir behavior. As the amount of dolomite increases from 10 to 80%, regardless of original rock type, permeability decreases progressively. At 80–90% of dolomite, significant amounts of intercrystalline porosity and permeability occur. If the dolomite content exceeds 90%, porosity and permeability again decrease.

Rarely, recrystallization is also accompanied by anhydrite cementation and replacement, mainly in the transitions from carbonate rocks to evaporites. Then, megablasts of anhydrite do form.

Silicification was not observed in the Arab interval.

4.3.2.4. Production Data, Reserves, and Summary

Saudi Arabia's recoverable oil reserves were estimated at about 26×10^9 m^3 early in 1987, those of the Ghawar Field at about 7×10^9 m^3 (1970). Saudi ARAMCO has produced more than 50×10^{12} m^3 of crude since 1938, when commercial production began in Saudi Arabia. The field is characterized by a very high per-well yield requiring a small number of wells per field. However, it produces a considerable amount of associated gas.

The crude oil is undersaturated, with saturation pressure decreasing to the south of the field. Gravities range from 36° API in the north to 33° API in the south. The sulfur content ranges from 2 to 8%. All producing wells in the field are flowing wells and none of them requires pumping.

The former ARAMCO (Arabian American Oil Company) concession is now that of a state-owned company (PETROMIN). ARAMCO still operates as a service company for the state.

The giant oil accumulations of Saudi Arabia result from an exceptional combination of conditions necessary for hydrocarbon generation, entrapment, and preservation:

– Long history of quiet and almost continuous marine sedimentation;
– Widespread development of a thermally mature, organic-rich source-rock facies underlying or juxtaposed with carbonate-evaporite shelf deposits containing extensive, very thick reservoir-rock units with high porosities and permeabilities;
– A depositional cycle terminating with a major regression in which the intrashelf basins and adjacent shelves are covered by efficient and extensive evaporite seals;
– Multiple source, reservoir, and seal developments;
– Tectonic activity sufficient to create numerous immense anticlines of extraordinary closure area, but not strong enough to disrupt oil migration paths or the evaporite caps.

This prolific habitat is explained by the wide horizontal extension of the basins, of the structural traps, and of the lithological units (Murris 1980, Beydoun 1988). Horizontal hydrocarbon migration was efficient. Traps with access to rich source rocks are completely filled because they drain large areas of source rocks.

Today, Saudi Arabia produces and exports more oil and gas and has greater reserves than any other area in the world. Petroleum exploration in Saudi Arabia will be further intensified in the near future; new discoveries are anticipated. Landsat satellite imagery and remote sensing supplement exploration to a large extent.

4.3.3. U.S.A. – DEVONIAN GAS SHALES

4.3.3.1. General Data

The so-called unconventional gas sources are gas occurrences which are not explored by conventional geological methods, such as normal geophysical exploration methods, and which are hard to evaluate (long-term testing). They include gas from: (1) tight

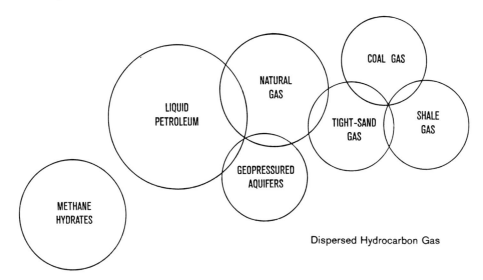

Fig. 115.   Venn diagram of different hydrocarbon sources as seen under genetic and reservoir-mechanical views (after Rose and Pfannkuch 1982).

formations, (2) methane-rich coal beds, (3) Devonian shales, (4) geopressure brines, and (5) methane hydrate deposits. The Venn diagram (Figure 115) demonstrates the overlap of different hydrocarbon sources according to their origin and reservoir mechanism. Liquid petroleum and conventional natural gas intersect as associated gas and gas condensates. Geopressured aquifers behave as a liquid with associated gas. Tight-sand gas is transitional between conventional and shale-gas reservoirs. Methane-rich coalbeds resemble tight-sands and shale-gas reservoirs in permeability and production mechanisms. The ice-like methane hydrates stand apart from the continuum of conventional and unconventional hydrocarbon sources.

The Eastern gas-bearing shales underlie an area of approximately 518 000 km^2 in the Appalachian, Michigan, and Illinois Basins (Figure 116). The most important Eastern oil-shale areas are in Kentucky, Tennessee, Indiana, and Ohio, where almost flat-lying beds of oil shale, averaging 9 to 12 m thick, are exposed at the surface in an outcrop belt nearly 1 600 km long. A closer-up view of the Appalachian Gas Province is shown in Figure 117.

The gas-bearing shales of the Appalachian Basin, mainly of Middle and Late Devonian age, are unique in a geological sense, being both source and reservoir at the same time. Since the fuel shortage in the 1970s, they became again attractive as an unconventional gas source. Potter, Maynard and Pryor (1982) summarized their geological setting and development.

During the Middle and Upper Devonian, the Appalachian Basin bordered the Old-Red Sandstone Continent. The gas-bearing shales were deposited in a large delta complex which prograded cratonward into a euxinic basin containing black shales. This delta

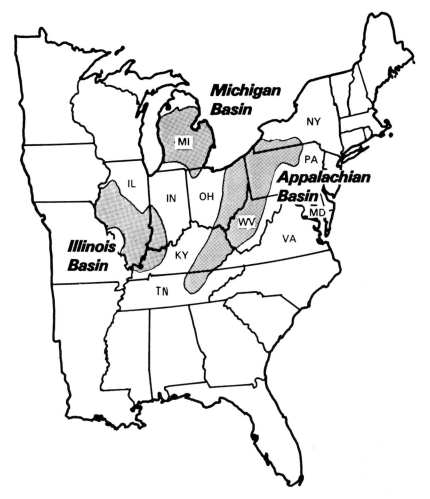

Fig. 116.   Eastern gas-bearing shale deposits in the Illinois, Michigan and Appalachian Basins, U.S.A. (Anonymous 1980).

complex had a westward-flowing paleoriver and marine dispersal system. A westward thinning wedge with a transition from delta to shelf to slope to basin formed. The main depositional environments of the basin model include from E to W: (1) an alluvial plain with a high sedimentation rate, (2) a combined delta plain and shelf, (3) a slope with turbiditic siltstones and sandstones, and (4) a euxinic basin with a very slow sedimentation rate. According to sedimentological evidence, i.e. occurrence of marine algae and presence of coarser silt grains, the black shales were everywhere deposited below the dysaerobic-anaerobic boundary which occurred at a water depth ranging from 100 to 200 m.

In the euxinic basin shales predominate. The black shale sequences are subdivided by

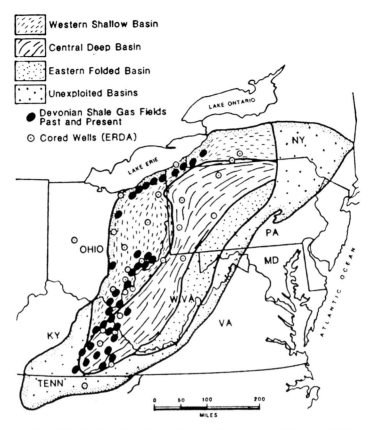

Fig. 117.   Geology of the Appalachian Gas Province (after Potter *et al.* 1982).

means of scintillometer surveys in outcrops and gamma-ray well logs based on the uranium distribution. Each black-shale unit is found at the base of a coarsening-up cycle. Some of the thin units can be traced for hundreds of kilometers along the depositional strike. A total of seven major black-shale units, which are distinguished by color (greenish-gray, gray, or black), occur in the basin (Figure 118). The organic carbon content correlates with the colors, and radioactivity as determined by gamma-ray logs with the organic carbon content.

The black, bituminous shales are laminated; bioturbation is absent. The greenish-gray shales are either laminated or bioturbated. Color, lamination, and bioturbation are interrelated. The black shales are characterized by a restricted fossil association of low diversity: brachiopods, crinoids and molluscs, fish bones, conodonts, and styliolina. Plant remains include coalified wood and phytoplankton.

Abundant *Tasmanites* is a characteristic constituent throughout the black-shale facies. All fossils are from the upper aerobic and middle dysaerobic parts of the water column. Brachiopods, crinoids, and molluscs are allochthonous (Figure 119). Nonmarine micro-

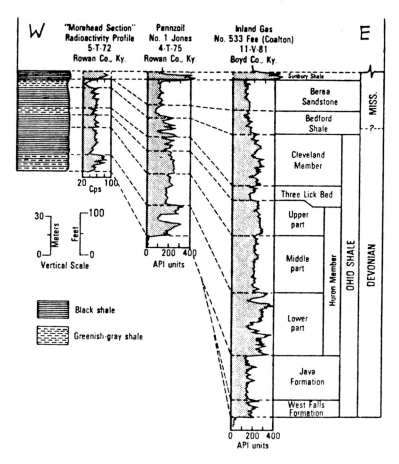

Fig. 118.   W–E cross-section through the Appalachian Basin, U.S.A., displaying major lithology, gamma-ray logs, and subdivision in 7 units (after Potter *et al.* 1982).

fossils, e.g. plant spores or minute wood fragments, are abundant in the black shales, being brought into the basin as detrital particles. Marine fossils in the greenish-gray shale include numerous benthic specimens.

The black shales, studied best by X-ray diffractometry, X-ray photography, and thin-section petrography (Figure 120) are ordinarily composed of clay minerals (65%), quartz (25%), calcite (±5%), and pyrite (< 5%). Illite and chlorite amount to approximately 60–80% of the clay-mineral fraction; kaolinite, illite-smectite mixed-layer, and illite-chlorite mixed-layer minerals are subordinate constituents. The mineral composition varies in a lateral and vertical direction reflecting facies-distribution and paleocurrent patterns. Illite crystallinity of the shales, however, seems to reflect provenance rather than diagenesis. The Devonian shales are characterized by a considerable amount of trace elements (Table 27). Pyrite contents of 5–10% are present with associated arsenic, zinc,

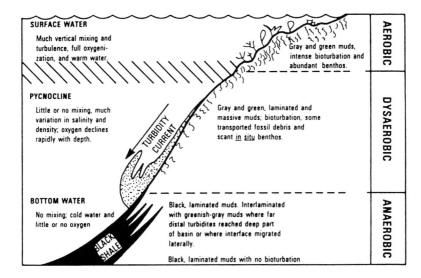

Fig. 119.   Environmental setting of the Devonian black shales (aerobic, dysaerobic and anaerobic), Appalachian Basin (after Potter *et al.* 1982).

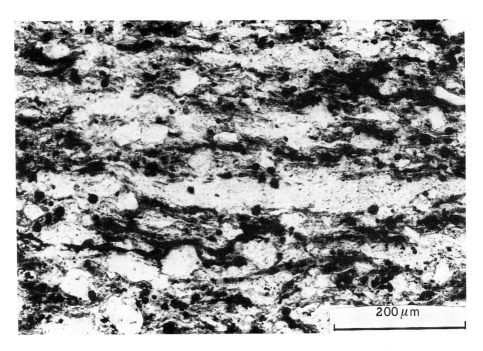

Fig. 120.   Photomicrograph of the Devonian black-shale standard sample (USGS Ohio Shale, SDO 1).

Table 27. Chemical composition of black-shale sample (SDO-1) Base Ohio Shale, Upper Devonian of Eastern Kentucky (from Maynard 1983).

	Percent		ppm		ppm
SiO_2	49.8	As	80	Ni	100
Al_2O_3	12.6	Ba	440	Rb	140
Fe_2O_3	9.3	Cd	2	Sc	15
MgO	1.6	Ce	120	Ag	<1
CaO	1.0	Cr	70	Sm	50
K_2O	3.3	Co	50	Sr	100
Na_2O	0.42	Cu	67	U	50.4
TiO_2	0.67	Ga	19	Th	9.7
P_2O_5	0.12	La	50	Sn	400
MnO	0.045	Pb	30	V	200
		Li	30	Zn	60
C	10.5	Hg	<0.2	Zr	150
H	1.31	Mo	160		
N	0.35				
S	5.31				

Source: Average of determination by various contractors of the U.S., Department of Energy, Eastern Gas Shales Project.

and mercury. The sources of the uranium and other metals were probably extensive beds of volcanic ash that were leached (Leventhal and Kepferle 1982). In the basin the shales contain normally less than 100 ppm of uranium. The total amount of uranium, however, is high, because the volume of black shale is large.

4.3.3.2. Source Rocks, Migration, and Trap Types

The Devonian Gas Shales are firstly source rocks. Their organic carbon content reaches up to 10% in the western portion of the basin. The amount of organic carbon controls chemical and physical properties of the dark shale, such as color, density, resistance to weathering, radioactivity, sulfur content, and, to a certain degree, elasticity and fracture density. The carbon-isotope ratio of organic matter suggests two sources of carbon, a heavier terrestrial and a lighter marine carbon. The sulfur-isotope ratio of the pyrite nodules is thought to reflect the relative sedimentation rates of the shale. Phosphorite nodules only occur at the top of sections characterized by extremely slow sedimentation. Thus, phosphorite formation is considered to be favored by slow sedimentation far from terrigenous sources. X-ray photographs reveal sedimentary structures and some detail in mineral distribution at the scale of a hand specimen.

The amount of organic carbon determines the gas and oil potential. Thermal maturity, as defined by vitrinite reflectance, increases regionally towards the SE from a vitrinite

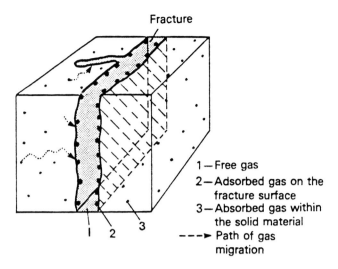

Fig. 121.   Fractured black shale with natural gas concentration (steep-dipping fractures without significant mineralization), Eastern gas-bearing shale (after Vinopal *et al.* 1982).

reflectance value of 0.5% near the western outcrop to 2% in western Pennsylvania. Simultaneously, conodont colors darken with increasing depth of burial and augmenting vitrinite reflectance.

Sedimentary fabric, fracture systems, and/or facies distribution influence the gas production. Laminae as well as fractures are normally pathways for natural gas. Four types of sedimentary fabric were observed in the black shales (Vinopal, Nuhfer and Hohn 1982): (1) thinly laminated, (2) lenticularly laminated, (3) banded, and (4) homogeneous layers. Natural fractures which cut these black shales can be classed as: (1) low-angle slickensided fractures, (2) steep-dipping heavily mineralized fractures, (3) steep-dipping fractures without significant mineralization, and (4) steep-dipping fractures with coarse hackle and offset. Optimum gas producing conditions exist in thinly laminated shales cut by fracture type (Figure 121).

Structure analysis (Shumaker 1980, 1982) showed that detached deformation and basement deformations influence markedly the development of fracture permeability within the shales. A porous fracture facies of regional extent within a black shale unit partially relates to specific physical properties of the organic-rich shales. However, another important factor for widespread gas production are fractures caused by differential shortening of sediments above a detachment surface in the shale unit. Mineralized, unidirectional, and slickensided fractures as well as increased fracture intensity within the organic-rich shales perpendicular to the stress caused by the Allegheny orogeny support this interpretation. The porous fracture facies is most permeable. Beyond the region of major tectonic transport only erratic permeabilities occur. Trends of abnormally high final open flows in the producing areas correspond to trends of intensely fractured black shale. The gas migrated updip along open fractures placing the best wells slightly updip along the fracture trend or on the flank of adjacent low-relief flexures. Thus, these unique reservoirs form their own source and seal.

The gas seems to be adsorbed rather than freely held in pores and fractures. Gas in the shales is presumably held in three forms, namely in open pores, adsorbed on clays, and dissolved in organic matter (Schettler 1979). Dissolved gas in organic matter may be the main reason for the longevity of production from gas wells in shale. In all shales the interval surface area, which is a measure of matrix porosity, appears to be a key factor for gas production. The surface area decreases continuously with depth of burial and with augmenting carbonate content.

4.3.3.3. Diagenetic History

The Late Paleozoic black shales of the Eastern Gas Province, normally composed of illite/muscovite and chlorite, are laminated and hard due to compaction and tectonism. The shales are in an advanced diagenetic alteration stage. The degree of parallel orientation of clay minerals as seen under crossed nicols is likely a measure of the amount of clay mineral alteration during compaction. Increasing perfection of lamination causes increase in tensile strength, acoustic velocity, and hardness. The original clay-mineral composition is unknown; however, a primordial smectite composition is presumed based on the general geology and the presence of tuff intercalations. Such a clay-mineral composition enhanced easy transformation of minerals and of the organic matter. The heavy minerals present such as pyrite, leucoxene, zircon, tourmaline, and rutile are the relics of a more unstable heavy mineral suite degraded by intrastratal solution. The internal surface area, a key factor for gas production, decreases steadily with depth of burial and with increasing carbonate content. Thermal maturity, ranging between vitrinite-reflectance values of 0.5 to 2%, is a critical parameter for gas-sourcing as well as fracturing for gas-releasing. The most productive gas shales are those with maturity values of R_0 between 0.5 and 1.5%. This makes the decisive difference to Devonian shales of the Hercynian Mountains of central Europe, which are characterized by much higher vitrinite-reflectance values of R_{min} 3% to R_{max} 7%.

4.3.3.4. Production Data, Reserves, and Summary

The first well drilled specifically for natural gas in the U.S.A. was in the Appalachian Basin. It produced gas from a Devonian shale sequence in 1821, 38 years before Colonel Drake discovered oil in Pennsylvania.

The different shale units extend from outcrops to depths down to 2 600 m. An estimated 76×10^9 m^3 have been produced from this area (about 9 600 producing wells). Up to 25.5×10^{12} m^3 of natural gas are estimated to be contained in the fractures and matrix of the gas shales but only a small percentage is being recovered. Natural gas currently provides about 27% of the U.S. energy needs. In 1978 this amounted to almost 570×10^9 m^3 of gas. Of the major primary-energy forms, natural gas accounted for 19.6 quads of energy in 1977.

Estimates of the Devonian Shale gas vary; they range from 8×10^{12} m^3 to 25.5×10^{12} m^3 for the total as-in-place and from 283×10^9 m^3 to 14.7×10^{12} m^3 for the economically

Table 28.   Natural gas production from conventional and unconventional sources, U.S.A. (Anonymous 1980).

| | U.S. Natural Gas Production | | | |
	Current Production (10^9 m³/year)	Estimated Recoverable Reserve	Estimated Well Life (years)	Production Per Well
Conventional				
On-shore	425×10^9 m³	$4\,248 \times 10^9$ m³	20–30	142×10^6 m³
Off-shore	113×10^9 m³	$1\,416 \times 10^9$ m³	20	$142–282 \times 10^6$ m³
Unconventional				
Western Sandstones	28×10^9 m³	$1\,416–9\,911 \times 10^9$ m³	20–30	14×10^6 m³
Coalbeds	3 (vented)	$57–14\,158 \times 10^9$ m³	10–20	
Devonian Shale	3	$283–14\,725 \times 10^9$ m³	30–40	

recoverable shale gas (Table 28). Potentially productive shale zones may reach a thickness of over 300 m. In many areas the gas-bearing shale deposits are shallow and only about 10% of the gas-productive black and brown shales lie at depths below 2 500 m. The recoverable gas can be increased by developing effective exploration techniques to locate fractured reservoirs and by use of more efficient stimulation technologies.

Optimum gas production comes from the thinly laminated black-shale type cut by steep-dipping fractures with coarse hackle and offset. Appropriate knowledge of the interval stratigraphy of the shale sequences is needed to identify all productive horizons and to stimulate different gas-shale horizons if necessary.

In summary, the most productive gas shales are thick and rich in organic matter (above 3%) with maturity values of Ro between 0.5 and 1.5. Other factors, not yet fully understood, comprise the significance of interlaminated and interbedded siltstones, the proportion of interlaminated gray shale, and the spatial density of fractures.

4.3.4.   U.S.A. – MINERAL CEMENTATION, PERMIAN LYONS SANDSTONE

4.3.4.1.   General Data

The productive Lyons Sandstone of Permian age (Leonardian) crops out almost continuously along the eastern side of the Colorado Front Range (Figure 122); it subcrops in the Denver Basin, Colorado, U.S.A., which is an asymmetrical northward-trending syncline. This basin produces oil in several fields.

The Lyons Sandstone ranges in thickness from 0 to over 70 m in the subsurface. It thins to zero in areas adjacent to the evaporite basins. Based on subsurface correlations and on directional data from surface exposures, a NE trending eolian-sand sea is postulated across the Denver basin. The sandstone appears to be derived from the northeast. Evaporites were deposited in depressions adjacent to the sand sea. Two depositional systems merge

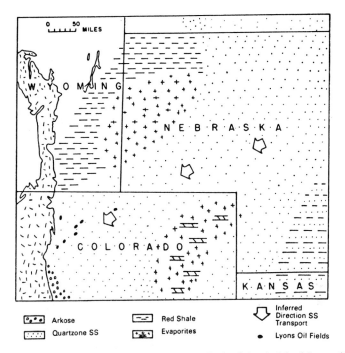

Fig. 122. Lithofacies map of the Lyons Sandstone, Denver Basin, Colorado (after Momper 1963).

along the margin of the Front Range, eolian sandstones being transported to the west and south intertongue with arkoses being transported by streams flowing eastward from the source area (Momper 1963). Most of the oil production from the Lyons Sandstone comes from anticlinal structures. Major oil fields producing from the Lyons Sandstone occur in isopach thick areas or paleostructural low areas.

The sandstone is quartz-rich with about 75% quartz; individual beds vary in composition from orthoquartzites to subarkoses. The intergranular matrix and the cements comprise clay and chlorite (1–10%), iron oxide (traces to 10%), secondary quartz (0–28%), solid organic matter (up to 25%), anhydrite (0–25%), and carbonate cement (calcite and dolomite). Pyrite occurs in minor quantities as cubes and small patches.

4.3.4.2. Source Rock, Migration, Trap Types, and Seal

The source beds for oil from the Lyons Sandstone have not yet been identified. Several theories exist concerning source beds for the production found in the Denver basin from the Lyons Sandstone. The source bed for oils found in the Lyons Sandstone is regarded to be one of the following: (1) adjacent Pennsylvanian and Permian shales, (2) adjacent Satanka Formation, (3) overlying Cretaceous source beds, or (4) the Phosphoria Formation of W Wyoming (Sonnenberg and Weimer 1981). Crude oil produced from the Lyons Sandstone

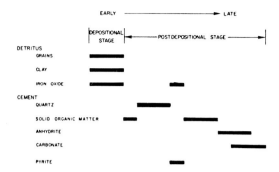

Fig. 123.   Paragenetic sequence of early and late mineral cements in the Lyons Sandstone, Denver Basin, Colorado (after Levandowski *et al.* 1973, reprinted with permission).

resembles crudes produced from Paleozoic reservoirs in the Powder River Basin. Also Pennsylvanian and Permian radioactive black shales in NW Nebraska and E Wyoming have excellent source capability (Wilson 1980).

Anticlinal closure is the trapping mechanism in the Black Hollow, Pierce, New Windsor, Fort Collins, Keota, and Berthoud Fields. Diagenetic trapping appears to be responsible for trapping of oil in the Loveland Field and partially in the New Windsor Field. Thus, trapping in the Lyons Sandstone appears to be structural with minor diagenetic control.

The anhydrite-rich 'Upper Satanka' provides the impermeable seal.

4.3.4.3. Diagenetic History

In their classical study on the Lyons Sandstone Levandowski *et al.* (1973) demonstrated the important role of mineral cementation in reservoir evolution, especially in respect of migration and accumulation of hydrocarbons in sandstones. Cementing minerals normally reduce porosity and permeability of reservoir sandstones. In the present study a distinction was made between early, late, and differential cementation. The paragenetic sequence of the matrix and the cements is shown in Figure 123. Solution and overgrowths on detrital grains are absent where organic-matter, iron-oxide, and clay coatings are thickest, and, thus, have prevented chemical interaction between detrital grains and solutions. Anhydrite is paragenetically associated with secondary quartz and dolomite, but rarely with calcite replacing quartz.

The regional distribution of the important mineral cements is shown in a SW–NW cross-section across the Denver Basin (Figure 124). The cross-section depicts those parts of the Lyons Sandstone in which various cementing minerals are observed, i.e. quartz, solid organic matter, carbonate, and/or anhydrite and in which present porosity occupies more than 50% of the minus-cement porosity. The distribution of red and gray iron oxide-bearing sandstones is shown in Figure 125.

The distribution of quartz cement is independent of depth, but tends to increase towards the W (Figure 126). This accumulation of quartz cement accounts for a permeability barrier in the subsurface. In the eastern part of the basin, detrital quartz grains are strongly etched,

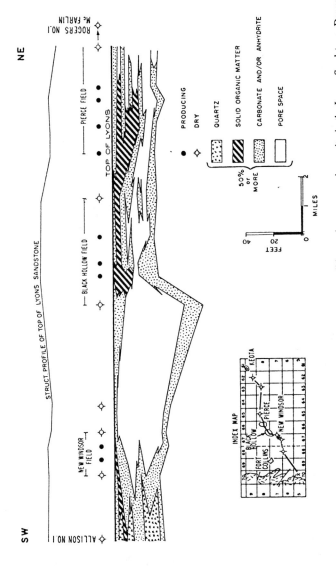

Fig. 124. Vertical and lateral distribution of mineral cements, solid organic matter, and porosity in the Lyons Sandstone, Denver Basin, Colorado (after Levandowski *et al.* 1973, reprinted with permission).

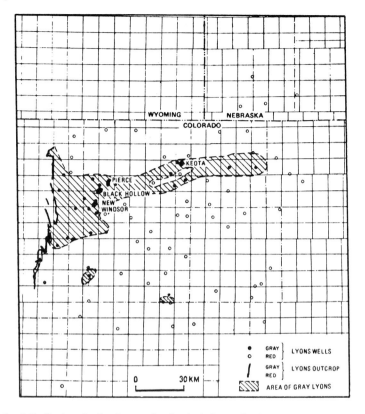

Fig. 125.   Areal distribution of red and grey coloration in the Lyons Sandstone, Denver Basin, Colorado (after Levandowski *et al.* 1973, reprinted with permission).

but little or no deposition of secondary quartz took place. These features indicate removal or loss of silica. The solid organic matter has a wider distribution than oil; it is closely associated with the gray variety of the Lyons Sandstone. It much resembles the solid organic matter in the gray Rotliegende sandstones of NW Germany. The carbonate and anhydrite cements are concentrated near the top of the Lyons Sandstone, below the organic interval and below intervals containing a relatively large amount of quartz. The general build-up of the carbonate and anhydrite cements in the west, just east of the concentration of quartz cement, is evident.

Spectrochemical analysis of anhydrite cement demonstrated a regional Sr content variation in anhydrite, as a substitute of Ca^{2+}. The Sr content has been used as hint at the anhydrite type or origin: (1) Primary anhydrite, present in bedded evaporite sequences, tends to have a nearly constant $Sr \times 10^2/CaSO_4$ ratio, because it formed in equilibrium with brines of similar composition. (2) 'Secondarily redistributed' anhydrite has a Sr content that suggests repeated mobilization. The Sr content of the normal anhydrite is lowered notably by intimate contact with formation waters. (3) 'Secondarily introduced'

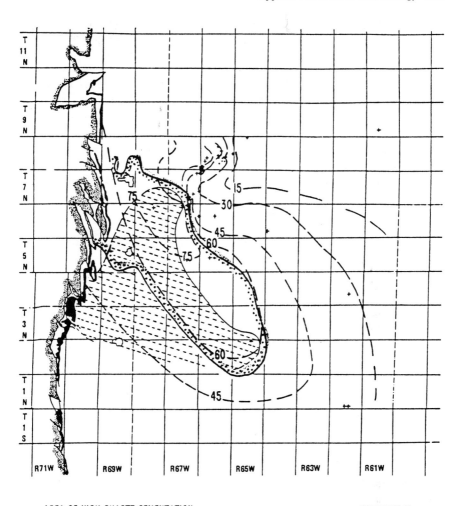

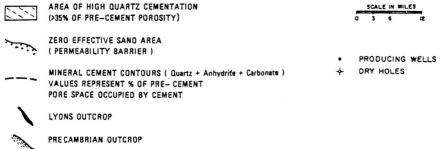

Fig. 126.   Regional distribution of quartz cementation in the Lyons Sandstone, Denver Basin, Colorado (after Levandowsky *et al.* 1973, reprinted with permission).

anhydrite from veins and solution cavities is distinguished by higher than normal Sr content. Anhydrite is enriched in a NW–SE trending zone, roughly paralleling the zone of high silica cementation. It is associated with the known Lyons Sandstone oil fields and the greatest enrichment occurs near these fields. The beds underlying the Lyons Formation and much of the Lyons Sandstone to the east appear to contain leached anhydrite. The anhydrite of the cap rock overlying the Lyons Sandstone is of uniform composition. It is a normal anhydrite, not affected by solution.

Based on these regional distribution patterns, the formation waters seem to have moved upward into the Lyons Sandstone as well as laterally toward the edge of the basin because of the impermeable cap rock.

The formation of cements depends on pressure and temperature and, thus, on the burial history of the sedimentary basin. The presence of fresh detrital plagioclase with clay and chlorite coatings suggests that burial metamorphism was absent or minor. The coatings are the results of diagenesis.

Silica for the secondary quartz cement was considered to be derived from: (1) solution of detrital quartz grains, (2) deposition from sea water, (3) precipitation from connate water, and (4) deposition from fluids derived from the adjacent shale beds. The fluids of types 3 and 4 are likely associated, wholly or partly, with compaction fluids.

Extremely high pressure, usually not common in sedimentary rocks, is required to appreciably increase the solubility of quartz. Levandowski *et al.* (1973) speculated that granulation and submicroscopic fracturing at grain contacts may produce some amorphous silica, similar to quartz ground-up in the laboratory (Siever 1962). Interstitial waters would then preferentially dissolve the amorphous silica. As the amorphous silica is removed in solution, pressure along grain contacts would produce new amorphous silica. Under natural conditions other parameters in addition to the pressure would have to be considered, e.g. temperature and composition of pore fluids.

Moreover, anhydrite and carbonate replacement of quartz releases silica to solutions being then available for cementation. Silica solubility increases with augmenting temperature, whereas $CaCO_3$ solubility below $120°C$ and at constant P CO_2 decreases with increasing temperature. If intrastratal fluids saturated with anhydrite, $CaCO_3$, and $CaSO_4$ may precipitate with concomitant dissolution of silica.

The observation that silica was moved from the basin center towards the margin where it was precipitated indicates, that the composition of the intrastratal fluids also influenced regional variations in cementation. Calcite-quartz replacement reactions depend on pH, e.g. dissolved silica tends to precipitate below a pH of 9.8. Therefore, the formation fluids in the basin center were presumably highly alkaline because of the presence of evaporites. High alkalinity and existing pressure and temperature tended to increase the solubility of the quartz. At late-diagenetic burial and compaction the silica-enriched waters moved horizontally underneath the tight anhydrite-rich unit. The basinal compaction fluids mixed with less alkaline meteoric waters from the basin margin. The pH dropped and silica precipitated.

Carbonate and sulfate cements formed last in the paragenetic sequence as it is common in cementation sequences. Calcite was evidently the latest cement. The vertical and horizontal regional distribution of the last cements was at least partly controlled by the distribution of the earlier formed intergranular pore fillings (e.g. quartz cement). Trace-element composition of the anhydrite indicates that the solutions moved from below and migrated laterally to the basin margins underneath the tight anhydrite cap rock. This migration took place during compaction.

Pore-water geochemistry controlled diagenetic mineral cementation. The Lyons Sandstone sequence contained highly alkaline formation fluids with a large amount of sulfate. During compaction the upward and, subsequently, laterally moving solutions migrated towards the basin margins, where anhydrite and possibly some calcite precipitated due to pressure decrease. The concept of ionic impedance, first introduced by Fothergill (1955), was used to explain the concentration of carbonate and anhydrite cement near the secondary quartz barrier that presumably acted as a semipermeable membrane behind which ions were concentrated on the high-pressure, basinward side. As compaction progressed, the formation fluids on the basinward side may have been subjected to higher pressures than on the landward side. Behind the semipermeable membrane, ionic impedance progressively increased the concentration of Mg^{2+}, Ca^{2+}, and Sr^{2+} ions, until the solubility products of the respective minerals were reached. Anhydrite and calcite were precipitated as open-space fillings. The sequence of late cementation presumably was Sr-enriched anhydrite, anhydrite, and calcite.

The various stages of mineral cementation can be summarized as follows:

— Sand deposition with some detrital clay and iron oxide in the matrix. Basinward the sands interfingered with a thick evaporite series (lime muds, dolomites, anhydrites, and clays).
— Accumulation of the anhydrite-rich unit 'upper Satanka' providing a rather impermeable seal.
— Progressive accumulation of younger sediments increased the overburden, thus pressing out the connate water as compaction fluid. Migration was mainly lateral towards the basin margin within the sandstone underneath the anhydrite cap. The high alkalinity (pH) of the water caused silica dissolution and its subsequent reprecipitation as cement when the pH was lowered.
   Silica solution predominated along a zone of basinward sandstone-evaporite interbedding. On the other hand, near the western basin margin, the mixing of the alkaline connate waters with meteoric waters lowered the pH causing silica precipitation.
— Oil, generated in the basinal shales and carbonates, migrated into the Lyons Sandstone prior to complete compaction. The oil moved laterally because of buoyant forces and the pressure gradient resulting from compaction.
— With increasing burial, critical overburden pressure favored solution processes such as stylolite formation, associated with further expulsion of connate waters. These fluids dissolved anhydrite and carbonates in underlying horizons and reprecipitated anhydrite while migrating upwards and laterally.

In summary, the reconstruction of a geological model based on petrographic, isotopic, and spectrochemical analyses demonstrates that mineral cementation controlled migration

Table 29.   Reservoir and production characteristics, Lyons Sandstone, Denver Basin, Colorado (after Sonnenberg and Weimer 1981).

Field	Trap	Porosity % (avg)	Perm. (md/ft) (avg)	Oil Gravity ° API	Oil Zone Thickness	Area (km^2)	Com. Prod. (m^3)
Black Hollow	Anticl. Clos.	12	88	36.0	26	6.27	1.61×10^6
Pierce	Anticl. Clos.	11.6	44	36.0	27	8.09	1.56×10^6
New Windsor	Anticl. Clos. red. of ϕ cement	9.0	21	41.1	13	0.97	133 074
Fort Collins	Anticl. Clos.			26.9	36	?	37 136
Loveland	ϕ reduction updip	4.4	0.9	?	40	0.16	24 296
Keota	Anticl. Clos.	?	?	?	32	?	2 038
Berthoud	Anticl. Clos.	?	?	?	?	?	41 042
La Porte	Anticl. Clos.	?	?	?	?	?	22.4

Modified from Levandowski 1973. Cumulative production figures from Colorado Oil and Gas Conservation Commission statistics, 1979.

and accumulation of oil in the Lyons Sandstone to a very large extent. In this respect the cementation of the Lyons Sandstone resembles that of the coeval Rotliegende sandstones in NW Europe.

4.3.4.4. Production Data, Reserves, and Summary

Lyons Sandstone production was established in the Keota Field in 1953. Additional fields, all on the western flank of the Denver Basin, now produce from depths between 1 500 and 2 900 m, generally structurally controlled. Cumulative production from all Lyons fields was 3.4×10^6 m^3 of oil as of December 1979. Porosities in the productive sandstones range form 1 to 26% while permeabilities range from 1 to 1 000 mD. The bulk of the production comes from the Black Hollow and Pierce Fields. Reservoir and production data from Lyons fields are summarized in Table 29.

Water injection of the Lyons Sandstone fields started in 1958. Available data indicate technically successful projects.

Oil production from the Permian Lyons Sandstone could be extended to other adjacent areas. Porous and permeable sandstones in the Lyons Formation exist over a broad area.

4.3.5. MEXICO – CRETACEOUS TAMABRA LIMESTONE

4.3.5.1. General Data

Mexico occupied the fifth place in the world's hydrocarbon production in 1987. The major oil-producing basin of Mexico is the Tampico Embayment, the so-called 'Faja de

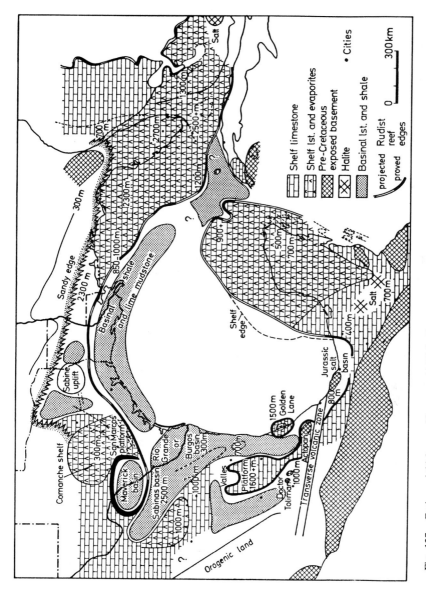

Fig. 127.   Facies and thickness of Lower and Middle Cretaceous strata around the Gulf of Mexico (after Wilson 1975, p. 326).

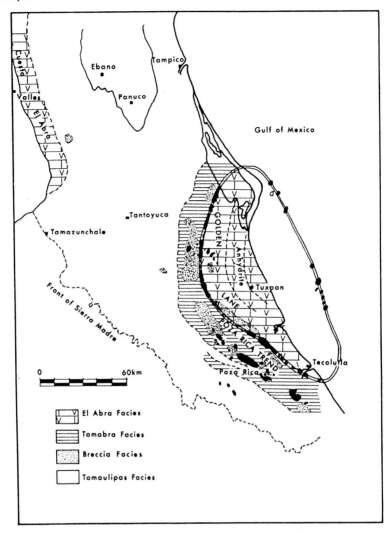

Fig. 128.   Facies and distribution of oil fields in the Middle Cretaceous Golden Lane and Poza Rica trends on the Gulf coast of Mexico (after Wilson 1975, p. 332).

Oro' (Golden Lane) discovered in 1908, that extends from the central Mexican platform, near the front of the Sierra Madre Oriental fold belt, to the continental shelf of the Gulf of Mexico (Figure 127). Most of the production in the Tampico Embayment comes from Albian-Cenomanian limestones. As one of the great oil fields of the world exists along the edges of the Golden Lane, the facies of these rocks have been studied in detail by Mexican and U.S.A. geologists recently, hoping that the vast production of Mexican fields can be extended into offshore areas around the margins of the Gulf and into Texas.

These limestones form three regional facies belts (Figure 128): (1) the Tamaulipas

Limestone – a fine-grained basinal limestone, (2) the El Abra Limestone – a shallow-water platform facies with a reef, back-reef, and lagoonal facies, and (3) the Tamabra Limestone – a wedge of coarse skeletal debris and breccia at the basin margin (Figure 129). The present review focusses on the Tamabra Limestone, the term 'Tamabra' being a corruption of Tamaulipas and Abra, and refers to the facies that is intermediate between the El Abra reefal facies and the dense micritic Tamaulipas equivalent.

The Lower/Upper Cretaceous Tamabra Limestone (Albian-Cenomanian) is well developed in the subsurface of the Tampico Embayment, close to the Golden Lane platform ('atoll' in El Abra Facies). Oil production comes mainly from the greater Poza Rica Field that is located at the NE flank and nose of a broad anticline which plunges SE (Figure 130). The limestone will ultimately produce more than 365×10^6 m^3, mainly from secondary porosity in probable basin-margin mass-flow deposits (Enos 1988). The Golden Lane platform developed on an igneous basement-high of sialic composition which probably lies over a structural culmination of the Late Paleozoic circum-Gulf orogenic belt (Wilson 1975).

The facies has been described by Barnetche and Illing (1956) from the Poza Rica Field. It contains mixed and interbedded biota of pelagic lime mudstone and shallow-water benthos. The thick Tamabra Limestone section in Poza Rica is interrupted by numerous thin shale units, presumably bentonites. The principal carbonate-rock types of the Tamabra Limestone are in order of abundance (Enos 1977, Enos and Moore 1983, Moore 1989): (1) skeletal-fragment grainstones and packstones, (2) breccias, (3) rudist-fragment wackestones, (4) pelagic-microfossil wackestones, and (5) dolomites. These lithologies, with the exception of dolomites, are summarized in Table 30.

Regional trends are to finer-grained lithologies upward and into the basin. Grainstones/packstones predominate in the greater Poza Rica Field; breccias are more common to dominant in other parts of the field.

The skeletal-fragment grainstones and packstones are porous and productive with porosities ranging from 5 to 25%. Massive beds prevail. Rudists are the dominant skeletal fragments. Primary depositional features include fine/coarse interbeds, presumable imbrication, long-axis particle orientation, and locally graded bedding. Grains are commonly coarse (> 1 cm) and rounded.

The sedimentary breccia intervals, conspicuous rock types within the Tamabra Limestone, range from less than one meter to about 100 m in thickness. Breccia clasts are angular and range generally between 1 and 10 cm in size. They include skeletal-fragment grainstones and packstones, pelagic-microfossil wackestones, and rarely crystalline dolomites but not of specific Tamabra lithology. Platy clasts are embedded at high angles, even perpendicular to the bedding. Sutured grain contacts are common. The amount of matrix within the breccia varies considerably; unaltered breccia matrix normally contains pelagic microfossils, e.g. globigerinid foraminifera. The porosity of the breccia is mostly low (Table 30), because of the micritic matrix and of sutured grain contacts.

The rudist-fragment wackestones are rare and form intervals 3 m or thicker. The rudists are rarely in growth position. Incomplete internal sediment fillings show geopetal structure. The micritic matrix contains minute, unidentifiable skeletal fragments. The porosity is normally poor (< 5%). Moldic, rarely vuggy porosity develops by leaching

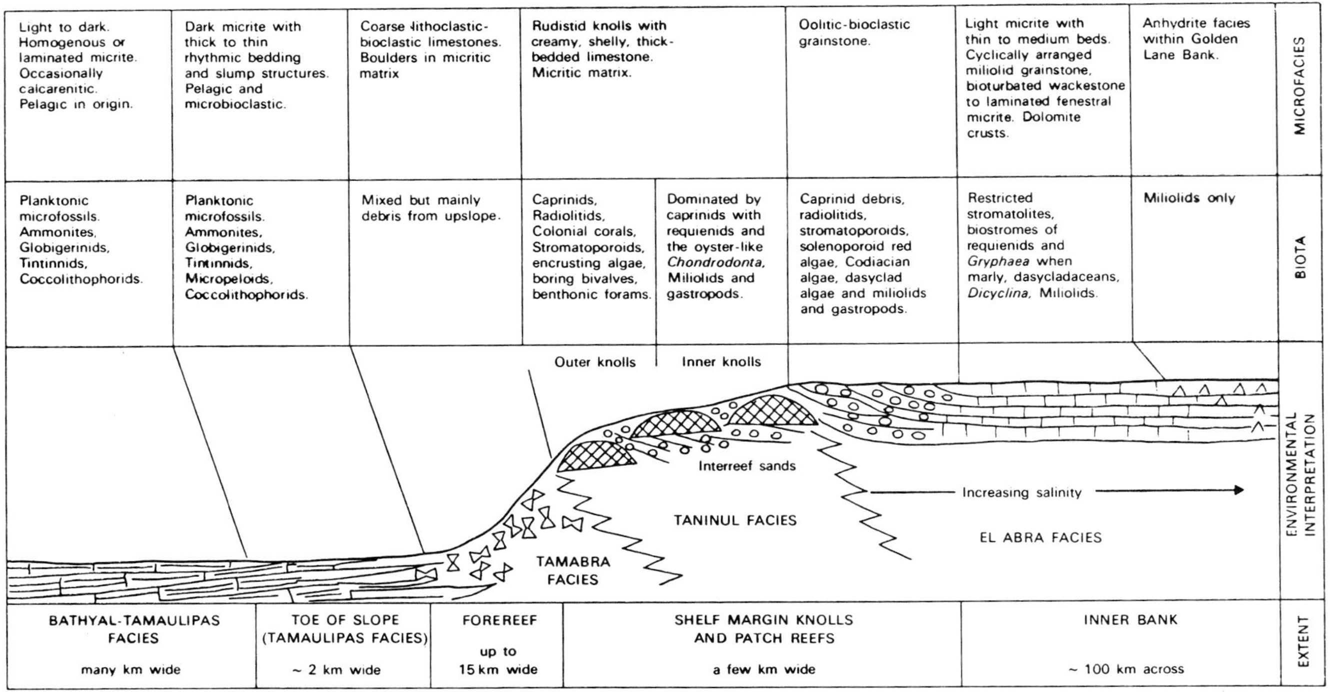
MICROFACIES
Light to dark. Homogenous or laminated micrite. Occasionally calcarenitic. Pelagic in origin.
Dark micrite with thick to thin rhythmic bedding and slump structures. Pelagic and microbioclastic.
Coarse lithoclastic-bioclastic limestones. Boulders in micritic matrix
Rudistid knolls with creamy, shelly, thick-bedded limestone. Micritic matrix.
Oolitic-bioclastic grainstone.
Light micrite with thin to medium beds. Cyclically arranged miliolid grainstone, bioturbated wackestone to laminated fenestral micrite. Dolomite crusts.
Anhydrite facies within Golden Lane Bank.
BIOTA
Planktonic microfossils. Ammonites, Globigerinids, Tintinnids, Coccolithophorids.
Planktonic microfossils. Ammonites, Globigerinids, Tintinnids, Micropeloids, Coccolithophorids.
Mixed but mainly debris from upslope.
Caprinids, Radiolitids, Colonial corals, Stromatoporoids, encrusting algae, boring bivalves, benthonic forams.
Dominated by caprinids with requienids and the oyster-like Chondrodonta, Miliolids and gastropods.
Caprinid debris, radiolitids, stromatoporoids, solenoporoid red algae, Codiacian algae, dasyclad algae and miliolids and gastropods.
Restricted stromatolites, biostromes of requienids and Gryphaea when marly, dasycladaceans, Dicyclina, Miliolids.
Miliolids only
ENVIRONMENTAL INTERPRETATION
Outer knolls
Inner knolls
Interreef sands
Increasing salinity
TANINUL FACIES
EL ABRA FACIES
TAMABRA FACIES
EXTENT
BATHYAL-TAMAULIPAS FACIES
many km wide
TOE OF SLOPE (TAMAULIPAS FACIES)
~ 2 km wide
FOREREEF
up to 15 km wide
SHELF MARGIN KNOLLS AND PATCH REEFS
a few km wide
INNER BANK
~ 100 km across

Fig. 129. Biofacies distribution across the margin of the Golden Lane ‘Atoll’, Mexico (modified from Wilson 1975, after Sellwood 1986).

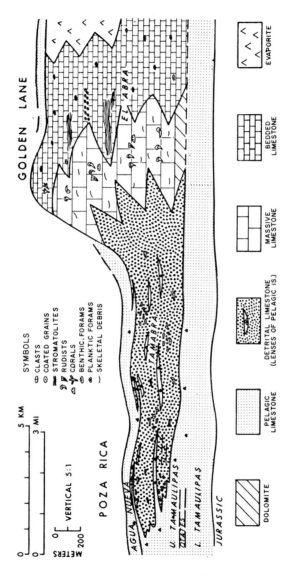

Fig. 130.   Lithofacies distribution in a SW–NE cross-section from Poza Rica to Golden Lane, Mexico (after Enos 1977, p. 286, Fig. 10).

Table 30.   Petrographic characteristics of major Tamabra Limestone lithologies, Cretaceous Poza Rica trend, Mexico (after Enos 1977, p. 287).

Lithology	Skeletal-Fragment Grst/Pkst	Rudist Wackestone	Pelagic-Microfossil Wackestone	Breccia	
Abundance	very common	very rare	rare	common	
Color	dark brown or grey	light brown, cream	white, light grey, rarely dark grey	mottled white, brown and grey dark matrix	
Oil Staining	common, heavy	rare	lacking	patchy	
Interval Thickness	may exceed 100 m	generally thin	thin	may exceed 100 m	
Bedding	vague, massive	vague, massive?	thin	massive	
Sedimentary Structures					
Graded Beds	vr	–	m	vr	
Coarse/Fine Layering	m	vr	r	vr	
mm Lamination	–	–	vr	–	
Imbrication	f	vr	–	vr	
Long-Axis Orientation	m	–	–	–	
Bioturbation	–	–	m	–	
Soft-Sediment Deformation	–	–	vr	f	
Reoriented Geopetals	r	m	vr	f	
Grains					
Whole Skeletons	r	c	c	vr	
Skeletal Fragments	ab	c	f	m	
Pelletoids	vr	f	m	–	
Lithoclasts	vr	r	–	ab	
Intraclasts	vr	f	r	f	
Matrix	r	ab	ab	ab to r	
Grain Size	coarse–fine	coarse–med.	v. fine	coarse–v. coarse	
Roundness	r.–subang.	ang.	–	ang.	
Sorting	good to poor	poor	good to poor	poor	
Fossils				matrix	clasts
Caprinids	ab	ab	–	vr	ab
Radiolites	c	c	–	–	c
Other Mollusks	c	c	vr	r	c
Coral	m	f	–	–	f
Red Algae	r	r	–	–	r
Green Algae	f	f	–	–	f
Echinoderms	c	c	f	–	c
Bethonic Forams	–	vr	–	vr	vr
Pelagic Forams	–	vr	ab	ab	–
Calcispheres	–	r	ab	ab	–
Radiolaria	–	–	r	r	–
Cement				matrix	clasts
Micritic	–	r	ab	ab	c
Fibrous (ragged)	r	–	–	–	c
Blocky	ab	c	r	–	c
Silica	r	vr	f	–	vr
Recrystallization	extensive	moderate	slight	slight	moderate
Dolomitization	vr	r	c	ab	r
Pyrite	vr	vr	ab	c	vr
Stylolites	f	f	c	ab	c
Porosity	fair to excellent	poor to good	nil	nil to fair	
Primary Intergr.	c	r	–	r	
Primary Intragr.	c	r	vr	vr	
Moldic	ab	m	–	m	
Vuggy	r	r	r	r	
Fracture	r	f	r	f	
Intercrystalline	–	r	r	c	vr

Symbols: – = lacking; vr = very rare; r = rare; f = few; m = many/much; c = common; ab = abundant.

(Table 30).

The pelagic-microfossil wackestones form thin beds (< 2 m thick). Contacts against grainstones or breccias are in general stylolytic. Internal sedimentary structures include vague burrows, local microlamination, and microgradation. Pelagic foraminifera, calcispheres, calcified sponge spicules, calcified radiolaria, ostracods, molluscs, echinoid fragments, and presumable pelagic algae are characteristic fossils. Pelletoid 'ghosts' are common. The well preserved microfossils are suited best to date the Tamabra Limestone. Locally some layers and nodules of black, translucent replacive chert with silicified calcispheres and foraminifera occur. Euhedral dolomite rhombs replace the calcitic matrix to a minor degree. No porosity is perceptible in the pelagic-microfossil wackestones (Table 30).

Dense and structureless crystalline dolomites occur in thin intervals either as massive vuggy and fossil-moldic dolomites, or as breccias with dolomite matrix and clasts, or as dolomitic skeletal grainstones and packstones, and as dolomitic microfossil wackestones. The 'average' porosity of the dolomites is poor. Massive dolomites may locally reach good porosities with 'relict moldic' and vuggy porosities.

The previous interpretation that the Poza Rica trend is a true shelf edge and the Golden Lane a faulted backreef-lagoon sequence, was rejected by Enos (1977) because of the stratigraphic thickening and the evident lack of fault structures at the Golden Lane escarpment. Geometry and sedimentology of the carbonate rocks indicate that the Golden Lane is a true shelf edge and the Tamabra-Limestone facies with its reef-derived debris was deposited in the basin. Most of the debris of the Tamabra Limestone seems to have been derived from reefs at the escarpment of the Golden Lane platform. The skeletal grainstones/packstones were transported grain-by-grain by sediment-gravity flows (turbidity current, debris flow, grain flow) and deposited at the edge of the basin. The breccias are also interpreted as debris-flow deposits. Graded beds are characteristic sedimentary structures; other structures indicate deposition in a current-dominated environment. No rudists in growth position were observed in the Tamabra Limestone. Fossils were only mechanically deposited with long axes in a vertical position.

4.3.5.2. Source Rocks, Migration, Trap Types, and Cap Rocks

Basinal pelitic sediments of the Upper Jurassic Taman Formation and, to a lesser extent, in the Upper Jurasic Pimienta Formation, which are calcareous or terrigenous and rich in organic matter, are supposed to source the Tamabra reservoirs. Migration is of short distance only. Stratigraphic trapping in the Poza Rica Field is caused by the westward pinchout of the porous Tamabra Limestone into the impermeable basinal Upper Tamaulipas Limestone. Cap rock is a dense basinal limestone of Upper Cretaceous age.

4.3.5.3. Diagenetic History

Diagenetic processes within the Tamabra Limestone comprise recrystallization, dissolution, leaching, and dolomitization. In the skeletal-fragment grainstones and packstones recrystallization obscured nature and amount of the original rock matrix. Interparticle

Table 31.   Porosities of Tamabra-Limestone lithologies, Cretaceous, Poza Rica trend, Mexico (after Enos 1977, p. 291).

Lithology	'Average' Porosity*	Maximum Porosity (%)	Principal Porosity Type	Other
Skeletal-fragment grainstone and packstone	good	25	moldic	intergranular
Rudist wackestone	poor	10	moldic	vuggy
Breccia	poor	8	intercrystalline	fracture, moldic
Pelagic wackestone	nil	1	intercrystalline	fracture
Dolomite	poor	25	intercrystalline	vuggy, 'relict moldic'

* Good 10–15%, fair 5–10%, poor 1–5%, nil <1%.

porosity and local leaching also contribute to interconnect the pore network. Grain-to-grain and bed-to-bed contacts are commonly stylolitic. Dolomitization is only subordinate and local; the average porosity of dolomites is poor.

Typical reservoir porosities are about 10% (Table 31); permeabilities average 2 mD and rarely exceed 100 mD. Porosity is largely the result of selective dissolution of rudist fragments which were originally composed of aragonite. A relatively simple diagenetic history is observed. The primary porosity was reduced through lithification of matrix mud and initial cementation by clear, non-ferroan calcite. Later dissolution produced extensive skell-moldic and minor vuggy porosity. Subsequently, non-ferroan calcite cement reduced porosity prior to the emplacement of hydrocarbons. The initial phase of cementation and presumed lithification of mud greatly reduced porosity. Dissolution produced secondary-porosity which even exceeded that of the initial sediment in some grainstones. Calcite cementation and local multiphase quartz cementation and dolomitization reduced porosity to present average values of 8–12% in grain-supported rocks and 3% in mud-supported rocks. The greater persistence of primary porosity and, therefore, of permeability in grain-supported rocks probably accounts for their greater secondary-porosity development and ultimate reservoir quality. Permeabilities range from 0.17 mD in wackestones to 3.85 mD in dolomites. Permeability increases with porosity in all lithologies. According to Enos (1988) the agent for early cementation and development of secondary porosity was presumably meteoric water. Subaerial exposure appears to be ruled out because of a basin-margin depositional environment and continuous burial beneath Upper Cretaceous pelagic sediments. Early exposure to meteoric water can be explained by a hydrologic head developed during penecontemporaneous exposure that produced cavernous porosity in the adjacent Golden Lane trend. Descending meteoric waters likely emerged as submarine springs along the Tamabra trend. Analogous freshwater circulation is found in Florida nowadays.

Table 32. Production, proven reserves and API gravity of Tamabra
Fields, Cretaceous, Poza Rica trend, Mexico (after Enos 1977, p. 276).

Field	Tamabra Fields Production (10^3 t)	Proved Reserves (10^3 t)	API Gravity
Tres Hermanos	6 358	14 294	21
Moralillo	1 652	1 155	21
Soledad	588	?	33
Miquetla	1 567	6 931	35
Castillo de Teayo	41	?	32
Jiliapa	1 957	3 167	34
Zapitalillo	13	?	
Nuevo Progresso	637	55	31
Pital y Mozutla	–	131	23
Poza Rica	127 548	148 397	35
Cerro del Carbon	67	?	–
San Andrés	(main pay Jurassic)		29
Totals	140 431	174 112	29.4

4.3.5.4. Production Data, Reserves, and Summary

Production and proven reserves from the Tamabra Limestone exceed 365×10^6 m^3; more than 320×10^6 m^3 from the greater Poza Rica Field alone. The oil fields producing from the Tamabra Limestone are listed in Table 32. Secondary recovery operations were applied to the Poza Rica Oil Field, an old field with declining oil production discovered early in the 1930s. 90% of the present production is due to secondary water flooding.

The oil fields of Mexico and Texas belong to the Extracontinental-Borderland Type 4 C open downwarps according to the typification of petroliferous basins (Klemme 1981).

The prolific oil province of the Tampico Embayment is caused by the following factors:

–   Wide extent of the geologically young carbonat-reservoir rocks;
–   Weak diagenetic alteration and tightening of the reservoir rock on the one side, and
    solution porosity on the other side;
–   Effective source rocks directly underneath the carbonate reservoirs, short migration
    paths, and efficient sealing.

Reservoirs at carbonate-basin margins, like the Tamabra Limestone, seem to offer new plays in oil exploration. Shelf-margin models can be attractive since the shelf edge may be recognizable from seismic records. Shelf-edge slope and relief are considered to be the best guides to basin-margin debris deposits in early stages of exploration for Poza Rica-type reservoirs.

4.3.6. C.I.S. – W SIBERIAN HYDROCARBON PROVINCE

C.I.S. hydrocarbon production comes mainly from two regions: W Siberia and Volga-Urals. Most efforts in the decades to come will be focused on W Siberia because large amounts of gas are piped over a long distance to W Europe. The giant W Siberian basin will remain for many years the major oil and gas producer in the C.I.S. Several aspects specific of hydrocarbon exploration in W Siberia are stressed in the following:

– Sandstone reservoirs of the giant Urengoy Gas Field;
– Exploitation of bituminous shales of the spectacular Bazhenov Formation;
– Influence of permafrost on the reservoirs.

4.3.6.1. The Urengoy Gas Field

The super-giant Urengoy Gas Field in W Siberia (northern Tyumen), extending over an area of about 6 000 km^2, is one of the world's largest gas fields (Figure 131). It is three to five times larger than all gas reservoirs of Mexico, Algeria, Canada, Great Britain, and the Netherlands together. In 1985 the Urengoy Field reached a production capacity of about 314×10^9 m^3/year. The reserves of the Urengoy Field were estimated to be 8.1×10^{12} m^3 (1991). In 1989 the Urengoy and Yamburg Fields accounted for more than 50% of the C.I.S. gas production.

4.3.6.1.1. General data. The Cenomanian reservoir sandstones of the Urengoy Field are found in depths between 1 000 and 1 200 m (Figure 132). Sandstones, siltstones (aleurolites in Russian terminology), and claystones are irregularly interbedded. The thickness of the claystone intercalations amounts to about 20% of the total sequence. The sandstones and siltstones are poorly sorted. They contain considerable admixtures of silt. The siltstones contain much sand admixture. Permeability is primarily associated with the fine-grained sandstones. The siltstones and sandstones are acidic arkoses composed of quartz (50–70%), feldspars (25–35%), and rock fragments (up to 10%) with traces of muscovite and biotite. Heavy minerals are leucoxene, magnetite/ilmenite, garnet, epidote, sphene, and zircon (Yastrebova 1976). Primary clay matrix predominates in the pores. Carbonate rocks, composed mainly of calcite and rarely of siderite, are sparse.

4.3.6.1.2. Source rocks, migration, trap types, and cap rocks. The gas is methane (98%) with an insignificant amount of ethane, nitrogen, and carbon dioxide (see also Bazhenov Formation, Section 4.2.6.2). The gas migrated into the traps during Neogene. The gas is confined to large structures extending north-south. The structures are of synsedimentary origin; structuring terminated in the Late Oligocene–Neogene. The structural closure within the Urengoy anticline is 235 m. The vertical gas column is 216 m thick.

Essential for the formation of this giant gas accumulation is a large artesian-pressure system with aquifers of considerable thickness, carrying ground water rich in hydrocarbons (Kortsenshteyn 1974). The total volume of gas dissolved in the ground water of this system is ten to a hundred times greater than the total reserves of the potential gas accumulation. The dissolved methane content of the ground water in the Cretaceous strata is of the order

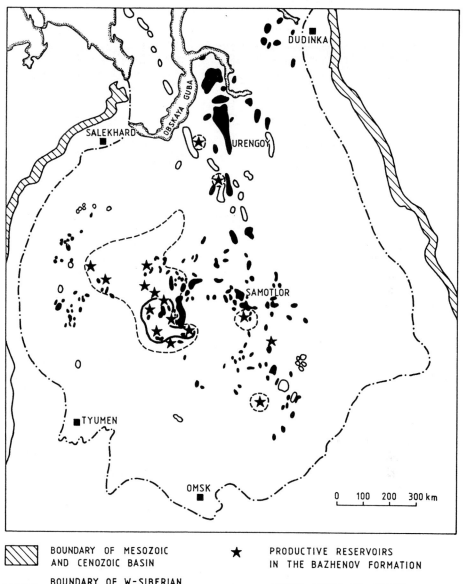

Fig. 131. Major oil and gas fields in W Siberia (redrawn from Bois and Monicard 1981, p. 857).

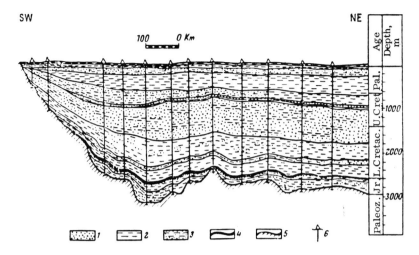

1) sand, sandstone; 2) clay, argillite; 3) sandstones and shales; 4) bituminous argillites of the Bazhenovsk formation (Volga stage); 5) Paleozoic rocks (granites, volcanic-sedimentary formations); 6) boreholes.

Fig. 132.   SW–NE cross-section through Mesozoic-Cenozoic sediments including the bituminous Bazhenov Formation and Lower Cretaceous reservoir sandstones, W Siberia (after Pluman 1975).

of 1.5 to 2.5 liter/liter. Heavier hydrocarbons contain 1 to 2% nitrogen and less than 1% carbon dioxide. The composition of gases dissolved in the ground water and of free gases in the reservoir is the same. The ground water is undersaturated in gas under the existing thermodynamic conditions; bed pressures exceed saturation pressure. These specific conditions are induced by the polar climate of W Siberia. The last glaciation not only reduced the mean annual temperature but also the secular temperatures down to a depth of 2 to 3 km. The Cenomanian artesian system lies in this depth range with temperatures 10 to 15°C below the present temperature (permafrost). This low temperature substantially increased gas solubility. Thus, by decreasing temperature, the system was transformed from a system saturated with gas into a system undersaturated in gas. In the relatively autonomous Cretaceous system, gas reserves are of the order of 400×10^{12} m^3.

Thus, the ground-water system and the gas reservoir are interrelated; the former being the primary and the latter the secondary accumulations of hydrocarbons.

The hydrodynamics of the Cenomanian artesian system controlling Urengoy and other reservoirs of the region are shown in Figure 133. The Urengoy reservoirs, located in the Cenomanian aquifer zone, are washed by a stream that filters from N to S. The pressure gradient of this stream is of the order of tenths of a meter per kilometer. Although the pressure gradient is not large, the slope of the gas-water interface (GWI) in the gas reservoirs reaches 20 m along the long axis of the gas-bearing belt.

The brines in the Cenomanian rocks have very low salinity, are almost sulfate-free, and are predominantly of the calcium-chloride type, less commonly of sodium-bicarbonate composition. Total salinity usually ranges from 16 to 20 g/liter (600 to 700 mg–eqv/liter).

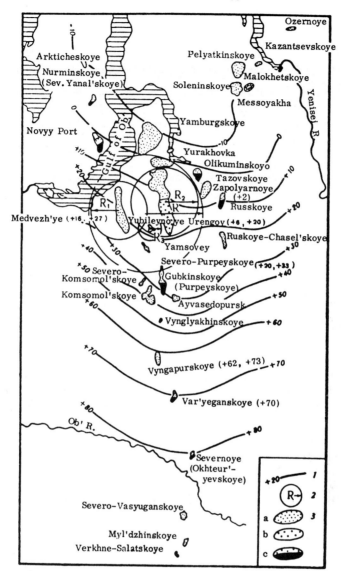

Fig. 133. Cenomanian artesian basin of W Siberia (after Kortsenshteyn 1974, p. 228). 1: Isopiestic lines (m) of the Cretaceous aquifers (calculated). 2. Inferred radii (km) of source areas of natural gas reservoirs: R1 – Medvezh'ye, R2 – Urengoy, R3 – Yubileynoye, R – radius of the three reservoirs together. 3. Hydrocarbon reservoirs: a – natural gas, b – condensate, c – gas-petroleum.

Chlorides dominate slightly over alkaline earths, whose content ranges from 290 to 315 mg–eqv/liter. The concentrations of alkaline earths are not high: Ca: 10–20, Mg: 5–12 mg–eqv/liter. Brines from the Cenomanian reservoirs contain characteristic trace elements (mg/liter):

Total salinity	Cl^-	Br^-	J^-	Ca^{++}	Mg^{++}	Br^{++}	NH^+	K^+
Concentrations	300–350	40–50	15–20	200–400	60–140	10–20	20–30	30–40

The geothermal setting of the gas reservoirs, evidenced by the low temperatures in the GWI zone, is unusual. In most deposits lying above $-1\ 000$ m the temperatures range from 13 to 30°C. Permafrost induced the low temperatures in the Cenomanian sequence.

4.3.6.1.3. Diagenetic history. Diagenesis is rather weak in the shallow-seated sandstone reservoirs of the Urengoy Gas field.

The 'immunizing' effect of oil against diagenetic alteration processes is a well known phenomenon important to oil exploration. Such electrolytic processes, associated with the pore water as an ion-exchange medium, slow down or stop further mineralization. On the other hand, data on the 'immunizing' effect of gas are sparse. For instance, gas-saturated rocks from productive horizons of N Sakhalin do not notably differ from water-saturated rocks in respect to diagenesis (Yurkova 1971). A comparable study of reservoir rocks was undertaken in the Urengoy Field based on the alteration of heavy minerals, clay minerals, and detrital biotite (Ovchinnikov 1976). The majority of the heavy minerals (magnetite, ilmenite, garnet, zircon, and leucoxene) in gas-saturated sandstones show the same frequency range as those in water-saturated sandstones. However, the sphene and epidote content in the gas-saturated sandstones is several times higher than in water-saturated sandstones (Table 33). Near the gas-water contact the behavior of sphene and epidote is more complex.

The clay minerals of permeable gas-saturated sandstones of the Urengoi and Medvezh'e gas fields, studied by X-ray diffraction, electron diffraction, and electron microscopy, do not differ in composition or crystallographic habit from that of the water-saturated sandstones. Kaolinite predominates in the matrix of gas- and water-saturated sandstones (Table 34). Characteristic is the low degree of crystallinity. The kaolinite particles are small (up to 5 μm), isometric-lamellar, and of irregular shape. Euhedral crystal habit of pseudohexagonal shape was not observed.

Quartz, detrital rock fragments, and muscovite did not undergo noticeable diagenetic alterations. The feldspars remain fresh; some of them are significantly altered by sericitization and kaolinization. Thus, the water-saturated rocks do not differ markedly from the gas-saturated rocks with respect to degree and quantity of feldspar alteration. Fresh biotite is rare; alteration processes comprise a decrease in pleochroism, swelling, widening of cleavage planes, sideritization, and kaolinization. For comparison of the gas- with water-saturated rocks, the following altered biotite types were distinguished: discolored, weakly pleochroic flakes (not swollen and foliated), swollen or foliated in a fan-like pattern; biotite partially or completely replaced by kaolinite. The number of altered biotite types varies considerably. However, the water-saturated rocks do not differ notably from the gas-saturated rocks as far as biotite alteration is concerned.

Thus, gas-saturated reservoir sandstones are not different from water-saturated reservoir sandstones in mineral composition. Only the pronounced decrease of sphene in sandstones outside the gas reservoirs proper and of epidote outside the gas reservoir and below the

Table 33. Heavy-mineral content of the reservoir rock in the Urengoy Field, W Siberia (after Ovchinnikov 1975). Frequency range and average (numbers in parenthesis) expressed in grain percent of total heavy minerals minus leucoxene: Leucoxene given as % of total heavy minerals.

Sample Type	Number of Samples	Sphene	Epidote	Magnetite and Ilmenite	Garnet	Zircon	Leucoxene*
Reservoir rocks	64	4–15 (10)	5–24 (12)	40–70 (53)	4–24 (14)	4–6 (5)	10–39 (22)
Rocks below the gas–water contact	10	1–10 (6)	1–5 (3)	54–75 (61)	5–23 (15)	3–5 (4)	8–30 (21)
Rocks outside the reservoir	12	1–7 (3)	1–4 (2)	50–60 (54)	10–23 (19)	4–7 (5)	10–22 (18)

* Frequency range and average in grain percent of total heavy minerals minus leuxocene; leuxocene given as % of total heavy minerals.

Table 34. Composition and characteristics of the clay-mineral fraction of permeable sandstones, Urengoy and Medve'zhe gas field, W Siberia (after Ovchinnikov 1975).

Sample Site	Clay Mineral Composition	Kaolinite Structure		Kaolinite Morphology	d(001) for glycerine-saturates Montomorillomite
		Cell	Lattice constant along c axis		
Sandstones overlying the Kuznetsov Formation	M[KHC]			not idiomorphic	17.60
Reservoir sandstones	K(M)[HC]	'M'	not constant	not idiomorphic	17.80–18.00
Outside gas-bearing	K(M)[HC]	'M'	not constant	not idiomorphic	17.80–18.00
Sandstones below the gas-water contact	K(M)[HC] rarely K[MHC]	'M'		not idiomorphic rarely idiomorphic	17.80–18.00

Legend: K = kaolinite; M = montmorillonite; H = hydromica = illite; C = chlorite. Predominating mineral indicated without brackets, secondary minerals in parentheses, mineral impurities in square brackets. 'M' – pseudomonoclinic unit cell.

gas-water contact evidently documents the retarding effect of gas on intrastratal solution processes of minerals unstable during diagenesis. Like oil, but not to the same extent, gas preserves reservoir sandstones against diagenetic alteration. Unlike oil, gas does not completely displace water from pore spaces. Water remains as a thin film on the mineral grains. Electrolytic interaction between detrital minerals and pore water may therefore continue, even after gas filled the traps. However, electrolytic activity decreased considerably with the decrease of pore moisture, which acts as the ion-exchange medium. Of all the minerals examined in the Urengoy Gas Field only the behavior of epidote seems to reflect the retarding effect of gas on diagenetic processes.

4.3.6.1.4. Production data, reserves, and summary. The Urengoy Gas Field is said to have proven reserves totalling 7.5×10^{12} m^3 (1982), it accounted for about 75% of W Siberia's gas output in 1985. Gas from Urengoy is being piped to Central Europe.

The very large oil and gas fields of W Siberia are located in an Intracontinental-Cratonic Type 2 A complex basin according to the typification of petroliferous basins (Klemme 1981). The major hydrocarbon accumulations occur in the Jurassic and Lower Cretaceous with minor reservoirs in the Upper Cretaceous. This basin type represents about 25% of the world's basin area and contains a similar proportion of the present world hydrocarbon reserves.

The formation of the Urengoy Gas Field as well as of the other major fields in W Siberia was favored by the following geological parameters:

- Gentle and wide-spanned structuring of the Siberian Shield and absence of severe tectonism and faulting;
- Relatively young age of source rocks, reservoirs, and structures;
- Permafrost enhancing perfect sealing;
- Weak diagenesis due to shallow reservoiring, low heat flow, and minor mineral cementation.

4.3.6.2. The Bazhenov Formation

The Baszhenov Formation, mainly of Upper Jurassic age (Volzh stage) (Krylov and Korzh 1984), subcrops over an area more than 1 000 km long from N to S and several hundred km wide in an E-W direction in the Irtysh-Ob river area (Figure 134). It is one of the few examples of primary oil-bearing pelite reservoirs rich in organic matter that sourced the oil simultaneously. Such a unique reservoir type is termed in the C.I.S. Bazhenite (Nesterov 1979, Nesterov *et al.* 1991). The oil-saturated fractured shales of the Bazhenov Formation in W Siberia were discovered based on the low energy of reflected waves ('dim spots') (Bobrovnik and Isakov 1978). This new type of reservoir has been found in W Siberia not only in the bituminous argillaceous sediments of the Bazhenov (J3v-K1b), but also in the Tutleym (J3v-K1v), Mulym'in (J3v-K1h) Formations, and partly in the Alyasov (K1v-K1h) Formation (Figure 135).

During diagenesis and catagenesis the shales do not act as reservoir. Under rather specific conditions the organic matter is converted to oil and in this process porous

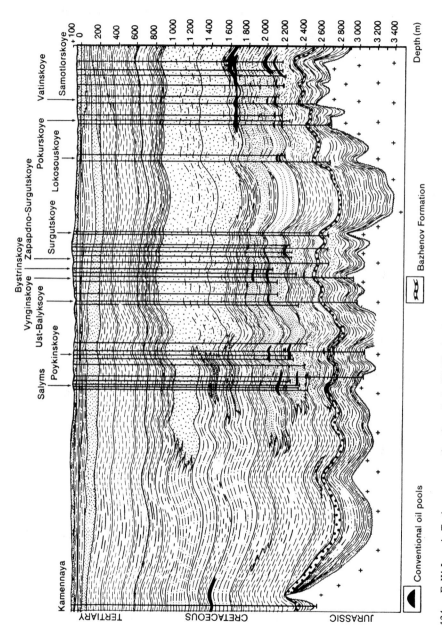

Fig. 134. E–W Jurassic-Tertiary cross-section through the W Siberian Basin showing conventional oil pools and the Bazhenov Formation (English version from Bois and Monicard 1981, p. 860).

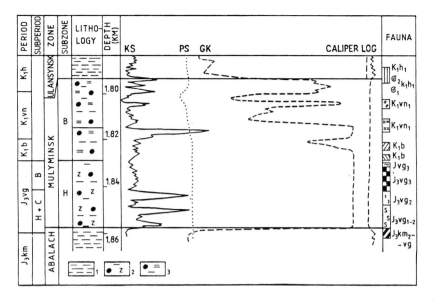

Fig. 135. Stratigraphy and log characteristics of the Bazhenov Formation, W Siberia (modified from Vysemirskij 1986).

reservoirs are formed.

Individual wells yield 700 tons of OPD. Comparable oil-bearing pelite reservoirs are known from California, western Venezuela, Angola, and Gabon.

4.3.6.2.1. General data. The thickness of the Bazhenov Formation ranges, in general, between 10 and 100 m; it increases to 200 m toward NE. With increasing thickness the organic matter content decreases considerably. Bazhenite occurs at depths between 1 and 4 km. In the central part of the W Siberian Basin it is most abundant and 20–40 m thick (average 32 m). The lower part of the bazhenite (Figure 136) consists predominantly of calcareous shales with thin beds of marls and limestones. The middle part is a massive, highly bituminous shale. The upper part is composed of thin-bedded shales. The top of the Bazhenov Formation is the main seismic reflection horizon in W Siberia. The higher the productivity of wells drilled on this formation, the lower is the amount of energy of the seismic waves reflected from it. The maximum apparent electrical resistivity of the Bazhenov Formation is 100–300 Ω m; in oil pools it reaches 3 000–4 000 Ω m.

The dark gray to black bituminous shales, in general, are horizontally bedded, pyritic, and contain up to 5–6% silt. Thin intercalations of clayey siltstones, bituminous clayey limestones, marls, and radiolarites occur. Locally, thin shale layers are calcareous containing microcrystalline calcite and glauconite. The shales are characteristically silicified which is evidenced by round chalcedony clusters, by cryptocrystalline quartz disseminated within the clay matrix, and by replacement of siliceous sponge spicules and radiolarian tests. Biogenic chalcedony, opal, and carbonate are preserved in tests that contain organic matter, too. The thorough silicification does presumably account for the ability of this

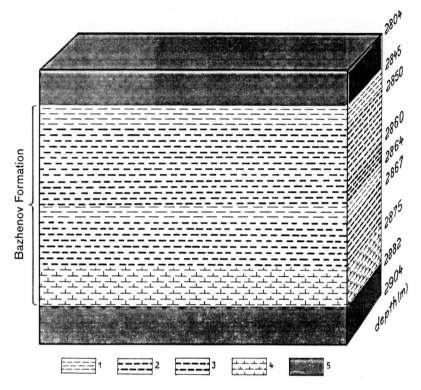

Fig. 136. Lithological profile through the Bazhenov Formation in the Salym Field, W Siberia (after Bois and Monicard 1981, p. 858). 1 = paper shale, 2 = compact shale, 3 = compact shale rich in organic matter, 4 = marlstone with lime intercalations, 5 = shale as seal.

rock to maintain fractures open, down to depths of 2 800–3 000 m. The content of organic matter ranges from 3 to 12% with an average of about 7%; the bitumen content can reach 20% in highly bituminous shales. Locally, organic matter makes up more than 50% of the rock volume, forming lenses about one meter thick. The particulate organic matter, 0.01–0.06 mm in size, is of irregular or elongate shape. Thin chips of the rock burn with a bright flame turning brick red.

Bazhenite is composed of clay minerals (35–50%), chalcedony and biogenic opal (15–20%), organic carbonate (10–15%), quartz and feldspar (4–6%), pyrite (5–10%), and organic matter (10–20%). The clay-mineral fraction consists of montmorillonite (15–20%), dioctahedral hydromica (10–15%), chlorite (6–8%), and kaolinite (4–7%). Common rock types are clayey-siliceous bazhenite, clayey bazhenite, and radiolarian bazhenite, all rich in organic matter.

Chemically, bazhenite contains silicon (40–50%), aluminum (8–12%), iron (4–6%), calcium (5–15%), magnesium (1–3%), potassium (1–2%). Trace elements are (in 10^{-4}%): 130–150 barium, 300–500 vanadium, 150–200 nickel, 150–200 strontium, 100–150

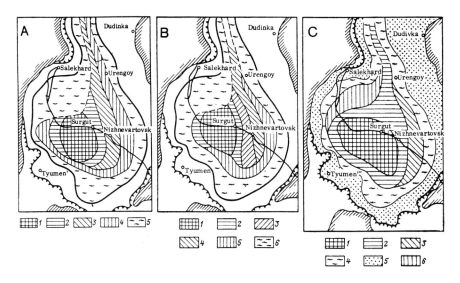

Fig. 137. Composition and regional prospectivity of the Upper Jurassic/Lower Cretaceous bituminous Bazhenov Formation, W Siberia (compiled from Ushatinskiy 1982). A. Organic-matter content. 1: > 10%, 2: 8–10%, 3: 7–8%, 4: 5–7%, 5: 2–5%. B. Amount of biogenic silica. 1: > 25%, 2: 22–25%, 3: 20–22%, 4: 15–20%, 5: 10–15%, 6: < 10%. C. Reservoir prospectivity. 1: extremely favorable, 2: relatively favorable, 3: less favorable, 4: hardly favorable, 5: unfavorable, 6: not studied in detail.

chromium, 120–150 copper, and 100–150 molybdenum.

Average chemical composition, C_{org} content, and various SiO_2 modifications of the bituminous shale in W Siberia are summarized in Table 35. The evaluation of the regional prospectivity of the Bazhenov Formation is depicted in Figure 137. Moreover, the proximity of gas-rich waters and petroleum production is shown in Figure 138. Russian geochemists published analyses of formation waters from the Lower Cretaceous collected during early wildcat drilling of the W Siberian Lowlands (Kartsev 1963, Hunt 1979). The waters were analyzed for organic carbon, dissolved gases, and several organic and inorganic compounds. The highest methane content of the formation waters was in a 500 km² area west of the Ob River. To the east and south, the methane content decreased with an increase in CO_2 and nitrogen in the dissolved gases. Other geochemcial data such as the presence of organic carbon also indicated that this was a favorable area for exploration. About ten years after these data were published, several giant oil and gas fields were found in the favorable area.

The bituminous shales, with an uranium content ranging from 1×10^{-3} to 7×10^{-3} ppm and a considerable pyrite content, were deposited in an anoxic environment at the sea floor, but oxygenated above as evidenced by the rich fauna of ammonites, belemnites, pelecypods, foraminifera, ostracods, and radiolarians. The relief of the basin floor was rather smooth; the source area of the clay minerals was subjected to intense chemical weathering.

Table 35. Average chemical composition of bituminous rocks from the Bazhenov Formation, W Siberia (after Ushatinskiy 1982). Producing regions in the W Siberian Trough: V = Krasnoleninsk, VIII = Tazovsk, XI = Gubinsk, XIV = Surgut, XVI = Vengapur, XVII = Salym, XX = Kaymysovka.

Components	Region							Through Area as a Whole
	V	VIII	XI	XIV	XVI	XVII	XX	
SiO_2	52.85	52.15	54.35	58.63	58.30	52.44	59.37	55.61
TiO_2	0.67	0.92	0.77	0.47	0.52	0.53	0.63	0.66
Al_2O_3	14.53	16.68	14.36	9.72	10.20	10.03	11.49	12.98
Fe_2O_3	6.30	5.34	3.75	4.50	4.88	4.84	4.88	4.87
FeO	2.53	4.14	3.09	1.51	1.44	1.54	1.25	2.13
CaO	3.85	2.10	2.95	4.21	3.08	6.38	2.93	3.18
MgO	1.12	1.09	1.72	1.12	0.62	1.97	1.22	1.33
MnO	0.05	0.10	0.06	0.05	0.07	0.03	0.05	0.05
Na_2O	0.60	1.63	1.77	0.82	1.10	0.78	1.01	1.05
K_2O	1.45	2.52	3.49	1.60	2.12	1.37	1.99	2.20
Calc. loss	16.71	14.24	14.09	18.01	17.10	20.76	15.85	16.48
P_2O_3	0.36	0.16	0.33	0.35	0.30	0.43	0.19	0.26
S_{tot}	3.45	1.68	3.11	2.85	2.33	3.34	2.14	2.83
C_{org}	10.24	2.61	8.37	9.05	6.01	10.97	7.19	7.22
SiO_{2cl}	26.73	41.61	31.22	29.36	31.24	21.68	31.41	31.98
SiO_{2qz}	4.80	8.24	7.51	4.35	6.10	5.07	7.13	6.97
SiO_{2ch}	20.57	1.92	15.21	24.05	20.60	24.72	20.11	16.10
SiO_{2op}	0.75	0.38	0.41	0.67	0.36	0.97	0.72	0.56
$\dfrac{SiO_{2aut}}{SiO_{2ter}}$	0.68	0.05	0.40	0.73	0.56	0.96	0.54	0.47
$\dfrac{SiO_{2aut}}{SiO_{2cl}}$	0.79	0.06	0.50	0.84	0.67	1.18	0.66	0.56
$\dfrac{SiO_{2aut}}{SiO_{2qz}}$	4.35	0.28	2.08	5.43	3.44	5.07	2.92	2.76
Fe + Mn/Ti	16.01	12.76	11.01	15.58	14.77	14.36	11.63	12.74
N	65	25	39	78	64	136	96	793

S_{tot} – total sulfur; C_{org} – organic carbon; SiO_{2cl}, SiO_{2qz} – silica associated with clay minerals, quartz and feldspars (terrigenous); SiO_{2ch}, SiO_{2op} – silica associated with chalocedony and opal (authigenic).

The capacity of the Bazhenov bituminous shales as hydrocarbon reservoirs is attributed to fracture and leaching porosity formed during diagenesis. No porous sandstone reservoirs of the conventional type are present. Porosities of the Bazhenov bituminous shales range between 2 and 16%, with an average of 8%; the permeability is 2.4 mD. Horizontal fractures show a density of 0.5–1.5 per cm, vertical fractures a density of 0.15–0.20 per cm. The fractures remain open; they are not mineralized due to the absence of formation water. Secondary dissolution porosity comprises the dissolution of fossil shells and tests,

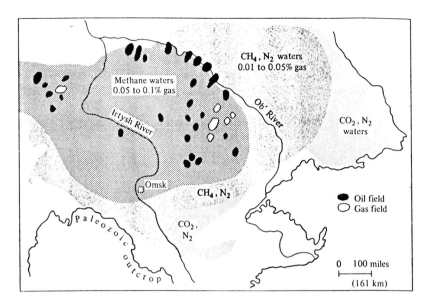

Fig. 138.   Gas-rich Lower Cretaceous formation waters as hint at oil and gas fields, W Siberia (from Kartsev 1963).

leaching of the fracture walls, and partial leaching of amorphous silica disseminated in the clay matrix. Leaching provides most of the effective porosity and fracturing causes the permeability. The pore throats as determined under the scanning electron microscope range from 1 to 5 μm. The density of bazhenite at a depth of 1.8–3 km is 2.14–2.15 g/cm^3. The density of the rocks above and below is 0.2–0.3 g/cm^3 greater.

4.3.6.2.2. Production data, reserves, and summary. In 1980, 25 oil fields produced oil from the Bazhenov bituminous shale. The giant Salym Field, discovered in 1965, and the giant Samotlor Field are both located within the Bazhenov bituminous shale belt. In the Salym Field the Bazhenov bituminous shale is sealed by underlying massive, highly bituminous shales, marls, and limestones of Kimmeridgian age (Georgriyev Formation) and by overlying thick shales of Berriasian to early Valanginian age (Achimov Formation). The anomalously high formation pressure of the Salym Field is due to gas generated from the organic matter by maturation processes during deeper burial and due to the hydrodynamic sealing of the Bazhenov Formation. The reservoirs of the Salym Field are characterized by low permeability. Estimates of reserves of the Salym Field are as high as 1.6×10^9 m^3. If recovery techniques are improved, the Bazhenov Formation is potentially very productive. The liquid fractions of petroleum of the Upper Jurassic bituminous shales of W Siberia are estimated totalling 2×10^{12} tons. No estimates are known of how much of these materials might be recovered, but the source-rock potential is significant.

Presumably, the underground nuclear exploration, registered in October 1979 in the vicinity of the Salym Field, has demonstrated the attempt of Russian earth scientists to stimulate oil production from relatively tight bituminous shales by extensive, nuclear-induced fracturing.

4.3.6.2.3. Permafrost as cap rock. Primary migration terminates where oil and gas migrating from a source rock reach an adjacent permeable bed. Further movement of hydrocarbons within potential reservoir rocks may lead to an accumulation of gas and/or oil, provided that favorable conditions such as structure or stratigraphic trap and sealing caprock exist.

A common belief is that a caprock should be impermeable both to oil and natural gas as well as to water. This assumption, especially as it relates to water, is not entirely correct. True water-impermeable rocks in sedimentary basins are permafrost and evaporites only.

The thickness of permafrost, the so-called cryolithosphere, developed in Pleistocene glacial climates, can be very high, up to 1 500 m in some areas of the Arctic zone of the C.I.S. Permafrost is the perfect seal for stopping the dispersion to the surface of the gas accumulated at relatively shallow depth in the impressive cluster of giant gas fields discovered in the northern area of the W Siberian Platform (e.g. Urengoy Gas Field). As the permafrost cover is impermeable to·water, the water displaced by the gas during its accumulation in the producing layers was evidently discharged either in southern areas with no permafrost or local uplift areas kept at a temperature above 0°C by the flow of the deep warm water.

Profound chilling of the sedimentary sequence has led to the formation of an anomalously low geothermal field which has in turn affected the hydrocarbon and non-hydrocarbon fluids. Scale and consequence of the effects of such chilling have not been studied extensively. Differentiation of the depth of the cryolithosphere may be explained by varying heat-flow magnitudes derived from basement and sedimentary cover rocks of different ages. Nature, development, and structure of underground ice in permafrost regions are discussed by Williams and Smith (1989).

4.3.7. CHINA – CRETACEOUS LAKE DELTAS

4.3.7.1. General Data

China is eighth in world's oil production (1985). In 1982 oil production was about 100×10^6 t.

In 1955 petroleum exploration started in the Songliao Basin, the largest basin in northeastern China. As a first step, comprehensive gravity, magnetic, electrical, and seismic surveys, as well as stratigraphic drilling were executed in the entire basin. In September 1959 the giant Daqing Oil Field, an anticlinal structure, 120 km long from south to north and 15–20 km wide from east to west with a maximum closure of 524 m covering an area of about 2 000 km^2, was discovered. Subsequently, oil production in China increased rapidly. With an annual crude production of 55.5×10^6 t (1985) the field accounts for 50% of the total production in China.

The NE–SW striking Songliao Basin is located in the so-called Cathaysian Rift System of NE China (Figure 139). Basins such as the Songliao Basin developed from grabens or half grabens into large rift basins. With the further extension of the crust, graben-horst

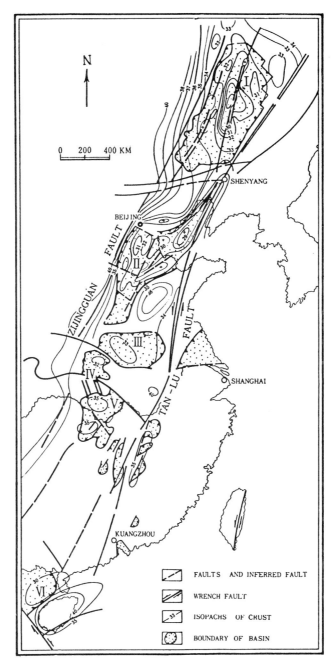

Fig. 139.   Location of the Songliao Basin in the Cathaysian Rift System, NE China (after Liu Hefu 1986, p. 382, reprinted with permission).

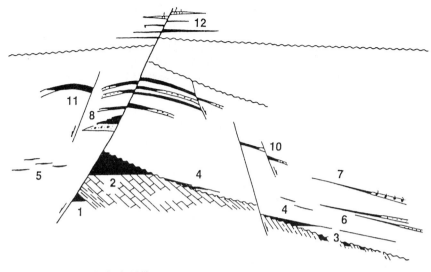

(1) Reservoir within buried hill
(2) Weathering-body reservoir of buried hill
(3) Unconformity of reservoir of buried hill
(4) Stratigraphic overlap reservoir
(5) Turbidity sand-body reservoir
(6) Stratigraphic pinch-out reservoir

(7) Bioherm (oolitic) limestone reservoir
(8) Pyramidal conglomerate reservoir
(9) Anticlinal reservoir
(10) Fault-block reservoir
(11) Rollover anticlinal reservoir
(12) Shallow secondary reservoir or gas reservoir

Fig. 140.   Types of stratigraphic oil and gas accumulations in rift basins of China (after Zhai Guangming *et al.* 1982).

structures developed with internal structures similar to those of the Basin and Range province in the western United States. As a result of further subsidence, a rift basin formed. In the middle Cretaceous it subsided to form a large-scale sedimentary basin filled with lacustrine sediments. The basin ceased to subside during the Late Cretaceous when the Daqing anticline formed.

The principal productive horizons are the Saertu, Putaohua, and Gaotaizi Formations of Lower Cretaceous age. The Putaohua Formation is the major producing unit. It is composed of clastic sediments deposited on the erosional surface of the Qingshankou Formation and can be divided into two parts: PI_3, the lower part, and PI_{1-2}, the upper part. The gross thickness of the Putaohua reservoir is 40–60 m. The sand is 5–20 m thick and is intercalated as a lens into the thick black shales of this area.

Rift basins commonly contain accumulations of petroleum in unconformity traps and pre-unconformity fault traps, typical in early Cretaceous (Figure 140). The rift basins are initiated by normal growth faulting, forming grabens which accumulate continental and marine sediments, usually in a dominantly regressive sequence. The Songliao Basin was one of the largest taphrogenic lake basins in Asia in Cretaceous time. It covers a total area of about 260 000 km². Oil and gas accumulations in such continental basins have their own nature, own structural, depositional, source and reservoir models. The formation model of a giant nonmarine oil field by Yang Wanli (1983) applies to these Cretaceous lake deltas.

Subsequent to the Yenshan orogeny a number of small, separate fault basins, filled with volcanic rocks and coal measures, formed in the Jurassic. During early and middle Cretaceous times, the basin gradually evolved from a faulted into a large depressional state in which thick lacustrine deposits (up to 5 000 m) were laid down. NNE-trending tensional faults controlled the early stage of basin development.

Paleogeomorphologic analysis revealed two long river systems flowing into the lacustrine basin at the southern and northern ends and paralleling its longitudinal axis. These river systems continually supplied sand, mud, and organic matter, and, thus, eventually formed a large quiet lake with fluvio-lacustrine delta sediments and lacustrine organic mudstones.

Argillaceous sediments in the rapidly subsiding basins, rich in organic matter and deposited in a deep-water lake environment, offered excellent conditions for oil formation.

In Albian time, the basin turned rather rapidly into a large lacustrine basin (150 000–200 000 km^2). Large quantities of sand were deposited at the inlets into the lake. Constructive deltaic deposition progressed as the lake retreated. The delta extended from the Heiyupao swamp southward to Putaohua, to Qinggang in the east, and to Wuyuer in the west. It is 150 km wide from east to west and 220 km long from south to north.

The delta is 500–600 m thick and covers a total area of nearly 30 000 km^2. It is composed of more than 200 elongate, tongue-like or bird-foot-shaped lobes. The distance from the piedmont to the lake shore of the southernmost lobe was about 200 km in middle Albian time. The paleotopographic slope was less than 0.5 m/km along the fluvial plain area, less than 0.11 m/km on the delta plain, and about 0.8 m/km at the delta-front zone. The center of the lacustrine basin was a zone of continuous subsidence in which 500–1 000 m of dark deep-lake muds, rich in organic matter, were laid down.

Three large fluvio-lacustrine deltas formed: The Xingshugang or Heiyupao delta, the Heidimiao delta and the Yingtai delta. They can be classified into two types: the Heiyupao and Heidimiao deltas developed along the longitudinal basin axis, i.e. long-axis deltas. The Yingtai delta, on the other hand, developed perpendicular to the longitudinal basin axis, i.e. short-axis delta, with only minor fluvio-lacustrine deltas (Table 36). The granitic detritus supplied into the Heiyupao delta came primarily from NE through high-energy rivers.

The deltas proper are composed of distributary-plain deposits and delta-front deposits. Flood-plain deposits, consisting of fining-upward cycles of sandstones and reddish-green mudstones, are of minor importance.

The distributary-plain facies is composed of fining-upward cycles of interbedded sandstones and mudstones. Linear channel sandstones are fine to medium-grained with cross-bedding and bottom scours.

The delta-front facies consists chiefly of distributary mouth bar and sheet sands. The fine to medium-grained, generally well-sorted string sands parallel the ancient river course. The delta-front deposits, accumulated in the turbulent part of the lake, do not contain many fossils.

Table 36.   Extension and morphology of fluvio-lacustrine deltas, northern Songliao Basin, NE China (after Xu Shice and Wang Hengjian 1981).

Name of delta	Direction of extension vs. longitudinal axis of basin	Length, km	Width, km	Shape	Volume of sandstone body, km³
Heiyupao	parallel	220	150	Tongue-like	2 900
Heidimiao	parallel	130	100	Tongue-like	910
Yingtai	perpendicular	Length along dip of deposition 40 km and along strike of deposition 130 km		Fan-like	770

The prodelta facies is composed of shallow to deep-lacustrine mudstones of gray-black color. The relatively monotonous mudstones are interbedded with bioclastic rocks, marls, and oil shales. They contain few species, but numerous individuals of fossils.

Large fluvio-lacustrine deltas are mostly constructive because of the relatively low coastal energy of lakes, regardless whether they are formed at the time of lake advance or lake retreat.

Individual shoestring sands are aligned in the direction of ancient river courses. Only a few of them parallel the ancient shore lines.

Fluvio-lacustrine deltas can be divided into three parts: the main body, the transitional zone, and the outer rim (Xu Shice and Wang Hengjian 1981).

In the transitional zone and the outer rim the delta sandstones interfinger with source rocks, thus favouring the accumulation of oil and gas.

The main body of the Heiyupao delta, made of cross-bedded, medium-grained sandstones (river-mouth bars), is more than 300 m thick, accounting for 60% of the total thickness. A single unit is 20–30 m thick.

The transitional zone comprises distributary-channel sands, delta-front bar and delta-front sheet sands, ranging from 200 to 300 m, with more mudstone. Medium-grained sandstones account only for 30–40%. Single beds may reach a maximum thickness of 10 m.

The outer rim is composed of delta-front sheet sands and isolated sand bodies.

In the large fluvio-lacustrine deltas the outer rim of the delta, rich in source rocks and with lenticular sand bodies, are important for oil and gas accumulations. In the main body of the delta, source rocks are not well developed; however, sandstones have good pore intercommunication.

The Daqing Oil Field (Figure 141) is located within the transitional zone of the Heiyupao delta complex, its long axis paralleling the depositional trend of the river-borne sediments. The principal reservoirs are overlain and underlain by source rocks. Moreover, its location within the transitional zone favors lateral migration of oil and gas. All these parameters contribute to the high petroleum reserves.

The Daqing Delta complex is a polycyclic delta composed of numerous overlapping delta lobes. It has a total thickness of 500 m and 38 cycles. Most of the cycles are 10–30 m

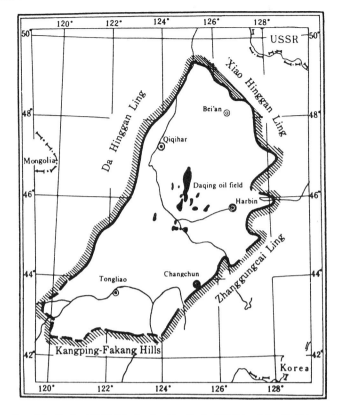

Fig. 141.   Location of the Daqing Oil Field in the Songliao Basin, NE China (after Xu Shice and Wang Hengjian 1981, Fig. 16-1).

thick, a few may reach 50 m (Figure 142).

The relationship of sand-body geometry with the heterogeneity of sandstones in the field was examined to optimize secondary recovery operations like water flooding as well as other stimulation operations within the oil field. Based on the study of minor sedimentary cycles the sandstones in the transitional zone were divided into four types: fluviatile-channel sandstones, distributary-channel sandstones, distributary-mouth-bar sandstones, and sheetlike delta-front sandstones.

The geometry of the fluviatile-channel sandstone bodies, distributed as strings of varying width on the delta plain, is characterized by a flat upper surface and an irregular scoured bottom. Petrophysical core data documents the heterogeneity of the fluviatile channel sandstones. Each time-unit is composed of numerous fining-upward cycles with a thickness ranging from 10 cm to 2 or 3 m. Horizontal air permeability and intercommunication fluctuate markedly. The more sedimentary cycles occur between producing and injection wells, the more noticeable are the differences in permeability.

The distributary-channel sandstones in the delta plains near the lake shoreline are

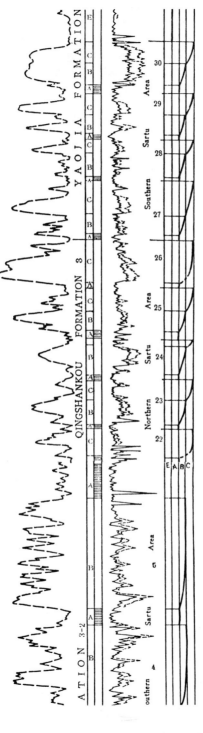

Fig. 142. Sedimentary cycles in the Daqing delta complex as shown by well logs, Songliao Basin, NE China (after Wu Chungyu 1982). A = Prodelta mud; B = Delta front; C = Delta plain; E = Littoral and shallow lacustrine facies.

formed by lateral and vertical accretion due to shifting river courses and shorelines.

The reservoirs are composed of sandstones and thin claystone beds of varying origin. Chief characteristics of these sandstones are fine grain size, moderate permeability, and persistent distribution of thin claystone intercalations within the thick producing sandstones. Reservoirs of this type give normally better yields; the injected water drive is more uniform. The exploitation of distributary-channel sandstones differs from that of fluviatile-channel sandstones.

The distributary-mouth-bar sandstones are clean, of uniform permeability and excellent intercommunication.

String-like as well as fan-like distributary bars are thicker, more permeable, and less clayey in the axial part than at the periphery. These reservoir sandstones are characterized by high yields of producing wells, late water cut, and slow increases of water cut.

The sheet-like delta-front sandstones were accumulated by floods beyond the distributary-mouth bars in the lake. Sand and clay interbeds of limited extent or thin sandstones intercalated between claystones formed.

Currents distributed the sediments over the delta front, removing fine-suspended components. Thin widespread blanket sands, reservoirs of uniform permeability, were laid down. These blanket sands are thin, multiple, fine-grained (generally less than 0.15 mm), and have low permeability. The higher the carbonte content, the lower the permeability. Close well spacing and high-pressure flooding operations are appropriate for effective development of these sands.

4.3.7.2. Source Rocks, Migration, Trap Types, and Cap Rocks

The total thickness of the lacustrine sediments is more than 1 000 m. The total organic carbon content is high (2.2–2.4%). The main center of oil generation is the Gulong depression, in which the area favorable for oil generation is about 50 000 km^2, and the thickness of the greyish black mudstone and shale suitable for source rock is 530 m in average and 1 000 m as the maximum. The average organic carbon content is 1%, total hydrocarbons 0.0868% and odd-even-preference (OEP) value 1–1.3. The sapropelic kerogen of the Daqing Oil Field mainly derived from fresh-water algae and the strongly bacteria-reworked higher plants, which are characterized by high hydrogen and low oxygen. The ratio of total hydrocarbons to organic carbon reaches 0.06–0.07.

Source-rock maturity is moderate (Figure 143). Oil generation generally starts at 65–95°C. The major properties of crude oil and natural gas are summarized in Table 37. The average depth of burial is about 1 000–1 800 m; it does not exceed 3 000 m.

Oil and gas may migrate from the center of oil generation over a distance up to 40 km and accumulate in the various types of traps along the flanks of the deep depressions. Thus, the distance of migration of oil and gas is relatively short. Oil and gas are mainly distributed in the oil generation area or its immediate neighborhood.

The major trap types in the Daqing Field are faulted anticlines as well as stratigraphic traps, the so-called structural, structural-lithologic, and lithologic pools (Figure 144). The lithologic pools are also termed subtle, i.e. difficult-to-detect oil pools (Ma Li 1985).

Table 37. Composition and physical properties of crude oil and natural gas in the Putschva oil zone, Daqing Oil Field, NE China (after Yang Wanli 1983, reprinted with permission).

Oil pool	Lamadran	Saertu	Xingshu gang	Taiping tun	Gaotaizi	Putaohua	Aobaota
Specific gravity	0.683	0.674	0.695	0.804	0752	0.777	0.8141
Methane (%)	87.63	85.25	82.12	73.93	73.77	68.54	72.75
Heavy hydrocarbon (%)	11.92	12.63	12.87	20.72	22.18	22.28	22.47
Specific gravity (D_4^{20})	0.864	0.858	0.852	0.848	0.853	0.839	0.856
Viscosity at surface (cp)	21.6	20.4	14.1	11.6	15.8	10.4	16.3
Pour point (°C)	26.2	25.0	26.0	30.0	26.5	26.0	25.5
Wax conent (%)	25.0	26.2	26.2	24.5	23.4	23.7	26.1
Resin content (%)	14.3	14.3	11.2	9.9	8.1	6.3	10.3
Total hydrocarbon (%)	81.9	78.8	84.8	88.9	87.1	89.0	–
Saturated hydro-carbon (%)	57.1	62.6	66.3	66.1	71.6	70.1	–
Aromatic hydro-carbon (%)	24.8	16.2	18.5	22.8	15.5	18.8	–
Non-hydrocarbon + bitumen (%)	18.1	21.1	15.2	11.1	12.8	11.0	–

The subtle oil pools of the Putaohua Formation in the Xingshugang delta are primary lenticular lithologic reservoirs. The expected reserves are about $300–500 \times 10^9$ t. All of the anticlinal oil traps in these formations have been drilled. Since 1975, search in the basin has been principally for subtle oil traps.

The reservoir characteristics of the Daqing Field are relatively simple. The structural configuration, pool type, fluid property, pressure system, and oil and water interface are almost identical in each oil reservoir. An E–W cross-section through the Songliao Basin (Figure 145) shows best the spatial relationship between source rocks, reservoirs, cap rocks, and fluid migration.

4.3.7.3. Diagenetic History

In the multilayered Daqing Oil Field, which was developed by separate-layer techniques, continuity, permeability, and geometric configuration of the different sand members in the same reservoir differ considerably from each other. The pore structure of the reservoir

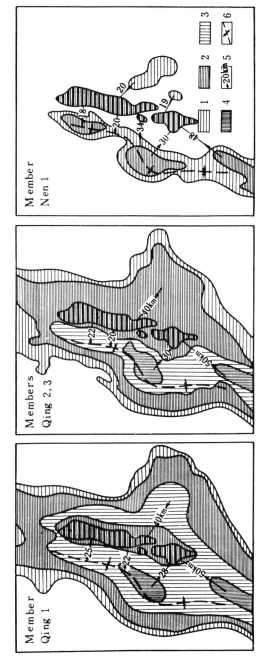

Fig. 143.   Maturation of hydrocarbon source rocks, Daqing Oil Field, Songliao Basin, NE China (after Yang Wanli 1985, Fig. 4, p. 1104, reprinted with permission). 1 – early low-maturation stage, 2 – late low-maturation stage, 3 – early high-maturation stage, 4 – late high-maturation stage, 5 – distance (km) from oil field to center of source rock depression, 6 – axis of subsidence. (The horizontal signature in the centers means maximum-maturation stage.)

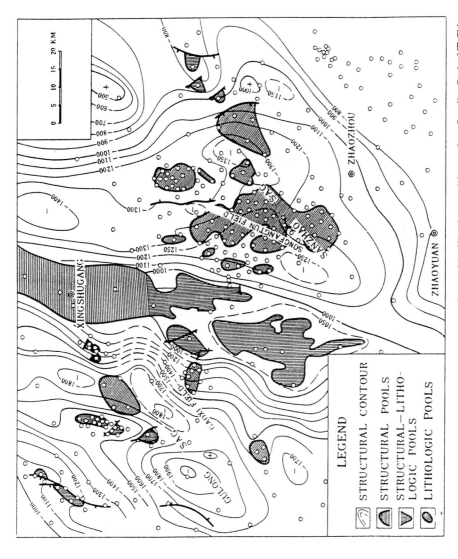

Fig. 144. Distribution of subtle pools in the Putaohua Group of the Xingshugang delta, northern Songliao Basin, NE China (after Ma Li 1985, Fig. 7, p. 1129, reprinted with permission).

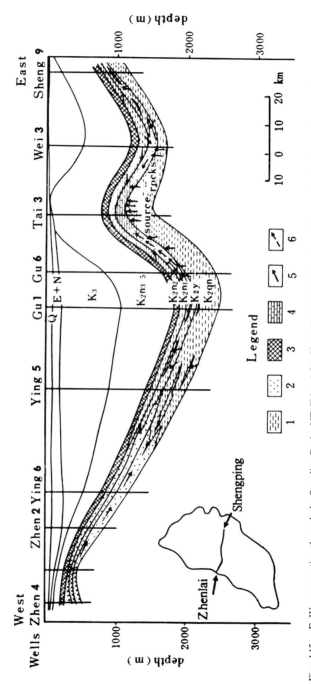

Fig. 145. E–W cross section through the Songliao Basin, NE China (after Yang Wanli 1985, Fig. 3, p. 1103, reprinted with permission). 1 = source rock, 2 = reservoir rock, 3 = cap rock, 4 = oil pool, 5 = direction of oil and gas migration, 6 = direction of subsurface-water flow.

sandstones is rather heterogeneous; the oil is normally moderately to highly viscous. The dominant clay mineral in the reservoir sandstones examined seems to be vermicular kaolinite which is normally located in the center of pores. Kaolinite is locally displaced in the pore space and easily mobilized during waterflooding. Microcrystalline clay of unknown mineralogical identity is located at the periphery of pores (Yang Puhua 1980, Jin *et al.* 1985, Mason and Dickey 1989). Low-permeability reservoirs, i.e. fine-grained, poorly sorted, and highly argillaceous reservoirs, are characterized by an increase of the curvature of porous intercommunication (tortuosity). The author is not aware of any integrated study of the entire mineral composition of the pore filling. The examination of undisturbed cores from flooded-out zones demonstrated that most of the pore surfaces changes from slightly oil-wet to slightly water-wet to water-wet as water injection proceeds. Scanning electron micrographs clearly reveal that the pore structure changed due to the long-term water flushing. Zhu Guohua (1982) showed that permeability increases with pore throat sorting.

4.3.7.4. Production Data, Reserves, and Summary

The huge multipay Daqing oil field, one of the largest in the world, was brought into production in 1960 with timely water injection. Geometry and internal structure of each oils and have been studied in detail. Techniques have been deviced to adjust the amount of injection water and production in separate layers at the bottom of the well, thus, maintaining a stable oil production for more than 20 years (Jin Yusun *et al.* 1985, Jin Yusun *et al.* 1989). In respect to the waterflooding behavior three kinds of thick sand bodies occur: bottom-, uniform-, and multizone-flooding. Flood-plain fluvial sands are bottom-flooding, river-mouth sand bars are uniform-flooding, and thick sand bodies with multistacked sedimentary-time units are multizone-flooding.

By 1976 the annual output of oil had reached 58×10^6 m^3. The average increase in water cut was from 2 to 3%. In 1981 the water cut was 61.7%, i.e. a high-water-cut stage of development.

In the Daqing Field distributary-mouth-bar sandstones rank highest in yield, followed by distributary-channel sandstones, fluviatile-channel sandstones, and sheet-like delta-front sandstones.

The geological and geochemical conditions for the formation of a large non-marine oil field like the Daqing Field are the following:
- Highly effective source rocks, not very large in volume;
- Reservoir rocks interfingering with the source rocks;
- High expulsion efficiency and favorable conditions for oil entrapment.

The lacustrine basins of China show the following primary sedimentary features, when compared with marine basins: smaller area, shallower water depth, weaker wave action, more moderate longshore currents, lack of tides, closer source area, more undulating relief, presence of more rapid rivers running into the lakes with a heavier load of clastic sediments, and, thus, greater influence of fluvial action. Consequently, lacustrine deltas are dominated by fluviatile sandstone sedimentation. Some rivers are relatively long and perennial, others are short to medium and seasonal.

4.4. Sedimentology as Applied to Well Treatment, Hydrocarbon Exploitation, and Enhanced Oil Recovery

Reservoir-quality analysis of sandstones and carbonate rocks also includes numerous sedimentological aspects. It applies to single wells or to oil fields, not to entire basins. It is aimed at optimizing hydrocarbon recovery of known hydrocarbon occurrences through all post-drill recovery operations. Well treatments are all measures to make hydrocarbon recovery of a single well most efficient. Enhanced oil recovery (*EOR*) are all operations, secondary as well as tertiary ones, which enhance the recovery of hydrocarbons in wells and fields.

The successful application of all secondary and tertiary recovery methods is normally hampered by primary reservoir heterogeneities which may cause bypassing of reservoir portions. Thus, reservoir characterization is crucial for a successful prediction of hydrocarbon-recovery operations (Lake and Carroll 1986). Heterogeneities such as shale discontinuities or permeability variations in cross-bedded sandstones considerably affect oil recovery.

Subsequent to drilling, reservoir-quality analysis helps to prevent drilling of additional wells when potential reservoirs do not exist. It helps in searching better reservoir qualities compared to those already discovered. The cost of a wasted exploration program can add up to millions of dollars.

A well drilled, completed, stimulated or treated incorrectly may not produce up to capacity or may even die altogether.

Reservoirs claimed by engineers as inadequate because well stimulation failed may turn out to be good reservoirs when properly treated based on the findings of a reservoir-quality study. Cooperation between petroleum engineers and geologists is essential in order to drill and treat uneconomic reservoirs or to avoid improper stimulation of good reservoirs.

4.4.1. THE GEOLOGICAL FRAMEWORK OF RESERVOIRS

The quality of a reservoir is controlled by: (1) depositional features, i.e. the primary sedimentology, (2) diagenetic processes, and (3) the tectonic framework.

Depositional features control reservoir quality primarily through their influence on textural properties and rock matrix. Grain size and sorting have a major effect on reservoir quality, while the influences of packing, fabrics (grain orientation), shape, and rounding are of varying importance. Grain size has no marked effect on porosity if shape is constant.

However, poor sorting significantly reduces porosity. Shape can be an important factor if grains with platy shapes, such as micas, or brick-like shapes, such as feldspars, occur in large amounts. Most grains in sands tend to have oval shapes. Although micaceous and feldspathic sandstones may have unusually high initial porosities as a result of bridging, they also are likely to loose porosity more rapidly than sandstones with grains of oval shape. Grains with plate or brick-like shapes can be compacted into arrangements with much closer packing than grains with oval shapes.

Permeability is influenced by both size and sorting, with size being the dominant influence. A decrease in grain size from coarse to very fine leads to a marked decrease in permeability.

Normally, very fine- and fine-grained sandstones are well-sorted, while coarse-grained sandstones are poorly sorted. Beach and dune sands have high initial porosities because of excellent sorting. For instance, Holocene sands of the Niger Delta show the following size-sorting interrelation: delta-mouth bar and delta-platform deposits have the best initial porosities (about 41%), but the beach, flood-plain, and tidal-bar deposits have significantly greater permeabilities.

Even moderate amounts of matrix clay dispersed throughout the pore system of a sandstone can severely reduce permeabilty.

Diagenesis commonly accentuates the influence and effects of the depositional environment. However, it may affect reservoir properties in a rather irregular manner, even reversing the effects of depositional features. Major mechanisms of porosity and permeability reduction are mechanical compaction (brittle-grain versus ductile-grain deformation), chemical processes such as dissolution and cementation or physico-chemical processes (pressure solution). In sandstones with abundant ductile grains, mechanical compaction is strong. Brittle-grain fracturing is another much overlooked process; it is primarily associated with faulting and/or intense folding.

The types of diagenetic processes and their effect on reservoir quality were briefly described in Section 3.4. Factors which influence diagenetic processes and control mechanisms of porosity and permeability reduction include temperature, effective stress, grain composition, pore-water composition, time, content of non-conductive materials in the pores, tectonic stress, and permeability.

In sandstone reservoirs, the composition of the detrital framework constituents, e.g. quartz, feldspar, and mica, has little effect on the chemical response of the reservoir to various drilling and completion fluids. The pore-lining or pore-filling diagenetic components of a reservoir, on the other hand, normally influence physical or chemical interactions. They are in direct contact with drilling, stimulation, and recovery fluids. Such chemical interactions may lead to a general degradation in reservoir quality. Even thin pore-linings can effectively block pore throats. A sandstone may have good porosity and zero permeability because of the presence of such a pore-lining. Thus, clay mineral compositon and pore-filling patterns control reservoir quality to a large extent. Sneider *et al.* (1984) show that discovery success is improved by integrating lithological and pore data on silica-rich clastic rocks with petrophysical laboratory data, log-derived data, and geology. Buttkus (1985) critically reviewed the methods for determining lithological parameters from seismic data in order to obtain better reservoir characterization.

A variety of clay minerals may be lined or filled in the pores of sedimentary rocks. Most of them can greatly reduce permeability, increase freshwater or acid sensitivity, totally alter the electric log response, and increase irreducible water saturations. Different composition of clay minerals will cause them to react in an undesirable manner with fluids which are introduced during drilling, completion, stimulation, and/or supplementary enhanced recovery operations. Noncommercial wells may result from low reservoir quality as well as from formation damage caused by an inadequately designed fluid system. Thus, fluids should be designed for the specific varieties of clay minerals present in the pores.

Clay minerals also possess a high-surface-area to volume ratio. Solitary clay particles are generally platy, flaky, and/or fibrous, and of minute particle size. They have an exceedingly large surface area as compared with quartz. Thus, clay minerals react readily and rapidly with fluids introduced into a reservoir, much more than detrital grains and cements of quartz, feldspar, and other silicate minerals.

Clay minerals found in sandstone and carbonate reservoirs can be subdivided into five major groups: smectite, mixed-layer minerals, illite, chlorite, and kaolinite (Figure 146). Each group contains several members, which can differ from each other, particularly as to their morphology and particle size. These members may also be different in chemical composition. Consequently, the reaction rate between clays and completion fluids may vary significantly. Thus, each group may cause different reservoir problems.

The smectite group includes clay minerals such as montmorillonite, beidellite, nontronite, saponite, and the mixed-layer composites of smectite and illite. Production problems in smectite-rich sandstones are numerous: (1) smectites are extremely water-sensitive and swell in freshwater, (2) smectite pore linings end to break loose and migrate during swelling, and (3) the crystal size and habit of smectites produce a very high surface-area/pore-volume ratio. Clay swelling caused by the introduction of freshwater into the pores will lead to sealing-off the pore-throats and to a loss of permeabililty. Diagenetic coatings of smectite tend to be destroyed by extensive swelling and to be mobilized within the pore system. The high surface-area/pore-volume ratio leads to high irreducible water saturation and high critical water saturation which can allow a well to produce water-free oil in the presence of high water saturation. The swelling problem associated with smectites can be overcome by using oil-base, potassium or ammonium chloride drilling, completion, and stimulation fluids. If swelling has already taken place within a reservoir, the damage may be corrected by acidizing with a weak mixture of hydrochloric acid and hydrofluoric acid (Table 38), provided injectivity is not totally lost.

Illite can form various crystal habits, e.g. a meshwork of long, hairlike crystals (Figure 146). The main engineering problem is that illite in a sandstone creates large volumes of microporosity and increases pore tortuosity considerably. Microporosity can bind water to the host grains and lead to high irreducible water saturation. With freshwater, illite fibers tend to clump, thus, further reducing permeability. 'Hairy' illites, not dissolved prior to production, may break loose during production, migrate to the pore throats, and act as a check valve. Illite can be dissolved using an acid mixture consisting of hydrochloric and hydrofluoric acid (Table 38).

Chlorite normally contains high amounts of Fe and Mg and is extremely sensitive to acid and to oxygenated water. It is readily dissolved in dilute HCl. When exposed to acid treatment it will dissolve and the iron, mobilized during dissolution, will reprecipitate as a gelatinous ferric hydroxide $Fe(OH)_3$, clogging the pores. Problems of iron precipitation can be avoided by adding appropriate chemicals to the acid (an oxygen scavenger and an iron-chelating agent).

Kaolinite is chemically very stable. It will react to acid similar to quartz. Thus, acid treatments have no real effect on kaolinite. Kaolinite is primarily responsible for the migration of fine particles found in many reservoirs. Large shear stress associated with high fluid-flow velocities can dislocate kaolinite booklets attached loosely to the surface of detrital host grains, most evidently in areas close to the borehole. This engineering

| | PARTICLE | | MORPHOLOGY | SPECIFIC SURFACE (m²/g) | | CATION EXCHANGE CAPACITY (m equiv. per 100 g) | SWELLING | FINE PARTICLE MOBILISATION |
	SIZE (μm)	SHAPE		PURE CLAY	SANDSTONE			
SMECTITE	2	PLATY TO FIBROUS	HONEY-COMB	85 – 100	0.5 – 2.0	80 – 140		
MIXED-LAYER MINERALS	2	PLATY TO FIBROUS	SIMILAR TO BOTH SMECTITE AND ILLITE			40 – 80		
ILLITE	2	PLATY TO FIBROUS	CURLED FLAKE w/ PROJEC. & FIBROUS MAT	90 – 115	1.5 – 10	20 – 40		
CHLORITE	2	PLATY	CARD-HOUSE ROSETTE	40 – 60	0.5 – 2.0	10 – 30		
KAOLINITE	30	PLATY, VERMICULAR	BOOKS FANS	15 – 18	0.05 – 0.2	3 – 15		

Fig. 146. Physical properties of major clay minerals and their implications for EOR processes (from Gaida *et al.* 1973, Wilson 1982, p. 115, Gaida *et al.* 1987).

Table 38. Reservoir characteristics of authigenic clay minerals (after Wilson and Pittman 1977).

	Morphology of Individual Flakes	Form of Aggregates	Relationship to Sand Size Detrital Grains	Thickness of Coating or Long Dimension of Aggregates (Microns)	Special Features
Kaolinite and Dickite	pseudohexagonal	stacked plates (book)	pore filling	2–2 500 (generally 2–20)	flakes notched or embayed (twinned?)
	pseudohexagonal	vermicule	pore filling	10–2 500 (generally 120–200)	flakes notched or embayed (twinned?)
	pseudohexagonal	sheet	pore filling	0.1–1	flakes notched or embayed (twinned?)
Chlorite	pseudohexagonal	plates (2-dimensional cardhouse)	pore lining	2–10	
	curled equidimensional with rounded edges	honeycomb	pore lining	2–10	
	equidimensional with angular or lobate edges	rosette or fan	pore lining and pore filling	4–150 (generally 4–20)	
	fan-shaped fibrous bundles	cabbagehead	pore lining and pore filling	8–40	
Illite	irregular with elongate spines	sheet	pore lining	0.1–10	bridging between sand grains
Smectite	not recognizable	wrinkled sheet or honeycomb	pore lining	2–12	bridging between sand grains
Mixed-layer smectite/illite	subequant with stubby spins	imbricate sheet to ragged honeycomb	pore lining	2–12	bridging between sand grains

problem in kaolinite-rich sandstones is easily resolved through the use of any of the clay-stabilization systems, as long as the treatment is carried out early in the lifetime of a well.

Problems caused by clay minerals in the reservoir vary from formation to formation, and from area to area. Individual formations may show regional trends in clay-mineral composition which may help predicting problems to be encountered in wildcat wells. Problems, induced by specific clay-mineral composition, are known in sandstone reservoirs of Rotliegende age in NW Europe, of Jurassic age in the North Sea area, and of Lower Cretaceous age in NW Germany. In the U.S.A. numerous Tertiary reservoir sandstones of the Gulf Coast, such as the Frio, Hackberry, Hosston, and Wilcox, suffer from sensitive clay-mineral composition.

4.4.2. RESERVOIR FRACTURING

In reservoir fracturing one has to distinguish between naturally induced fracturing and artificially induced hydraulic fracturing. Either fracturing is controlled by the geological setting of a reservoir and its lithological and sedimentological aspects, especially by its small- and large-scale anisotropies.

Naturally fractured hydrocarbon reservoirs occur in a variety of rock types including chert, chalk, shale, limestone, dolomite, siltstone, sandstone, and igneous and metamorphic rocks (Aguilera 1980). Fields that produce from fractured rocks are the Asmari limestone in SW Iran, the Aquila Field in Libya, the Lacq field in France, certain foothill fields of Alberta, Canada, the La Paz and Mara Fields in Venezuela, and others in Indonesia, Mexico, and China.

Hydraulic fracturing of potential producers or of producing wells is performed to generate cracks in a formation for easier flow of fluids or gas into the well. Fluids used to generate and extend the cracks generally carry aproppant. Purpose of the proppant is to prevent the fracture from closing after the hydraulic pressure is released at the completion of the treatment. Clean, well-sorted silica sand is used most often as proppant. Other granular materials have been used as proppants in fracturing treatments, too. The sand quality is defined by the silica content, silt content, acid solubility, roundness, cleanness, and strength.

Hydraulic frac jobs may cause formation damage due to problems with the proppant and/or the frac fluid. A common mistake is the use of standard frac sand at depths where a high-strength proppant is needed. Below 3 000 m sintered bauxite, a high-strength proppant in use, exhibits much less crushing. Frequently, the hydraulic frac fluid is the damaging element in the frac treatment. If mud acid forms the base for an acid frac, CaF precipitates can develop. Another cause of poor frac response is the use of fluid-loss additives which can plug pores in the frac face.

Hydraulic fracturing is executed by service companies like Halliburton, Dowell, and Otis.

Naturally occurring and artificially induced fractures greatly increase production, even from formations with low or variable matrix permeabilities.

Reservoirs with fracture-enhanced permeability can be recognized by their production characteristics and by the much greater yields when compared with adjacent parts of the same formation where no factures occur.

However, only certain fracture systems lead to a significant increase in production; i.e. fractures must be penetrative, interconnected, and open to enhance permeabilities markedly. At present two ways of approach are applied for analyzing fracture systems: (1) lineament studies are used to locate fracture traces at the surface, (2) high-resolution seismic reflection surveys are conducted to locate fracture traces at the surface, and (3) high-resolution seismic reflection surveys are conducted to locate structures in the subsurface. Moreover, geochemical surveying techniques also can be used to detect fracture systems.

A new approach to fracture analysis combines geochemical and structural techniques into a powerful tool to differentiate, interpret, and characterize fracture systems. Basic parameters for this approach are chemical and textural data obtained from vein and matrix mineral assemblages (Tillman and Barnes 1983).

In exploring and developing reservoirs with fracture-enhanced permeability it is necessary to carefully differentiate and characterize all fracture systems in order to identify which fractures are important to production.

In regions where artificially induced fracture permeability is necessary for significant production from low-permeability formations, these combined geochemical and structural techniques help to develop potentially productive reservoirs.

4.4.3. WELL TREATMENTS

Well treatments are applied to single wells in order to optimally drill wells without formation damage, to stimulate wells in tight reservoirs or wells of low porosity and permeability, to increase production rates, and to optimize hydrocarbon recovery. Well treatments comprise drilling proper, stimulation techniques such as acidizing and fracturing of tight reservoirs or reservoirs of low porosity and permeability (compare Section 4.2.2), and well-completion techniques. Wilson (1982) reviewed reservoir quality and formation-damage susceptibility in sandstones, Ely (1985) and Economides (1992) summarize data on well-stimulation treatments.

The well treatments are of purely mechanical and chemical nature. Formation damage during drilling can be caused by drilling-mud particulate invasion and by drilling-mud filtrate invasion. Formation damage during acidizing may be induced by iron-hydroxide precipitates, by sulfate and fluoride precipitates, and by migration of fines. Formation damage during fracturing may be produced by proppant embedding, water-loss additive particulate invasion, proppant crushing, and polymer residues. Formation damage during well completion may be triggered by migration and plugging by fines, iron-hydroxide precipitates, sulfate and carbonate scaling, and clay-mineral swelling.

Knowing lithology, mineral composition of detrital framework components and of mineral cements, and the physical properties of the reservoir rock as well as the spatial arrangement of clay minerals, formation damage can normally be avoided by selecting the appropriate drilling mud, the best acid for acidizing, and the most appropriate fracturing procedure.

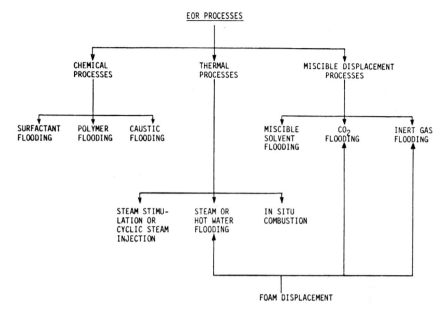

Fig. 147. Main types of EOR processes (after Donaldson *et al.* 1989, p. 6).

If well stimulations – or enhanced oil recovery operations – are designed without knowing the clay mineralogy of a given reservoir, they may fail or even damage the reservoir. Sometimes the damage is permanent. In other instances a new, properly designed stimulation procedure may produce marked increases in flow.

Preventive and remedial chemical and physical actions are tabulated in Table 39.

Improved well performance can be monitored best by modern logs such as the well-performance-simulation log (WPS), the flow-analysis log (FAL), or the mechanical properties log (MECHPRO).

Therefore, in designing waterflood projects, or enhanced oil recovery operations, it is vital to know which types of clay minerals occur in the pore system of the reservoir rock and what potential formation-damage mechanisms may inflict the reservoir.

4.4.4. ENHANCED OIL RECOVERY

Formerly, oil fields were produced by primary mechanisms until exploitation was no longer profitable. Then new wells were drilled for water injection (secondary recovery). When oil fields are exhausted by water flooding, they become candidates for tertiary recovery or enhanced oil recovery (EOR). Even water floods leave large amounts of oil in the reservoir, around 30% of the pore volume or 50% of the original oil-in-place. EOR may be started at any time during the exploitation history of an oil reservoir. It includes several very expensive and difficult, and, thus, very risky processes (Figure 147).

Modern EOR methods include solvent drive, carbon-dioxide drive, surfactant drive,

Table 39.   Formation damages introduced during drilling, well completion, stimulation (acidizing/fracturing), and enhanced oil recovery (modified after Wilson 1982).

	FORMATION DAMAGE	PREVENTIVE ACTION	REMEDIAL ACTION
	Drilling Mud Particulate Invasion	Mud Conditioning	Perforation – Production Clean-up; Fracturing – Surging
During Drilling	Drilling Mud Filtrate Invasion		
	Clay Swelling and Migration	Clay Stabilizers; Oil- or Polymer-Based Mud	Perforation – Fracturing
	Water Block	Reduce Mud Weight	Perforation – Fracturing; Production Clean-Up
During Well Completion	Migration and Plugging by Fines	Reduce Flow Rates – Clay Stabilizers; Avoid Acidizing Impure Carbonates	Surging (?) – Fracturing; Acidizing
	Iron-Hydroxide Precipitates	Sequestering Agents Maintain Low pH – Low Residence Time	Reacidize – Fracturing
	Sulfate Precipitates	Avoid HCl Acids – Sequestering Agents – Preflush	Fracturing
	Clay Swelling and Migration	Clay Stabilizer	Fracturing – Acidizing
During Stimulation Treatments	Fracturing		
	Proppant Embedment	Adjust Proppant Concentration	Re-fracture
	Water Loss Additive	Avoid Particulate Water Loss Additives	Production Recovery
	Particulate Invasion		
	Proppant Crushing	Monitor Proppant Delivered	Re-fracture
	Polymer Residues	Polymer Selection	Re-fracture
	Acid Sensitivity		
	Iron-Hydroxide Precipitates	Sequestering Agents Maintain Low pH – Low Residence Time	Reacidize – Fracturing
	Sulfate Precipitates	Avoid HCl Acids – Sequestering Agent – Preflush	Fracturing
	Fluoride Precipitates	Avoid HF Acids – Sequestering Agents – HCl Preflush	Fracturing
	Migration of Fines	Avoid Acidizing Impure Carbonates	Fracturing
During Enhanced Recovery	Temperature-Induced Clay Transformations		
	Kaolinite to Illite	Avoid KCl – Clay Stabilizer Preflush	Mud Acid Treatment
	Smectitic Clay to Illite	Avoid KCl – Acid Preflush	Mud Acid Treatment
	Kaolinite to Smectite	Acid Preflush?	Mud Acid Treatment
	Loss of Injected Surfactants and Polymers		
	Absorption on Clay Surfaces	Preflush with Sacrificial Agent – Mud Acid Treatment?	?
	Divalent Ion-Induced Precipitation	Preflush Cation Control	?

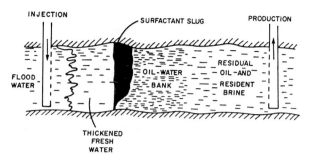

Fig. 148. The principal steps of chemical flooding (after Donaldson *et al.* 1989, p. 6).

micellar polymer drive, steam injection, and in-situ combustion. Each method requires specific sedimentological and mineralogical considerations to optimize application. The principal steps of the chemical flooding process are schematically illustrated in Figure 148.

When petroleum reservoirs contain low-gravity (less than 20° API), high-viscosity oil and high porosity, secondary recovery methods are not effective for displacement of oil. For such reservoirs, thermal processes have received the most attention. The injection of steam reduces the oil viscosity which causes an increase in the oil mobility. Depending on the way in which the heat is generated in the reservoir, the thermal processes can be divided into three categories: (1) in-situ combustion, (2) steam injection, and (3) wet combustion.

The fundamentals and analytical approaches to EOR, as well as the various methods used, are treated in texts by Donaldson, Chilingarian and Yen (1985, 1989) and Tillman and Weber (1987). Polymer flooding is summarized by Littmann (1988) and thermal recovery methods are dealt with by White and Moss (1983) and Garon and Jennings (1985).

Microbial enhanced oil recovery is discussed by Zajic *et al.* (1983), Donaldson, Chilingarian and Yen (1989) and Donaldson (1991). Microorganisms and their metabolic products are used to stimulate oil production. This technique involves the injection of selected microorganisms into a reservoir and the subsequent stimulation and transportation of their in-situ-growth products for further reduction of residual oil left in the reservoir after secondary recovery is exhausted.

The solvent drive means that light hydrocarbons – propane to hexane (LPG) completely miscible with the oil – are injected as gas or liquids into the reservoir. They dissolve in the oil and form a bank of light hydrocarbons pushing the oil towards the producing well. Other methods are to inject methane enriched with higher hydrocarbons (ethane, hexane) or dry gas, mostly methane, at even higher pressures.

The carbon-dioxide drive refers to carbon dioxide being dissolved in crude oil. It makes the crude to separate into two fractions: light and heavy. If CO_2 is injected at sufficient pressure, it will form a bank consisting of light hydrocarbons and CO_2. This CO_2 bank is then driven by water. A large-scale example is the Kelly-Snider Field, Texas, U.S.A.

The surfactant drive consists of a detergent added to the flood water. It lowers the interfacial tension at the interface between oil and water. Surface-active agents, called surfactants, have this property. Surfactants are immediately adsorbed out of the water.

The micellar-polymer drive consists of an emulsion of water in oil as driving medium; it gives a piston-like displacement. The emulsion is stabilized by high concentrations of a surfactant, which tends to surround the droplets of water in structures called micelles. The viscosity of the emulsion is increased by thickening agents, chemical compounds consisting of aggregates of smaller compounds called polymers. Surfactants are adsorbed on the rock and on oil-water interfaces. They are destroyed by metallic ions, such as calcium and magnesium, which occur in the pore water and are attached to the clay minerals. The chemicals needed are expensive. Like in all other secondary and tertiary recovery methods, reservoir heterogeneities cause bypassing. Tests did not reveal always reproducible patterns, i.e. similar or equal numbers of injection of producing wells.

Steam injection is applied to heavy oils with a gravity between 10 and 20° API (density 1.0 to 0.939 g/cm^3) which are highly viscous and contain very little gas. Original dissolved-gas drives are very ineffective, recovering only 10 or 20% of the oil-in-place. Steam injection into heavy oil sands has given excellent results, especially in California and Venezuela. Bypassing due to sand heterogeneities is less detrimental in the case of steam. However, geological parameters may also affect steam injection. The most successful method of steam injection is the steam soak or 'huff-and-puff' method. Steam injection is the only EOR process that has been commercially successful on a large scale. It is the only practical measure to exploit very heavy oil (10 to 20° API). Large-scale steam injection plants are being planned for large fields of heavy oil in Venezuela and Alberta. The sand horizons must be fairly thick (over 3 m). The well spacing has to be close.

In-situ combustion means burning oil in the reservoir by injecting large volumes of air or oxygen. Air passes through the burned region until it reaches the burning front. The hot combustion products are driven ahead. They vaporize the water and hydrocarbons, leaving behind non-volatile material called coke. This coke burns and provides additional heat. More than 100 field tests of in-situ combustion have been reported (Farouq Ali 1970). Most are small experimental projects, no really large-scale projects exist yet. The setting of an in-situ combustion field test in the Suplacu de Barcau Field, Romania, delimiting the burning fronts, is shown in Figure 149. Problematic and expensive is injecting enough air to keep the underground combustion going. Temperature in the combustion zone ranges from 500 to 800°C. It moves about 3 to 30 cm per day. Large increases in oil production usually occur, and the average oil recovery has been about 50 per cent of the oil in place. The lateral sweep efficiency is normally rather good, although the burning zone tends to move through the upper part of the sand. Many failures of in-situ combustion have disclosed the need for better appraisal of the geology prior to a fire flood. Sands have to be permeable to admit large amounts of air required. The sands should be sufficiently deep that injection pressures can be high enough to force the air into the sands. Marked heterogeneities in sand bodies caused detrimental bypassing.

The basic principles and the processes of EOR operations have remained unchanged over the years. Steam flooding remains the number-one process. However, the conventional EOR processes are being optimized with respect to performance and economy, mainly in order to reduce costs. In this respect, chemical methods are predominant, using

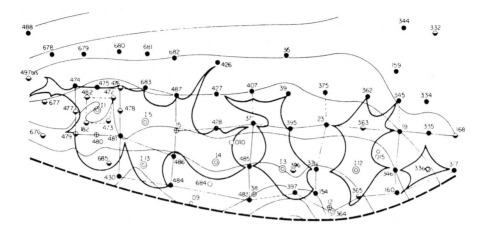

Fig. 149.   In-situ combustion field test showing the 1970 burning fronts, Suplacu de Barcau Field, Romania (after Latil 1980, p. 203) Double circles = new injection wells.

low-cost organic and inorganic chemicals as sacrificial agents. Low-cost organic chemicals are polyethylene glycols, lignosulfonates, carboxylic or quarternary ammonium salts. Inorganic chemicals are polyphosphates, borates or carbonates. These sacrificial agents, injected in front of or together with high-cost chemicals like surfactants and/or polymers used in chemcial flood processes, are preferentially adsorbed on the rock surface and prevent detrimental loss of high-cost flood chemicals. In steam floods as well as in gas floods (carbon dioxide, nitrogen) highly viscous foams, generated from surfactants and gas, are injected or produced in-situ preferentially in the highly permeable portions of the reservoir. The foam blocks the highly permeable streaks, prevents early break-through of steam or gas and, thus, improves sweep efficiency.

Conventional hydrocarbon exploitation, always the pre-stage of EOR, requires among other things sound sedimentological knowledge of the reservoir. Geological considerations as to the primary environment of deposition, diagenetic alteration, and the tectonic framework were discussed in Section 4.4.1.

Large-scale geological factors that affect reservoir performance are the distribution and attitude of the reservoir rocks and enclosing less permeable strata, the facies pattern within the reservoir interval, its stratification, the presence of shale 'breaks' within the reservoir interval, natural fractures, and the influence of a hydrodynamic pressure gradient on the movement of fluids within the reservoir. Small-scale features include the texture and mineralogy of the reservoir rock, the types and the arrangement of clays, and the geometry of the pore system.

Detailed sedimentological description, when combined with engineering or subsurface geological techniques, furnish reservoir models which may be used for reservoir management during field development and during secondary or tertiary EOR (Tillman and Weber 1987).

The framework for field description relies essentially on conventional cores. In addition, frozen and rubber-sleeve cores may be utilized. The integration of sedimentological examination with other techniques leads to synergism. Oriented cores or interpretation of dipmeter measurements may be incorporated into studies of shelf, delta, and fluvial sandstones. Oriented cores allow determination of flow directions for bottom currents. Studying reservoir heterogeneities and their scale is important. Heterogeneities range from as large as lithological and facies changes to as small as shale lenses and secondary pore-filling clays and cements. Present outcrop studies in De Chelley Canyon, Arizona, were used to predict the scale of features in the Permian Rotliegende Sandstone in the Leman field (North Sea). Also results from pressure buildups, pulse testing and interference tests can be integrated into geological-engineering models.

The preceding reservoir-specific parameters such as the geological framework of the reservoir as well as physical and chemical properties of the oil phase and of the flooding medium, reservoir dynamics, oil-in-place, wetting condition and actual saturation ratio, the technical field installation and the results of laboratory tests have to be evaluated before an EOR operation is applied successfully.

In addition, formation damage through EOR operations may be caused by physico-chemical interactions of the solid framework and the pore-filling clays with the surrounding displaced and displacing liquid phases. Such interaction processes influencing oil recovery are: adsorption of chemicals used, dissolution of soluble minerals, swelling of clay minerals, and cation exchange of divalent against monovalent ions. Cation-exchange capacity depends on particle size of clay minerals, on the substitution of quadrivalent against trivalent ions in the tetrahedral sheet, and trivalent against divalent ions in the octahedral sheet of the clay-mineral lattice.

Figure 150 demonstrates the influence of certain mineral cements in sandstones on texture and petrophysical parameters. The figure shows abundance of cementing minerals in sandstones. The texture of the cementing minerals in the pore space are symbolized: vermicular for kaolinite, flaky and fibrous for illite, honeycomb for smectite, and homoaxial overgrowths for silica cement. Clay minerals will affect EOR significantly when smectite and mixed-layer minerals are present, but only moderately if kaolinite occurs.

The effects of cementing minerals in sandstones on productivity and oil recovery are shown in Figure 151. Cation exchange, adsorption, and swelling are primary processes. They trigger secondary effects such as precipitation or particle mobilization. Surfactant floods would become ineffective by increasing interfacial tension. Polymer floods are affected by viscosity reduction caused by loss of chemicals due to adsorption. Caustic floods may be impaired additionally by dissolution of silicates, precipitation, and particle mobilization.

In CO_2-floods, in-situ combustion, and steam flooding, swelling and shrinking phenomena, dewatering, decomposition or mineral alteration influence EOR.

Iron-bearing minerals influence EOR significantly. Scaling problems in CO_2-floods, oxidation, corrosion, and sour-gas production result from physico-chemical interactions when in-situ combustion and steam flooding are applied. Soluble gypsum affects surfactant floods negatively as tests have shown; whereas after alteration of gypsum to less soluble anhydrite oil recovery in surfactant and polymer floods is less problematic.

The following factors must be considered in order to optimize EOR operations: (1) type

Cementing Minerals in Sandstones		Abun-dance	Texture early | late diagenetic	Petrophysics		
				Poros-ity	Permea-bility	Internal Surface Area
Clay Minerals	Illite	c-a	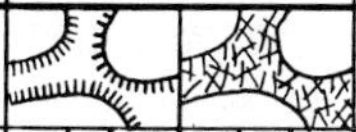	m	m	m
	Mixed-layer min.	r-c	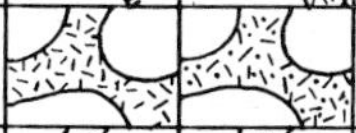	l-m	l	h
	Smectite	r-c	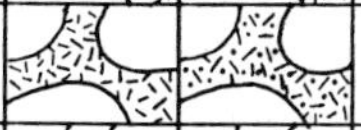	l-m	l	h
	Kaolinite	r-a	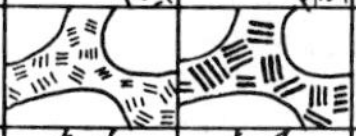	m	m	m
	Chlorite	r-c	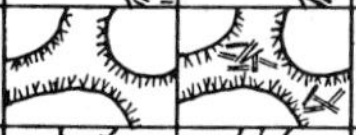	m	l	m-h
Silica	Quartz	c-a	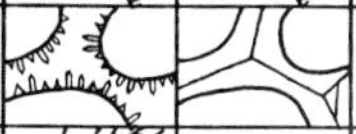	l-h	l-h	l
Silicates	Feldspars	c	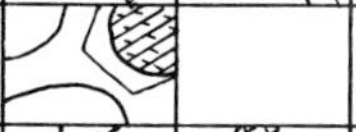	m	m	l
	Zeolites	r	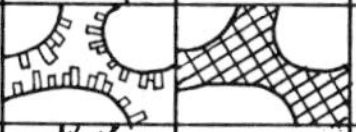	l-m	l-m	m
Fe Oxides and Hydroxides	Hematite	r	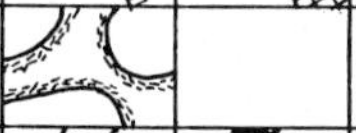	l-m	l-m	l
	Goethite	r	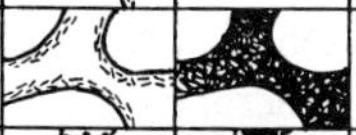	l-m	l-m	m
Fe Sulfides	Pyrite, Marcasite	c	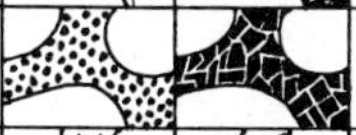	l-m	l-m	l-m
Carbonates	Calcite	a	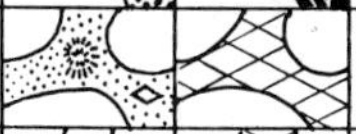	l-m	l-m	l
	Dolomite	c-a	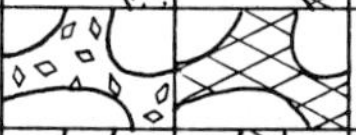	l-m	l-m	l
	Ankerite	r-c	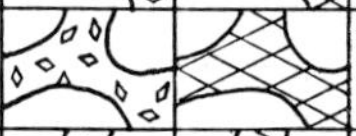	m	m	l
	Siderite	r-c	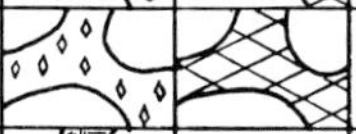	l-m	l-m	s/m
Sulfates	Gypsum	r	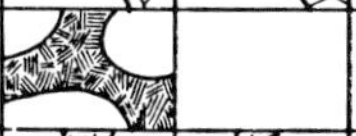	m	m	m
	Anhydrite	r-c	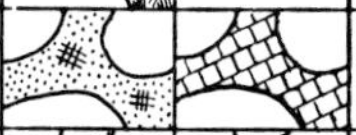	l-m	l-m	m
Phosphates	Apatite	r	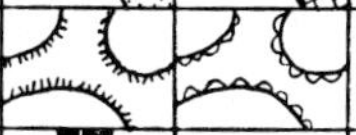	l	l	m
Organic Matter	Bitumen	r	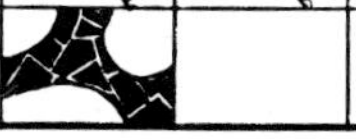	l	l	l

a — abundant, c — common, r — rare, h — high, m — moderate, l — low, s — small

Fig. 150. Influence of cementing minerals in sandstones on texture and petrophysical parameters.

Cementing Minerals in Sandstones		Enhanced Oil Recovery						
		Surfactant Petrol. | Ether Sulfonate	Polymer Hydrol. Polyacr. Fresh-Water	Polymer Poly-Sacchar. Brine	Caustic Solution	Foam/Misc./Immisc. CO₂Fl.	Foam/Steam Flooding	In Situ Combust.
Clay Minerals	Illite	CATION EXCHANGE/ADSORPTION/ PRECIPITATION/PARTICLE MOBIL./ INCREASE INTERFACIAL TENSION		ADSORPTION	CAT. EXCHA./PARTICLE MOBILIZ./ INCREASE INTERFACIAL TENSION	CATION EXCHANGE	SWELLING/ PART. MOBIL.	DEWATER-/ SHRINKING/ PART. MOBIL.
	Mixed-Layer Min.		SWELLING					
	Smectite					SWELLING		
	Kaolinite				SWELLING			DECO.
	Chlorite							
Silica	Quartz	CAT.EXCH./ADSORP./ PART. MOB./PRECIP./ INCR.INTERFAC.TEN.	ADSORPTION/RETENTION/INCREASE OF INTERFACIAL TENSION	ADSORPTION/VISCOSITY REDUCTION	DISSOLUTION/ PRECIPITATION			
Silicates	Feldspars							
	Zeolites							DEWAT.
Fe Oxides and Hydroxides	Hematite							OXIDATION
	Goethite						OXIDA.	
Fe Sulfides	Pyrite, Marcasite							OXIDAT/CORROS/ SOUR GAS PROD.
Carbonates	Calcite	CATION EXCHANGE/ADSORPTION/ PRECIPITAT/PARTIC. MOBILIZAT./ INCREASE INTERFACIAL TENSION			CATION EXCHANGE/ PRECIPITATION/ INCREASE INTERFACIAL TENSION	EQUILIBRIUM SHIFT/ DISSOLUTION/SCALING		DECOMPOSI./ OXIDATION
	Dolomite							
	Ankerite							
	Siderite							
Sulfates	Gypsum							DEWAT.
	Anhydrite							CORROSION
Phosphates	Apatite							
Organic Matter	Bitumen	CE/P/ PM/A/ IIT		ADSORPT/ VISC. RED.	CE/P/ PM/A/ IIT			COMBU.

Fig. 151.   Effects of cementing minerals in sandstones on productivity and oil recovery.

and distribution of the mechanical framework of the reservoir, i.e. size, distribution, and configuration of pores, (2) chemical constituents present in the reservoir and in the stimulation fluid, (3) potential chemical reactions which can occur between these chemicals. Service companies do furnish data on the chemical make-up of the fluid systems upon request.

Analysis of prospective reservoirs for reservoir quality and fluid sensitivity should comprise four steps (Almon and Davies 1981): (1) geological assessment, (2) petrographic examination, (3) X-ray diffraction analysis, and (4) scanning electron microscopy.

The geological assessment includes consideration of the local and/or regional geological and tectonic framework, i.e. dominating lithologies, primary sedimentary structures, and secondary fault patterns, especially in respect to vertical and lateral reservoir anisotropies. The petrographic examination should include megascopic core and thin-section inspection in order to determine textural features such as grain orientation, cross-lamination or intercalated laminae which might cause directional permeability. Point counting can provide accurate model analysis, i.e. the volumetric proportions of minerals present. X-ray diffraction analysis comprises bulk-rock analysis and identification of the fine fraction. Bulk-rock analysis is applied to control the overall composition of the sedimentary rock examined in thin section. It may detect minor amounts of opaque minerals (such as pyrite, hematite, etc.), difficult to identify in thin section. Identification of the fine fraction refers to clay minerals present and their detrital versus authigenic origin. Scanning electron microscopy reveals details on mineral distribution, pore configuration, size, and sorting as well as on the crystallographic habit of pore-filling clay minerals. The chemical mineral composition can commonly be determined by means of the microprobe or energy-dispersive X-ray analyzer, especially that of clay minerals. Such analytical procedures will lead the geologist/engineer team to design proper fluid systems which will be compatible with the reservoir geology and chemistry.

Case histories of optimizing hydrocarbon recovery operations through geological and sedimentological expertise are reported from the Leman Field in the southern North Sea and from the Morgan Field in the Gulf of Suez (Craig *et al.* 1977). In the Leman Field the resulting model has successfully predicted pressure for an additional two years. Such an accuracy of a model permitted proper planning of future platform locations. The clay mineral composition of the Lower Cretaceous Bentheim Sandstone in NW Germany was determined by sedimentological and petrographic techniques (Wittenhagen 1980) in order to prepare a surfactant pilot test in the Bramberge Oil Field (Figures 152 and 153). Detailed sedimentological and mineralogical investigations (Gaida *et al.* 1987) showed the importance of geologic and petrographic parameters of reservoir sandstones and their field-wide distribution for proper planning and operating an EOR project in one of the Middle Jurassic oil fields, the Hankensbüttel-Süd Field in the Gifhorn Trough, NW Germany (Figure 154).

In summary, EOR operations are dependent on and frequently limited by geological parameters of the reservoir. This intimate dependence has to be taken into consideration when planning successful EOR operations. The prime difficulty in EOR is the heterogeneity of reservoirs. Numerous reservoirs consist of channel and bar complexes with marked differences in permeability, both vertically and horizontally. Successful planning of EOR operations has to be preceded by detailed geological evaluation of cores and well logs, especially in respect to the depositional environment, predicting the geometry of the subreservoirs. Communication from one well to another has to be monitored by pulse tests. As projects progress, tracers may be injected in order to determine the preferred

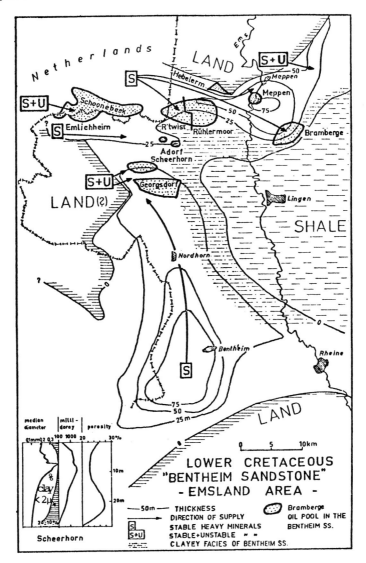

Fig. 152. Paleogeographic setting of the Lower Cretaceous Bentheim Sandstone (Emsland area, NW Germany) (after Wittenhagen 1980).

directions of flow. The well spacing for EOR is normally short because of the sand heterogeneities. Most EOR methods lead to the formation of an oil bank. However, the farther fluids go in inhomogeneous reservoirs, the greater the likelihood for fingering and bypassing which destroy the oil bank. In the case of surfactant floods, chemicals are more degraded or adsorbed by the reservoir rock as farther the fluids advance.

The types of reservoirs which are suitable for each of the different EOR methods are

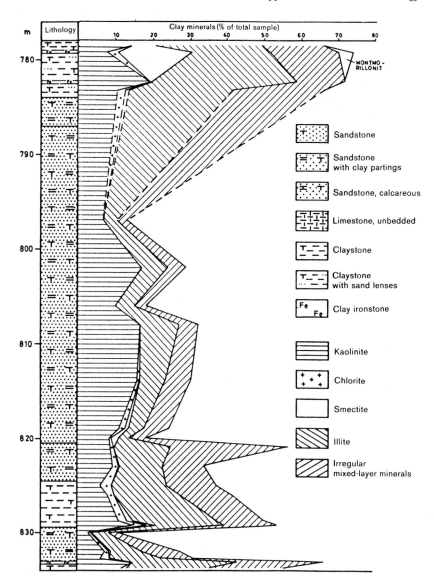

Fig. 153. Clay-mineral distribution in the Bentheim Sandstone, Bramhar 5 well, Bramberge Oil Field, NW Germany (after Wittenhagen 1980).

defined in Table 40, showing the limits on oil gravity and viscosity, reservoir porosity and permeability, and oil saturation. Intensified observation, improved technologies, and new concepts are needed to meet the final goal to recover more hydrocarbons from known reservoirs.

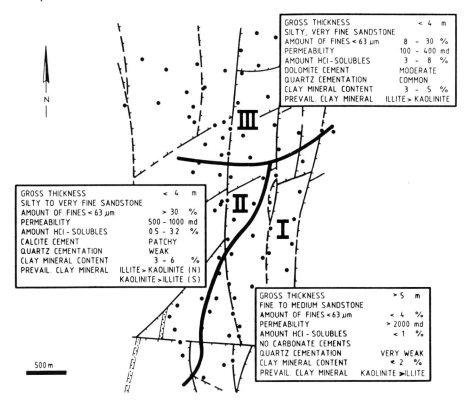

GROSS THICKNESS   < 4   m
SILTY, VERY FINE SANDSTONE
AMOUNT OF FINES < 63 μm   8 - 30 %
PERMEABILITY   100 - 400 md
AMOUNT HCl - SOLUBLES   3 - 8 %
DOLOMITE CEMENT   MODERATE
QUARTZ CEMENTATION   COMMON
CLAY MINERAL CONTENT   3 - 5 %
PREVAIL. CLAY MINERAL   ILLITE > KAOLINITE

GROSS THICKNESS   < 4   m
SILTY TO VERY FINE SANDSTONE
AMOUNT OF FINES < 63 μm   > 30   %
PERMEABILITY   500 - 1000 md
AMOUNT HCl - SOLUBLES   0.5 - 3.2 %
CALCITE CEMENT   PATCHY
QUARTZ CEMENTATION   WEAK
CLAY MINERAL CONTENT   3 - 6   %
PREVAIL. CLAY MINERAL   ILLITE > KAOLINITE (N)
                                         KAOLINITE > ILLITE (S)

GROSS THICKNESS   > 5   m
FINE TO MEDIUM SANDSTONE
AMOUNT OF FINES < 63 μm   < 4   %
PERMEABILITY   > 2000 md
AMOUNT HCl - SOLUBLES   < 1   %
NO CARBONATE CEMENTS
QUARTZ CEMENTATION   VERY WEAK
CLAY MINERAL CONTENT   < 2   %
PREVAIL. CLAY MINERAL   KAOLINITE > ILLITE

500 m

Fig. 154.   Reservoir properties, Lower Dogger-Beta Sandstone, Middle Jurassic, Hankensbüttel-Süd Field, Gifhorn Trough, NW Germany (Gaida *et al.* 1987).

4.4.5. Reservoir Simulation

One approach to predict performance of a hydrocarbon reservoir and to estimate the reserves is mathematical modelling or reservoir simulation (Aziz and Settari 1979). In such a model also heterogeneities of the reservoir must be taken into consideration. Reservoirs are normally subdivided into small cells which may be one-, two-, or three-dimensional. Dimensions and location of each cell are specified. An average value for porosity and permeability is assigned to each cell. Initial oil, gas, and water saturation and pressure are introduced into the simulation model. As the field is in production, changes in pressure and fluid volumes are calculated on a computer using appropriate programs. The actual production history is then compared with the computed history. One or several parameters may be changed to match the computed with the actual production. When the computed reservoir history coincides with the actual history, the future reservoir development may be predicted.

Table 40.　Reservoir types suitable for the application of EOR processes (classification after Donaldson *et al.* 1989a).

EOR Processes		Reservoire Types			Carbonate Reservoir
		Clean	Rich in Carb. Cem.	Rich in Clay	
Chemical	Surfactant	X		–	X
	Surfactant/Polymer	X	•	–	Multivalent ions less favorable
Processes	Polymer	X	X	•	X
	Caustic/Polymer	X	•	–	
	Caustic	X	X	–	
Thermal	Steam	X		–	
Processes	Steam/Foam	X	X	X	
	In-Situ Combustion	(+)			
Miscible/ Immiscible Processes	Miscible/Immiscible (CO_2, Methane)	X			
Immiscible Process	Inert Gas (Nitrogene)	X	X	X	
Microbic Process	Microbes	X	X		X
Technical	Profile Modification	X	X		
Processes	Water Shut-Off	X	X		

X = applicable; – = less favorable; • = restricted; (+) = under study

Most geological data required for reservoir simulation is furnished by geologists and sedimentologists. Sedimentology normally provides information on

- The reservoir configuration,
- Subdivisions of reservoirs by clay layers and faults,
- Size and shape of the aquifer,
- Porosity and its lateral and vertical variations, and
- Permeability and its variations.

On the other hand, the engineer normally provides information on

- Gas, oil, and water saturation,
- Pressure and pressure variations,
- Drill stem and other production tests, and
- Past production history.

Chapter 5

FUTURE TRENDS IN PETROLEUM SEDIMENTOLOGY

Sedimentology became an indispensable part of petroleum geology. Petroleum sedimentology is applied to petroleum exploration and exploitation. It comprises optimal sedimentological evaluation of well logs and seismic records; enhanced oil recovery of known conventional reservoirs; well treatment, stimulation and reservoir simulation; extraction of hydrocarbons from unconventional reservoirs (e.g. oil shales, tar sands, tight-gas sands and shales, in-situ coal gasification); and projecting underground storage of hydrocarbons.

Oil and gas forecasting becomes more difficult. Drew (1990) reflects on petroleum resource assessment and deals with forecasting oil and gas discovery rates and determining the distributional pattern of oil and gas fields of varying size. Halbouty (1981, 1986) outlined the future petroleum provinces of the world and pointed out the most effective exploration methods for discovering new oil and gas fields in mature exploration areas. Global-tectonic concepts like plate-tectonics, which considerably influenced hydrocarbon exploration in the past decades, will have to be further developed.

Improved photogeological and remote-sensing methods will gain increasing importance as cost-effective tools in future oil exploration and exploitation, especially in new and virgin exploration areas such as offshore areas (Figure 155). Remote sensing has a vast potential, even for petroleum sedimentology. For instance, the NASA unmanned Landsat system acquires images that are well suited for mapping fracture systems and viewing plate-tectonic features on the continents and examining their relationship to occurrences of hydrocarbons (Sabins 1980, Merin and Moore 1986, Morgan and Koger 1986, Cracknell and Hayes 1991). Scale, resolution, and broad coverage of the images are well suited for mapping such features as sedimentary basins, fold belts, and volcanic arcs. Local geological features associated with oil fields (folds, faults, and stratigraphic contacts) may also be mapped under favorable conditions. Halbouty (1980) demonstrated the geological significance of Landsat data from giant oil and gas fields. The 'Structural Atlas of Northwest Germany' is also based on Landsat imagery, besides the evaluation of well logs and geophysical surveys (gravity, magnetics, and seismic). Betz (1990) presented examples of Landsat-data interpretations to hydrocarbon exploration in frontier and mature basins of western Germany and Switzerland. To be effective, Landsat imagery analysis must be integrated with subsurface information, geophysical data, and a detailed

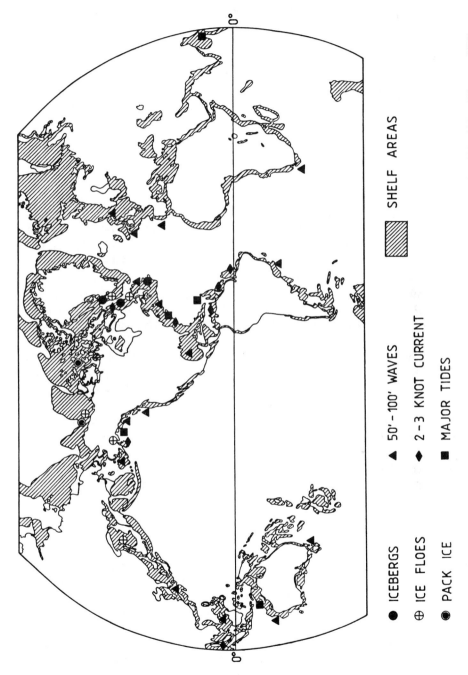

ICEBERGS

ICE FLOES

PACK ICE

SHELF AREAS

▲ 50' - 100' WAVES

◆ 2 - 3 KNOT CURRENT

■ MAJOR TIDES

Fig. 155.   Future hydrocarbon potential: shelf areas and offshore sedimentary basins (modified after St. John 1980, 1984 and Hammett 1980).

understanding of the geologic history of the area. Kramer (1992) gives a detailed survey and a short description of Earth Observing Missions and the sensors.

Sedimentological models will be increasingly applied for a better prediction of location and geometry of potential reservoir rocks during the exploration phase, and to aid exploitation and recovery of hydrocarbons during the development and production phase. Subsurface modelling procedures comprise three steps: (1) the determination of facies, their associates, and sequences, (2) the comparison of these facies parameters with published facies and environmental type models (i.e. generalized models), and (3) the establishing of a specific model. Modelling procedures and future models, however, shall rely on indirect methods such as facies-logging or seismic stratigraphy. But the specific calibration of each geological setting by studying rock samples from pilot wells will remain necessary. Consequently, the examination of the sediments proper will always be an integrated part of petroleum sedimentology, even for the times to come.

Since petroleum sedimentology is changing from a purely descriptive to the quantitative stage, both descriptive and measured data should be converted into a numerical form. Promising is the automation of routine analyses under the microscope using electronic image scanning systems and computerized pattern-recognition methods (Palaz and Sengupta 1992). Comprehensive knowledge of reservoir engineering and geophysics will be a requirement for future sedimentologists.

The concept of self-organization in geological systems, as demonstrated in geomorphology and diagenesis (Ortoleva 1990 and Korvin 1992), will stimulate new ideas. Sedimentary rocks represent complex systems of multifractal geometry. Bedding planes and sedimentary structures possess fractal properties such as the specific surface of a sandstone or the formation-resistivity factor (Fülöp 1991). Patterns may be generated by self-organization through the interplay of fluctuations, instability, and non-linearity. It is certainly opportune for sedimentologists to identify and investigate these phenomena. The development of such new ideas will complement more classical approaches to understanding ordered patterns in crystals, rocks, organic components, sediments, and landforms.

Explorative speculations must be tested thoroughly: the occurrence and potential of 'hot spots' of hydrocarbon generation and accumulation with multistoried source rock horizons, anorganic hydrocarbon generation, the hydrocarbon potential of deep-seated faults, the meaning of impact areas with cryptoexplosion structures for hydrocarbon exploration, and the multiresource exploration of hydrocarbon deposits (e.g. for helium, sulfur, precious metals, and other elements of economic interest).

Though new methods will continuously be developed, available methods must be put to full use. For example: mass spectrometer, microprobe, cathodoluminescence analysis, all at various stages of development, are to be applied to a much wider extent in petroleum sedimentology. And further research on the identity, provenance, and diagenesis of argillaceous sediments is indispensable to close the gap in knowledge of pelitic rocks.

Foremost, ideas and new exploration concepts will enhance further development of petroleum sedimentology. Because future hydrocarbon reserves are to be expected in non-structural traps, petroleum sedimentology will be of increasing importance.

In the long run oil industry will have to improve discovery rates, to stimulate flow from low porosity and permeability reservoirs, achieve more complete recovery of reserves, and evaluate accurately the potential reservoirs penetrated by the drill.

Principles, methods, and conceptual approach of petroleum sedimentology will not be restricted to oil exploration, but will find an even wider application in future exploration geology for any natural resources in sedimentary sequences.

PHOTOPLATES

Photoplate 1. Cathodoluminescence of sandstone with silica cement (courtesy U. Zinkernagel). Under the cathodoluminescence microscope the identical area of a polished sandstone thin section is shown under crossed nicols (A) and under cathodo-radiation (B).

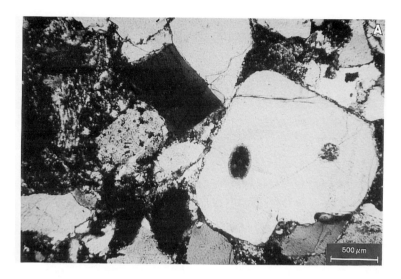

(A) The sandstone is composed of sand grains of different size which appear with various gray shades under crossed nicols. Note the large quartz grain with magmatic embayments. The different gray shades of the quartz grains, function of the varying optical orientation, irritate the observer and rather prevent a quick interpretation of the quartz-sandstone fabric as a whole.

(B) The detrital quartz grains display mainly two colors: bright blue (= dark gray) indicative of a high-temperature origin and dull reddish brown (= medium gray) indicative of lower-temperature regional-metamorphic provenance. Secondary quartz overgrowths show no cathodoluminescence. Thus, quartz grains of high-temperature rock provenance, regional-metamorphic provenance, and diagenetic quartz rims are easy to discriminate. The large quartz grain with magmatic embayments displays zonal growth under cathodoluminescence only.

Photoplate 2. Cathodoluminescence of carbonate rock (courtesy J. Schröder and U. Zinkernagel). Under the cathodoluminescence microscope the identical area of a polished carbonate thin section is shown under crossed nicols (A) and under cathodo-radiation (B).

(A) The carbonate rock, a dolomitized coral limestone, displays a sparry, dense mosaic of xenotopic dolomite crystals. The periphery of the view field is more pigmented than the clearer center.

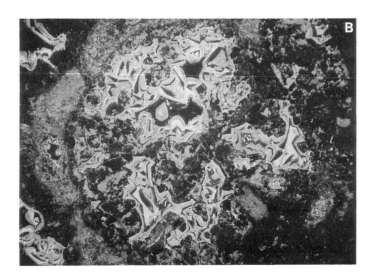

(B) The intensely pigmented periphery contrasts much with the clear center which displays zonal growth of the recrystallised dolomite crystals. Thus, in carbonate rocks cathodoluminescence helps to recognize palimpsest structures.

Photoplate 3. Petrography petroleum source rocks. Lower Jurassic (Lias Epsilon), Weseke 1015 well, North Rhine-Westphalia, W Germany – Source rock throughout Europe.

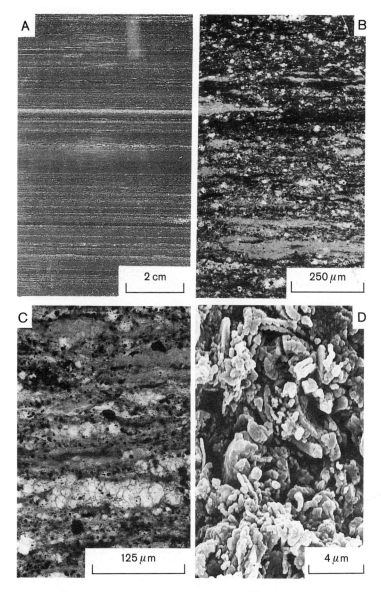

(A) Macroscopic view of cut and polished core: laminated bituminous marlstone with white laminae composed of fossil debris. (B) Thin section (general view): sequence of cryptocrystalline carbonate and bitumen laminae. (C) Thin section (close-up of B): elongate cryptocrystalline carbonate aggregates, i.e. deformed fecal pellets (medium gray), sparry biogenic carbonate (light gray), bitumen schlieren, and framboidal pyrite alternate with each other. (D) Scanning electron micrograph: coccolithic 'hash' of a deformed fecal pellet displaying some microporosity.

Photoplate 4. Petrography of petroleum source rocks. Lower Cretaceous (Berriasian), Schüttorf 3 well (63.5 m), Emsland, NW Germany – Source rock in northwestern Germany.

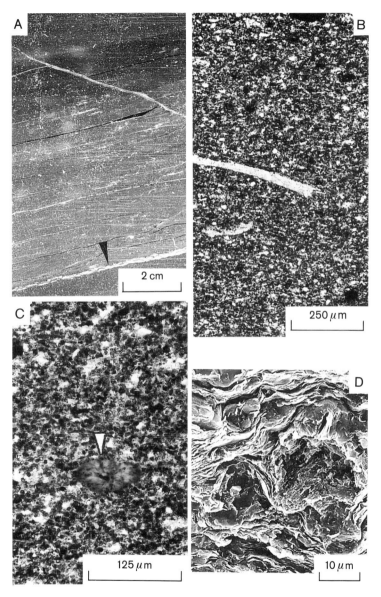

(A) Macroscopic view of cut and polished core: uniform and compact bituminous marlstone with a white fossil layer (arrow). (B) Thin section (general view): silty marlstone rich in organic matter. Note fabric of isometric particles. Two large shell fragments mark bedding. – Parallel nicols. (C) Thin section (close-up of B): *Botryococcus* colony (arrow) in isometric organic-rich marl matrix. – Parallel nicols. (D) Scanning electron micrograph: Slightly deformed hollow plant spherules in clayey matrix.

Photoplate 5. Petrography of gas source rocks. Coals from various locations in the Ruhr region, W Germany.

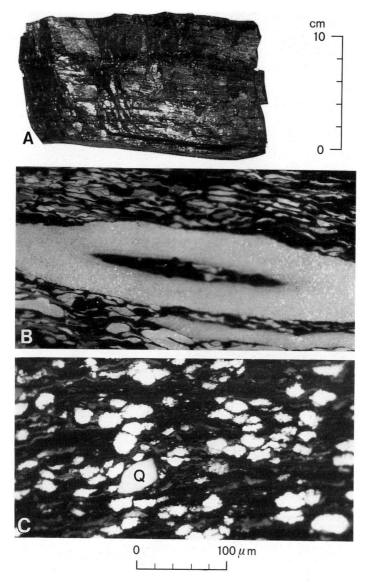

(A) Macroscopic view: gas coal hand specimen which is composed of durain coal, Mattstreifen coal, thin vitrain coal, and fibrous coal lenses (up to 3 mm). Zollverein 2 Seam, Essen Formation, Lower Westphalian B, Upper Carboniferous, Consolidation – Unser Fritz Mine (courtesy K. Burger). (B) Fluorescence photomicrographs perpendicular to bedding: fluorescent megasporinite and microsporinite cut at different orientations. The non-fluorescent matrix consists of macrinite, semifusinite, and collinite macerals (gas coal). Zollverein 6 Seam, Essen Formation, Lower Westphalian B, Upper Carboniferous. Core from Leven 2 well (courtesy G. Bieg). (C) Fluorescence photomicrograph perpendicular to bedding: strongly fluorescent alginite (white particles) with single subangular quartz grain (Q). The non-fluorescent portion consists of a cryptocrystalline clay matrix together with collinite, macrinite, and semifusinite macerals (flame coal). Nibelung Seam, Dorsten Formation, Middle Westphalian C, Upper Carboniferous. Core from Masskamp 1 well (courtesy G. Bieg).

Photoplate 6. Petrography of sandstone reservoirs: Upper Carboniferous (Stephanian), Fehndorf 2 T well, Emsland, NW Germany – Gas reservoir sandstone in northwestern Germany.

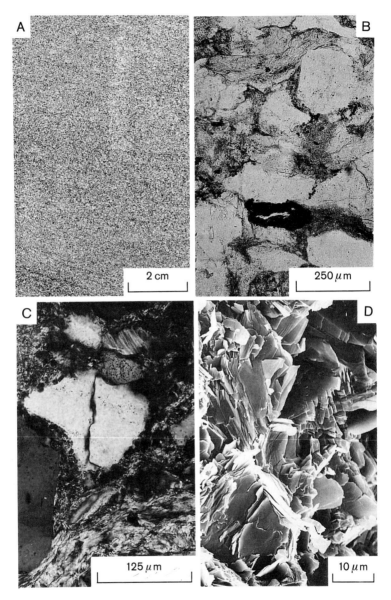

(A) Macroscopic view of cut and polished core: Uniform and compact fine-grained sandstone (graywacke) with faint indication of bedding. (B) Thin section (general view): Well sorted, fine-grained micaceous sandstone with deformed ductile micas and rock fragments. – Parallel nicols. (C) Thin section (close-up of B): Ductile micaceous sandstone matrix of low porosity and permeability. – Crossed nicols. (D) Scanning electron micrograph: Densely packed and well crystallized illite matrix with interspersed carbonate particles.

Photoplate 7. Petrography of sandstone reservoirs: Lower Permian (Upper Rotliegende) Wietzendorf Z 3 well, Lower Saxony, NW Germany – Gas reservoir sandstone in northwestern Germany.

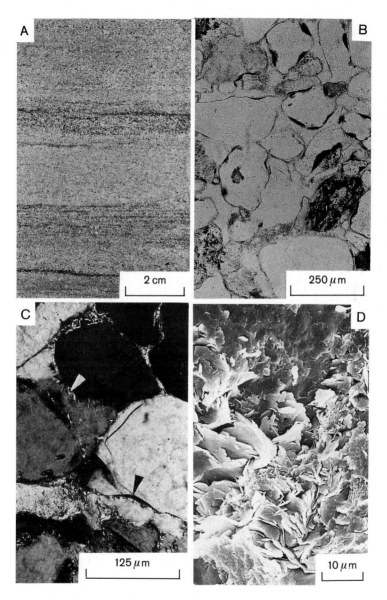

(A) Macroscopic view of cut and polished core: uniform, fine to medium-grained sandstone with marked bedding that causes reservoir anisotropy. (B) Thin section (general view): Well sorted, rounded, fine-grained sandstone composed mainly of detrital quartz. A few volcanic rock fragments (dark-pigmented grains) and feldspar. No detrital mica. – Parallel nicols. (C) Thin section (close-up): detrital quartz grains cemented by flaky illite growing on the grain surfaces (arrows) and by secondary quartz in the center of the pores. Low porosity and permeability. – Crossed nicols. (D) Scanning electron micrograph: flaky illite matrix encrusting a detrital grain that has been removed (in the micrograph seen from underneath).

Photoplate 8. Petrography of sandstone reservoirs: Middle Jurassic (Dogger-beta), Plön-Ost 102 well, Schleswig-Holstein, N Germany.

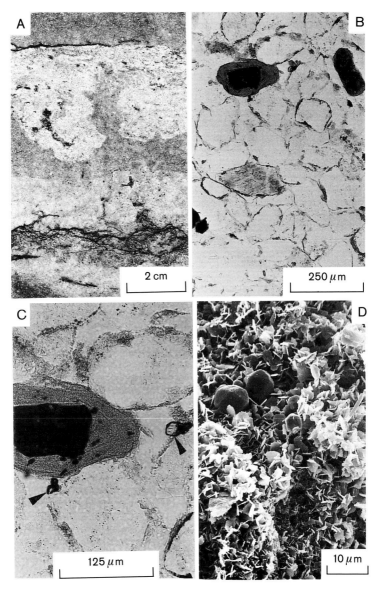

(A) Macroscopic view of cut and polished core: fine-grained sandstone with sedimentary structures: (1) dark lamina and layers rich in carbonaceous matter and (2) bedding deformation by burrowing. (B) Thin section (general view): well sorted, subrounded quartz sandstone with minor feldspar grains and single ooid. No carbonate. Good porosity. – Parallel nicols. (C) Thin section (close-up): subrounded quartz grains and single chamosite ooid with opaque core and thin chamosite crusts. Note minute authigenic brookite (arrows). Good porosity; no secondary quartz. – Parallel nicols. (D) Scanning electron micrograph: Pore space composed of flaky chamosite encrusting detrital grains underneath. Chamosite crusts impede secondary quartz precipitation.

Photoplate 9. Petrography of sandstone reservoirs: Lower Cretaceous (Valanginian), Bentheim Sandstone, Wettrup 11 well, Emsland, NW Germany).

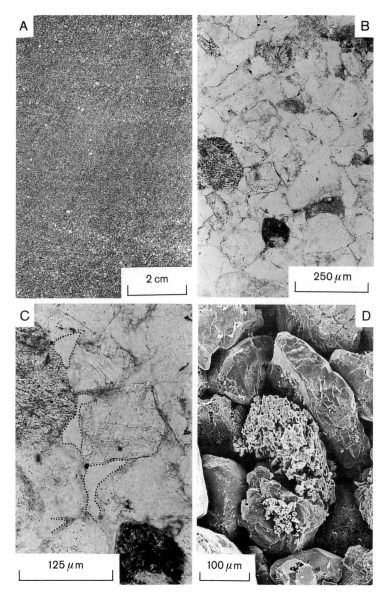

(A) Macroscopic view of cut and polished core: dark-colored, rather uniform, fairly sorted, fine to coarse-grained sandstone. Dark color is caused by thorough bitumen impregnation. (B) Thin section (general view): slightly dirty, porous, subangular, fine-grained quartz sandstone weakly cemented by clay coatings. – Parallel nicols. (C) Thin section (close-up of B): subangular quartz grains, pigmented instable grains (presumably volcanogenic), and fresh feldspar grain (center) with secondary overgrowth; loosely cemented by thin crusts of cryptocrystalline clay. Open triangular pores marked by dots. – Parallel nicols. (D) Scanning electron micrograph: good intergranular porosity of the friable sandstone. Note conchoidal fractures on detrital quartz, local idiotopic quartz overgrowth, and corroded feldspar grain (center).

Photoplate 10. Petrography of sandstone reservoirs: Lower Tertiary (Chattian), Breitbrunn C 8 well, Upper Bavaria, S Germany.

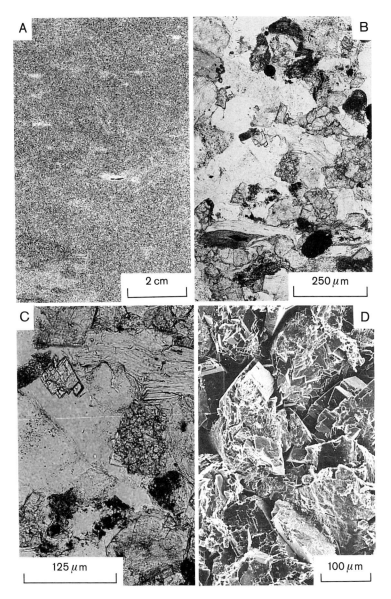

(A) Macroscopic view of cut and polished core: light-colored, fine-grained sandstone rich in carbonate debris. Note white halo around carbonaceous flake (center). (B) Thin section (general view): carbonate-rich sandstone (graywacke), an example of a Molasse sandstone. The detritus has been derived from the Calcareous Alps to the south. – Parallel nicols. (C) Thin section (close-up of B): heterogeneous assemblage of detrital quartz, light-colored mica, sparry carbonate, and undefined rock fragments. Pore space not visible. – Parallel nicols. (D) Scanning electron micrograph: carbonate-dominated pore space covered by plane idiotopic carbonate surfaces. Detrital carbonate grains prevail. Fair porosity.

Photoplate 11. Petrography of carbonate reservoirs: Middle Devonian (Givetian), Reef Limestone, Iberg Quarry, Harz Mts., Lower Saxony, NW Germany – An example of an old, completely oxidized oil reservoir within Paleozoic carbonate rocks.

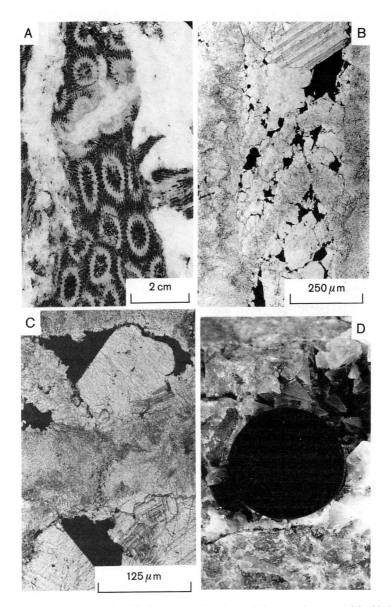

(A) Macroscopic view of cut and polished core: coral-bearing reef limestone impregnated by black solid bitumen (impsonite) and traversed by mainly vertical white calcite veins. (B) Thin section (general view): black bitumen within the intergranular space of the sparry reef limestone. – Parallel nicols. (C) Thin section (close-up): angular pores filled with black bitumen. – Parallel nicols. (D) Scanning electron micrograph: Macrophotograph: round droplet of black solid bitumen (500 μm in diameter) within a limestone vug.

Photoplate 12. Petrography of carbonate reservoirs: Upper Permian, subsurface northwestern Germany, Biohermal limestone.

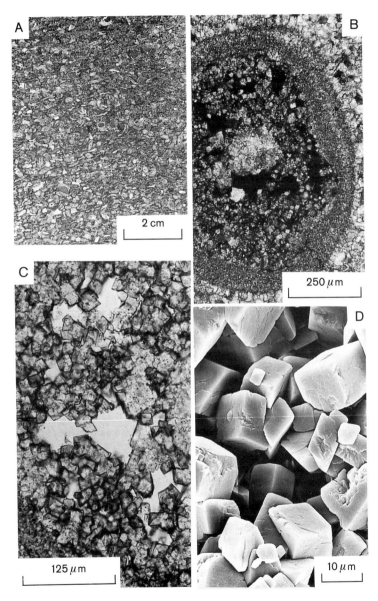

(A) Macroscopic view of cut and polished core: uniform, well-sorted oncolitic carbonate (biosparitic dolomite). (B) Thin-section (general view): oncolitic carbonate thoroughly replaced by subsparitic dolomite crystals. Pore space appears black. – Crossed nicols. (C) Thin section (close-up of B): intercrystalline pores of varying size (white) within a dolomitized oncoid. – Parallel nicols. (D) Scanning electron micrograph: distribution of porosity within a framework of idiotopic dolomite crystals.

Photoplate 13. Petrography of carbonate reservoirs: Upper Jurassic (Upper Malm 1), Messingen 5 well, Lower Saxony, NW Germany.

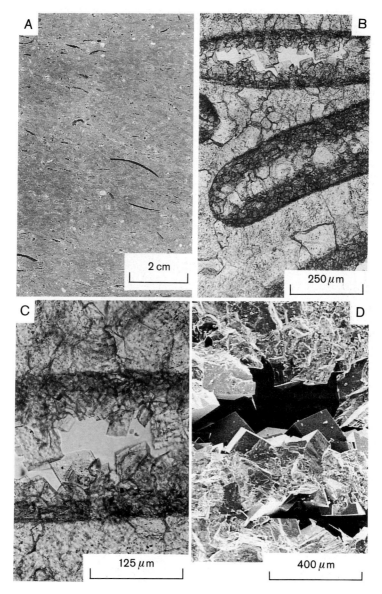

(A) Macroscopic view of cut and polished core: coarsely bedded oncolitic limestone displaying biomoldic porosity derived from the dissolution of shells. (B) Thin section (general view): recrystallized oncolitic biosparite with vuggy pores (white). – Parallel nicols. (C) Thin section (close-up): recrystallized biogenic carbonate with ghost of oncoid and central vuggy pores (white). – Parallel nicols. (D) Scanning electron micrograph: elongate vuggy pores (black) rimmed by idiotopic calcite crystals.

Photoplate 14. Petrography of carbonate reservoirs: Lower Tertiary *Lithothamnion* Limestone, Wolfersberg 2 well, Upper Bavaria, S. Germany.

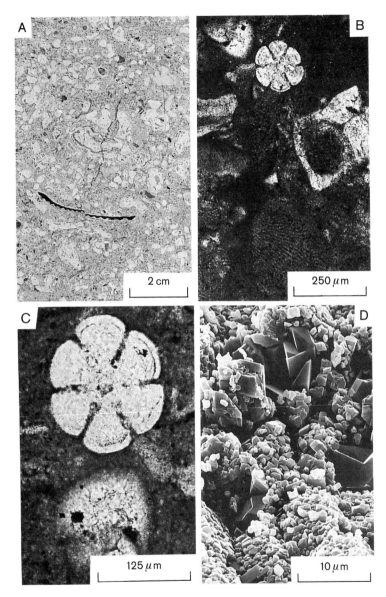

(A) Macroscopic view of cut and polished core: light-colored, poorly sorted algal limestone with a single shell fragment (black). (B) Thin section (general view): Algal micrite (dark gray) and a few echinoid fragments (light-colored). – Parallel nicols. (C) Thin section (close-up of B): cross section of echinoid spine (single calcite crystal) in dense micritic matrix. – Parallel nicols. (D) Scanning electron micrograph: carbonate matrix consisting of microcrystalline calcite particles and patches of coarser idiotopic carbonate crystals. Pore space is finely dispersed within the microcrystalline as well as fine-crystalline portions of the limestone.

Photoplate 15. Petrography of cap rocks from Germany. Macroscopic view of cut and polished cores.

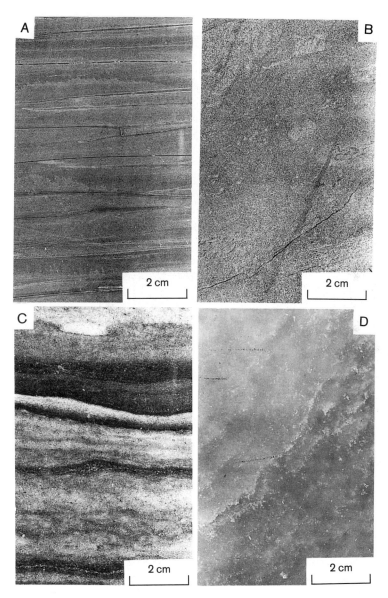

(A) Dark-colored fissile shale. Lower Jurassic (Lias epsilon), Wesendorf Ost 1 well, Lower Saxony. (B) Silty marlstone displaying some bioturbation. Upper Tertiary (Aquitanian), Inzenham W 11 well, Upper Bavaria. (C) Laminated anhydrite. Upper Permian (Zechstein), Lower Werra Anhydrite, Fehndorf 2 T well, Emsland. (D) Light-colored rock salt with undulose deformation structure. Upper Permian (Zechstein), Werra Rock Salt, Vierbaumerheide well, North Rhine-Westphalia.

REFERENCES

Aguilera, R. (1980): *Naturally Fractured Reservoirs*, 703 pp., 483 figs., 105 tables, appendix; Tulsa, Okla. (PennWell Books).

Aigner, T. and Bachmann, G. H. (1989): 'Dynamic Stratigraphy of an Evaporite-to-Red Bed Sequence, Gipskeuper (Triassic), Southwest German Basin', *Sediment. Geol.*, **62**, 5–25, 16 figs.; Amsterdam.

Akramkhodzhaev, A. M., Babadagky, V. A., and Dzumagulov, A. D. (1989): *Geology and Exploration of Oil-and Gas-Bearing Ancient Deltas*, 297 pp., 73 figs., 3 tables; Oxford (IBM Publ.).

Alexander, I. (ed.) (1982): 'Natural Gamma Ray Spectrometry Tool', 83–91, 12 figs.; in: Schlumberger, *Well Evaluation Development, Continental Europe*.

Allen, J. R. L. (1963): 'The Classification of Cross-Stratified Units. With Notes on Their Origin', *Sedimentology*, **2**, 93–114, 4 figs., 2 tables; Amsterdam.

Allen, J. R. L. (1970): 'Studies in Fluviatile Sedimentation: A Comparison of Fining-Upwards Cyclothems with Special References to Coarse-Member Composition and Interpretation', *J. Sediment. Petrol.*, **40**, 298–323, 18 figs., 12 tables; Tulsa, Okla.

Allen, J. R. L. (1982): 'Sedimentary Structures; Their Character and Physical Basis', *Developments in Sedimentology 30A*, **1**, 593 pp., 248 figs., 19 tables; *Developments in Sedimentology 30B*, **2**, 663 pp., 334 figs., 4 tables; Amsterdam (Elsevier).

Allen, P. A. and Allen, J. R. (1990): *Basin Analysis. Principles and Applications*, 451 pp., 342 figs., 27 tables, 1 appendix; Oxford (Blackwell).

Allman, M. and Lawrence, D. F. (1972): *Geological Laboratory Techniques*, 335 pp., 189 figs., 43 tables, 9 pls., appendices A–L; London (Blandford Press).

Almon, W. R. (1981): 'Depositional Environment and Diagenesis of Permian Rotliegendes Sandstones in the Dutch Sector of the Southern North Sea'; in: F. J. Longstaffe (ed.), *Short Course in Clays and the Resource Geologist*, 19–147, 13 figs., 4 tables; Toronto (Miner. Assoc. Canada).

Almon, W. R., Fullerton, L. B., and Davies, D. K. (1976): 'Pore Space Reduction in Cretaceous Sandstones through Chemical Precipitation of Clay Minerals', *J. Sediment. Petrol.*, **46**, 89–96, 7 figs., 3 tables; Tulsa, Okla.

Anderson, R. N. (1986): *Marine Geology. A Planet Earth Perspective*, 328 pp., 212 figs., color insert; New York (Wiley).

Anderton, R. (1976): 'Tidal-Shelf Sedimentation: An Example from the Scottish Dalradian', *Sedimentology*, **23**, 429–458, 19 figs., 3 tables; Oxford.

Anonymous (1980): *Eastern Gas Shales Project – Increasing U.S. Natural Gas Reserves from Eastern Gas-Bearing Shales. A Special Publication on Unconventional Gas from Shales*, rev. ed., 20 pp., unnumbered figs. and tables; Morgantown, West Virginia (Sci. Applications, Inc.).

Anstey, N. A. (1980): *Seismic Exploration for Sandstone Reservoirs*, 136 pp., 81 figs., 2 tables; Boston (Intern. Human Resources Develop. Corp.).

Aoyagi, K. and Asakawa, T. (1984): 'Paleotemperature Analysis by Authigenic Minerals and Its Application to Petroleum Exploration', *Amer. Assoc. Petroleum Geologists, Bull.*, **68**, 903–913, 11 figs., 2 tables; Tulsa, Okla.

ARAMCO (1959): 'Ghawar Oil Field, Saudi Arabia', *Amer. Assoc. Petroleum Geologists, Bull.*, **43**, 434–454, 8 figs., 2 tables; Tulsa, Okla.

Aswathanarayana, U. (1985): *Principles of Nuclear Geology*, 397 pp., 126 figs., 40 tables, appendix A–F; Rotterdam (Balkema).

353

Ayres, M. G., Bilal, M., Jones, R. W., Slentz, L. W., Tartir, M., and Wilson, A. O. (1982): 'Hydrocarbon Habitat in Main Producing Areas, Saudi Arabia', *Amer. Assoc. Petroleum Geologists, Bull.*, **66**, 1–9, 14 figs.; Tulsa, Okla.

Aziz, K. and Settari, A. (1979): *Petroleum Reservoir Simulation*, 475 pp., 143 figs., 7 tables, 2 appendices, with exercises; London, (Applied Science Publ.).

Bachmann, G. H. and Grosse, S. (1989): 'Struktur und Entstehung des Nordddeutschen Beckens. Geologische und geophysikalische Interpretation einer verbesserten Bouguer-Schwerekarte', 23–47, 19 figs., 1 table, Karte 'Verbesserte Bouguer-Schwere des Prä-Zechstein', 1 : 500 000; in: *Das Norddeutsche Becken – Geophysikalische Untersuchungen des tieferen Untergrundes. Niedersächsische Akad. Geowiss.*, Veröff. 3: 47 pp., 32 figs., 1 table, 3 maps; Hannover.

Bagnold, R. A. (1954): *The Physics of Blown Sand and Desert Dunes*, 265 pp., 16 pls., 84 diagrams; London (Methuen).

Bailey, E. H. and Stevens, R. E. (1960): 'Selective Staining of K-Feldspar and Plagioclase on Rock Slabs and in Thin Sections', *Amer. Mineralogist*, **45**, 1020–1025, 1 fig.; Menasha, Wisc.

Bally, A. W. (ed.) (1987): *Atlas of Seismic Stratigraphy, Vol. 1*, Amer. Assoc. Petroleum Geologists, Studies in Geology, Vol. 27, 125 pp., 179 figs., 1 table; Tulsa, Okla.

Banner, F. T., Collins, M. B., and Massie, K. S. (eds.) (1979): *The North-West European Shelf Seas; The Sea Bed and the Sea in Motion. I. Geology and Sedimentology*, Elsevier Oceanographic Series, Vol. 24A: 300 pp., 163 figs., 5 tables; Amsterdam (Elsevier).

Bard, J. P. (1980): *Microtextures des Roches Magmatiques et Métamorphiques*, 192 pp., 120 figs., 7 tables; Paris, New York, Barcelone, Milan (Masson).

Bark, E. van den and Thomas, O. D. (1980): 'Ekofisk: First of the Giant Oil Fields in Western Europe', 195–224, 26 figs., 3 tables'; in: M. T. Halbouty (ed.), *Giant Oil and Gas Fields in the Decade 1968–1978*, 596 pp., 514 figs., 38 tables; Tulsa, Okla.

Barnard, P. C. and Cooper, B. S. (1981): 'Oils and Source Rocks of the North Sea Area', 169–175, 4 figs.; in: L. V. Illing and G. D. Hobson (eds.), *Petroleum Geology of the Continental Shelf of North-West Europe*, 521 pp., 379 figs., 14 tables, 7 pls., 3 appendices; London (Institute of Petroleum).

Barnetche, A. and Illing, L. V. (1956): 'The Tamabra Limestone of the Poza Rica Oilfield', 38 pp., 14 figs., *20th Congr. Geol. Intern. Mexico 1956*; Mexico.

Bartenstein, H. and Teichmüller, R. (1974): 'Inkohlungsuntersuchungen, ein Schlüssel zur Prospektierung von paläozoischen Kohlenwasserstoff-Lagerstätten?', *Fortschr. Geol. Rheinld. u. Westf.*, **24**, 129–160, 17 figs. 1 table; Krefeld.

Barwis, J. H., McPherson, J. G., and Studlick, J. R. J. (eds.) (1990): *Sandstone Petroleum Reservoirs*, 583 pp., 443 figs., 20 tables, 48 color pls., appendix; New York (Springer).

Basan, P. B. (ed.) (1978): *Trace Fossil Concepts*, Soc. Econ. Paleontologists Mineralogists, Short course, No. 5, 181 pp., 176 figs., 7 tables, 1 chart; Tulsa, Okla.

Bates, R. L. and Jackson, J. A. (eds.) (1987): *Glossary of Geology*, 3rd ed., 788 pp.; Alexandria, Va.

Bathurst, R. G. C. (1975): *Carbonate Sediments and Their Diagenesis*, 2nd enlarged ed., 658 pp., 359 figs.; Amsterdam (Elsevier).

Beaumont, C. and Tankard, A. J. (eds.) (1987): *Sedimentary Basins and Basin-Forming Mechanisms*, Canad. Soc. Petroleum Geologists, Mem. 12, 527 pp., 375 figs., 32 tables, 5 appendices; Calgary. (Atlantic Geoscience Soc., Special Publ. No. 5: Halifax.)

Beckmann, H. (1976): 'Geological Prospecting of Petroleum'; in: H. Beckmann (ed.), *Geology of Petroleum*, Vol. 2, 183 pp., 110 figs.; Stuttgart (Enke).

Belderson, R. H., Kenyon, N. H., Stride, A. H., and Stubbs, A. R. (1972): *Sonographs of the Sea Floor. A Picture Atlas*, 185 pp., 163 figs., 1 table; Amsterdam (Elsevier).

Berg, R. R. (1975): 'Depositional Environment of Upper Cretaceous Sussex Sandstone, House Creek field, Wyoming', *Amer. Assoc. Petroleum Geologists, Bull.*, **59**, 2099–2110, 8 figs., 2 tables; Tulsa, Okla.

Berg, R. R. (1986): *Reservoir Sandstones*, 481 pp., 247 figs., 39 tables; Englewood Cliffs, N.J. (Prentice-Hall).

Berger, W. H. (1974): 'Deep-Sea Sedimentation', 213–241, 23 figs., 3 tables; in: C. A. Burk and C. L. Drake (eds.), *The Geology of Continental Margins*, 1009 pp., 636 figs., 79 tables; Berlin (Springer).

Berner, R. A. (1980): *Early Diagenesis – A Theoretical Approach*, 241 pp., 62 figs., 17 tables; Princeton, NJ (Princeton University Press).

Berner, R. A. (1981): 'A New Geochemical Classification of Sedimentary Environment', *J. Sediment. Petrol.*, **51**, 359–366, 1 table, 2 photos; Tulsa, Okla.

Berners, H. P. (1983): 'A Lower Liassic Offshore Bar Environment, Contribution to the Sedimentology of the Luxemburg Sandstone', *Ann. Soc. géol. de Belgique*, **106**, 87–102, 10 figs., 2 tables; Brussels.

Berners, H.P., Hendriks, F., Muller, A., and Schrader, E. (1983): 'Faktorenanalytischer Vergleich mesozoischer Sedimenttypen der Eifeler Nord-Süd-Zone und vom E-Rand des Pariser Beckens', *Jber. Mitt. oberrhein. geol. Ver., N. F.*, **65**, 143–166, 6 figs., 4 tables; Stuttgart.

Bernoulli, D. and Jenkyns, H. C. (1974): 'Alpine, Mediterranean, and Central Atlantic Mesozoic Facies in Relation to the Early Evolution of the Tethys', 129–160, 8 figs.; in: R. H. Dott and R. H. Shaver (eds.), *Modern and Ancient Geosynclinal Sedimentation*, Soc. Econ. Paleontologists Mineralogists, Spec. Publ. 19, 380 pp., 133 figs., 21 tables, 1 appendix; Tulsa, Okla.

Betz, D. (1990): 'Neue Maßtäbe durch geowissenschaftliche Hochtechnologie (New Dimensions in Earth Sciences with High Technology)', *Erdöl Erdgas Kohle*, **106**, 471–477, 11 figs.; Hamburg, Wien.

Beydoun, Z. R. (1988): *The Middle East: Regional Geology and Petroleum Resources*, 292 pp., 30 figs., 7 tables; Beaconsfield, Bucks., U.K. (Scientific Press).

Bhatia, M. R. (1983): 'Plate Tectonics and Geochemical Composition of Sandstones', *J. Geol.*, **91**, 611–627, 7 figs., 10 tables; Chicago, Ill.

Bigarella, J. J. (1972): 'Eolian Environments: Their Characteristics, Recognition and Importance', 12–62, 38 figs., 6 tables; in: J. K. Rigby and W. K. Hamblin (eds.), *Recognition of Ancient Sedimentary Environments*, Soc. Econ. Paleontologists Mineralogists, Spec. Publ. 16, 340 pp., 217 figs., 30 tables; Tulsa, Okla.

Bish, D. L. and Post, J. E. (eds.) (1989): 'Modern Powder Diffraction', *Reviews in Mineralogy*, **20**, 369 pp., 167 figs., 30 tables; Washington, D.C. (Mineral. Soc. America).

Bitterli, P. (1962): 'Untersuchung bituminöser Gesteine von Westeuropa', *Erdöl u. Kohle*, **15**, 2–6, 2 figs.; Hamburg.

Bjørlykke, K. (1989): *Sedimentology and Petroleum Geology*, 363 pp., 186 figs., 20 tables; Berlin (Springer).

Bjørlykke, K., Dypvik, H., and Finstad, K. G. (1975): 'The Kimmeridgian Shale, Its Composition and Radioactivity', Paper No. 12, 20 pp., 12 figs., 1 table; in: K. G. Finstad and R. C. Selley (eds.), *Jurassic Northern North Sea Symposium*, 1975, Proc., 510 pp., 194 figs., 6 tables, 4 pls.; Stavanger (Norwegian Petroleum Society).

Blatt, H. (1982): *Sedimentary Petrology*, 564 pp., 338 figs., 26 tables; San Francisco (Freeman).

Blatt, H., Berry, W. B. N., and Brande, S. (1991): *Principles of Stratigraphic Analysis*, 512 pp., 253 figs., 24 tables; Boston (Blackwell Scientific Publications).

Blind, W. (1964): 'Der Lithothamnienkalk der ostbayerischen Molasse und seine Eigenschaften als Trägergestein', *Erdöl u. Kohle*, **17**, 341–345, 13 figs.; Hamburg.

Blundell, D. J. and Gibbs, A. D. (eds.) (1990): *Tectonic Evolution of the North Sea Rifts*, 272 pp., 144 figs., 7 tables, 3 pls., Bouguer Gravity Anomalies 1 : 4 000 000; Oxford (Clarendon Press).

Bobrovnik, I. I. and Isakov, N. G. (1978): 'Method of Seismoenergetic Mapping', *Problems of Oil and Gas in Tyumen*, Proc. 39..

Boenigk, W. (1983): *Schwermineralanalyse*, 158 pp., 77 figs., 8 tables, 4 pls.; Stuttgart (Enke).

Boersma, J. R. and Terwindt, J. H. J. (eds.) (1983): 'Basin Analysis and Sedimentary Facies: Sedimentology at Various Scales', *Sediment. Geol.*, **34**, 83–265, 78 figs., 1 table; Amsterdam.

Boigk, H. (1981): *Erdöl und Erdgas in der Bundesrepublik Deutschland. Erdölprovinzen, Felder, Förderung, Vorräte, Lagerstättentechnik*, 330 pp., 109 figs., 51 tables, 2 folded pls.; Stuttgart (Enke).

Boigk, H., Hagemann, H. W., Stahl, W., and Wollanke, G. (1976): 'Isotopenphysikalische Untersuchungen. Zur Herkunft und Migration des Stickstoffs nordwestdeutscher Erdgase aus Oberkarbon und Rotliegend', *Erdöl u. Kohle*, **29**, 103–112, 16 figs., 5 tables; Leinfelden.

Bois, C. and Monicard, R. (1981): 'Pétrole: peut-on encore decouvrir des gisements géants', *La Recherche*, **124**, 854–861, 6 figs.; Paris.

Bortfeld, R. K. (1983): 'New Developments in Exploration Seismics', *Geol. Jb.*, **E 26**, 5–33, 23 figs.; Hannover.

Bouma, A. H. (1962): *Sedimentology of Some Flysch Deposits; A Graphic Approach to Facies Interpretation*, 168 pp., 31 figs., 17 tables, 8 pls., 3 enclosures; Amsterdam (Elsevier).

Bouma, A. H. (1969): *Methods for the Study of Sedimentary Structures*, 458 pp., 229 figs., 38 tables, 5 appendices; New York (Wiley-Interscience).

Bourgeois, J. (1978): 'Conglomerates', 183–186, 3 figs.; in: R. W. Fairbridge and J. Bourgeois (eds.), *The Encyclopedia of Sedimentology*, 901 pp.; Stroudsburg, Pa. (Dowden, Hutchinson & Ross Inc.).

Bowen, R. (1988): *Isotopes in the Earth Sciences*, 647 pp., 63 figs., 59 tables; London (Elsevier).

Boyles, J. M. and Scott, A. J. (1982): 'A Model for Migrating Shelf-Bar Sandstones in Upper Mancos Shale

(Campanian), Northwestern Colorado', *Amer. Assoc. Petroleum Geologists, Bull.*, **66**, 491–508, 17 figs., 2 tables; Tulsa, Okla.

Bradel, E. and Draxler, J. (1982): 'Erweiterte Computerauswertungen von Bohrlochmessungen in Rotliegend-sedimenten Nordwestdeutschlands (Ost-Hannover)', *Erdöl Erdgas Z.*, **98**, 16–24, 13 figs., 3 tables; Hamburg.

Bramkamp, R. A. and Powers, R. W. (1958): 'Classification of Arabian Carbonate Rocks', *Geol. Soc. America, Bull.*, **69**, 1305–1318, 2 tables; New York.

Braun, A. (1994): 'Zum Wert der Auflichtmikroskopie als sedimentpetrographisches Untersuchungsverfahren', *Mitt. Wiss. Technik*, **10**, 245–252, 18 figs.; Wetzlar (Leitz).

Brenchley, P. J. and Williams, B. P. J. (eds.) (1985): *Sedimentology. Recent Developments and Applied Aspects*, Geol. Soc. London, Spec. Publ. 18, 342 pp., 189 figs. (8 in color), 8 tables; Oxford (Blackwell Scientific Publ.).

Brenner, R. L. and Davies, D. K. (1973): 'Storm-Generated Coquinoid Sandstone: Genesis of High-Energy Marine Sediments from the Upper Jurassic of Wyoming and Montana', *Geol. Soc. America, Bull.*, **84**, 1685–1697, 14 figs., 2 tables; Boulder, Colo.

Brenner, R. L. and Davies, D. K. (1974): 'Oxfordian Sedimentation in Western Interior United States', *Amer. Assoc. Petroleum Geologists, Bull.*, **58**, 407–428, 25 figs., 1 table; Tulsa, Okla.

Brindley, G. W. and Brown, G. (eds.) (1980): *Crystal Structures of Clay Minerals and Their X-Ray Identification*, Mineral. Soc. Monograph No. 5, 495 pp., 113 figs., 104 tables, appendix with tables for the determination of d; London (Mineral. Soc.).

Brodzikowski, K. and Loon, A. J. van (1991): 'Glacigenic Sediments', *Developments in Sedimentology*, **49**, 674 pp., 321 figs., 4 tables; Amsterdam (Elsevier).

Bromley, R. G. (1990): *Trace Fossils – Biology and Taphonomy*, 280 pp., 157 figs.; London (Unwin Hyman).

Brongersma-Sanders, M. (1971): 'Origin of Major Cyclicity Evaporites and Bituminous Rocks: An Actualistic Model', *Mar. Geol.*, **11**, 123–144, 5 figs., 1 table; Amsterdam.

Brookfield, M. E. and Ahlbrandt, T. S. (1983): 'Eolian Sediments and Processes', *Developments in Sedimentology*, **38**, 660 pp., 371 figs., 27 tables; Amsterdam (Elsevier).

Brookins, D. G. (1988): *Eh-pH Diagrams for Geochemistry*, 176 pp., 98 figs., 61 tables; Berlin (Springer).

Brooks, J. and Welte, D. H. (1984): *Advances in Petroleum Geochemistry 1*, 344 pp., 126 figs., 5 tables; London (Academic Press).

Brooks, J. and Welte, D. H. (1987): *Advances in Petroleum Geochemistry 2*, 262 pp., 80 figs., 12 tables; London (Academic Press).

Bubb, J. N. and Hatlelid, W. G. (1977): 'Seismic Stratigraphy and Global Changes of Sea Level, Part 10: Seismic Recognition of Carbonate Buildups', 185–204, 17 figs., 1 table; in: C. E. Payton (ed.), *Seismic Stratigraphy – Applications to Hydrocarbon Exploration*, Amer. Assoc. Petroleum Geologists, Mem. 26, 516 pp., 420 figs., 12 tables; Tulsa Okla.

Budny, M. (1991): 'Seismisch-lithostratigraphische Interpretation permischer Reservoirgesteine (NW-Deutsches Becken)', 201–219, 13 figs., 1 table; in: M. Albertsen (ed.), *Der tiefere Untergrund des nordwestdeutschen Beckens: Sedimentologie-Tektonik-Kohlenwasserstoffe – Beiträge der DGG-DGMK-Gemeinschaftstagung Braunschweig 1989*, 388 pp., 128 figs., 9 tables, 13 pls.; Hamburg (DGMK).

Burrus, J. (ed.) (1986): *Thermal Modeling in Sedimentary Basins*, 1st IFP Exploration Research Conf., Carcans, France, June 3–7, 1985, Collection Colloques et Séminaires, **44**, 600 pp., 273 figs., 23 tables, 2 appendices; Paris (Technip).

Bushinsky, G. I. (1935): 'Structure and Origin of the Phosphorites of the U.S.S.R.', *J. Sediment. Petrol.*, **5**, 81–92, 12 figs., 3 tables; Tulsa, Okla.

Busson, G. (ed.) (1988): *Evaporites et Hydrocarbures*, Sciences de la Terre, Mém. Mus. nat. Hist. nat., (C), **55**: 139 pp., 50 figs., 5 tables; Paris.

Bustin, R. M., Cameron, A. R., Grieve, D. A., and Kalkreuth, W. D. (1985): *Coal Petrology: Its Principles, Methods and Applications*, Geol. Assoc. Canada, Short course notes, 3: 2nd revised ed., 230 pp., 127 figs., 40 tables, 18 pls.; St. John's, Newfoundland, Canada.

Butler, G. P., Harris, P. M., and Kendall, C. G. St. C. (1982): 'Recent Evaporites from the Abu Dhabi Coastal Flats', 33–64, 27 figs.; in: C. R. Handford, R. G. Loucks, and G. R. Davies (eds.), *Depositional and Diagenetic Spectra of Evaporites – A Core Workshop*, Soc. Econ. Paleontologists Mineralogists, Core Workshop 3, 395 pp., 190 figs., 5 tables, 1 appendix with core photographs; Calgary.

Buttkus, B. (1985): 'Seismic HC Reservoir Prediction: A (Critical) Review on the Determination of Lithological Parameters from Seismic Data', *Geol. Soc. Malaysia, Bull.*, **18**, 133–150, 12 figs. (partly in color); Kuala

Lumpur.

Cailleux, A. (1952): 'Morphoskopische Analyse der Geschiebe und Sandkörner und ihre Bedeutung für die Paläoklimatologie', *Geol. Rundschau*, **40**, 11–19, 6 figs. 4 tables; Stuttgart.

Carmichael, R. S. (ed.) (1982/1984): *CRC Handbook of Physical Properties of Rocks*, Vol. 1 (1982) *Mineral. Composition of Rocks – Electrical Properties of Rocks and Minerals – Spectroscopic Properties of Rocks and Minerals*, 404 pp., 175 figs., 185 tables; Vol. 2 (1982) *Seismic Velocities – Magnetic Properties of Minerals and Rocks – Engineering Properties of Rocks*, 345 pp., 43 figs., 65 tables; Vol. 3 (1984) *Densities, Elastic Constants, Inelastic Properties and Radioactivity Properties of Rocks and Minerals – Seismic Attenuation*, 340 pp., 81 figs., 41 tables; Boca Raton, Florida (CRC Press, Inc.)

Carozzi, A. V. (1989): *Carbonate Rock Depositional Models. A Microfacies Approach*, 604 pp., 556 figs., 3 tables; Englewood Cliffs (Prentice Hall).

Cartwright, J. A., Haddock, R. C., and Pinheiro, L. M. (1993): 'The Lateral Extent of Sequence Boundaries', 15–34, 9 figs.; in: G. D. Williams and A. Dobb (eds.), *Tectonics and Seismic Sequence Stratigraphy*, Geol. Soc. London, Special Publ. 71, 225 pp., 115 figs., 3 tables; London.

Carver, R. E. (1971): *Procedures in Sedimentary Petrology*, 653 pp., 188 figs., 72 tables; New York (Wiley-Interscience).

Cayeux, L. (1916): *Introduction à l'Etude Pétrographique des Roches Sédimentaires. Mémoires pour Servir à l'Explication de la Carte Géologique Détaillée de la France*, Text: 524 pp., 80 figs., tables; Atlas: 56 pls.; Paris (Imprimerie Nationale).

Cayeux, L. (1929): *Les Roches Sédimentaires de France: Roches Siliceuses*, 774 pp., 17 figs., unnumbered tables, 30 pls.; Paris (Imprimerie Nationale).

Chamberlain, C. K. (1975): 'Recent Lebensspuren in Nonmarine Aquatic Environments'; in: R. W. Frey (ed.), *The study of Trace Fossils*, 431–458, 10 figs., 2 tables; New York (Springer).

Chamberlain, C. K. (1978): 'Recognition of Trace Fossils in Cores'; in: P. B. Basan (ed.), *Trace Fossil Concepts*, Soc. Econ. Paleontologists Mineralogists, Short course, No. 5; 119–166, 130 figs. 1 chart; Tulsa, Okla.

Chamley, H. (1989): *Clay Sedimentology*, 623 pp., 243 figs., 65 tables; Berlin (Springer).

Chamley, H. (1990): *Sedimentology*, 285 pp., 177 figs., 17 tables; Berlin (Springer).

Chapman, R. E. (1983): *Petroleum Geology*, Developments in Petroleum Science, Vol. 16, 416 pp., 1890 figs., 17 tables; Amsterdam (Elsevier).

Chappell, J. (1980): 'Coral Morphology, Diversity and Reef Growth', *Nature*, **286**, 249–252, 4 figs.; London.

Charlet, J. M. (1971): 'Thermoluminescence of Detrital Rocks Used in Paleogeographical Problems', *Modern Geol.*, **2**, 265–274, 14 figs.; New York.

Chilingarian, G. V. and Wolf, K. H. (ed.) (1988a): *Diagenesis, I*, Developments in Sedimentology, Vol. 41, 591 pp., 297 figs., 85 tables, glossary; Amsterdam (Elsevier).

Chilingarian, G. V. and Wolf, K. H. (eds.) (1988b): *Diagenesis, II*, Developments in Sedimentology, Vol. 43: 268 pp., 128 figs., 29 tables; Amsterdam (Elsevier).

Choquette, P. W. and Pray, L. C. (1970): 'Geologic Nomenclature and Classification of Porosity in Sedimentary Carbonates', *Amer. Assoc. Petroleum Geologists, Bull.*, **54**, 207–250, 13 figs., 3 tables; Tulsa, Okla.

Chung-Hsiang P'an (1982): 'Petroleum in Basement Rocks', *Amer. Assoc. Petroleum Geologists, Bull.*, **66**, 1597–1643, 36 figs., 1 table; Tulsa, Okla.

Coleman, J. M. (1981): *Deltas: Processes of Deposition and Models for Exploration*, 124 pp., 83 figs., 4 tables; Minneapolis (Burgess).

Coleman, J. M. and Gagliano, S. M. (1965): 'Sedimentary Structures: Mississippi River Deltaic Plain', 133–148, 9 figs., 1 table; in: G. V. Middleton (ed.), *Primary Sedimentary Structures and Their Hydrodynamic Interpretation*, Soc. Econ. Paleontologists Mineralogists, Spec. Publ. 12, 265 pp., 159 figs., 23 tables, 15 pls., with glossary of primary sedimentary structures; Tulsa, Okla.

Coleman, J. M. and Wright, L. D. (1975): 'Modern River Deltas: Variability of Processes and Sand Bodies', 99–149, 29 figs., 2 tables; in: M. L. Broussard (ed.), *Deltas, Models for Exploration*, 555 pp., 333 figs., 51 tables, Appendix: Ancient deltas: Comparison maps; Houston (Geol. Soc. Houston).

Collinson, J. D. (1986): 'Alluvial Sediments', Chapter 3, 20–62, 53 figs.; in: H. G. Reading (ed.), *Sedimentary Environments and Facies*, 2nd ed., 615 pp., 564 figs., 15 tables; Oxford (Blackwell).

Colombo, U., Gazzarrini, F., Gonfiantini, R., Kneuper, G., Teichmüller, M. and R. (1968): 'Das Verhältnis der stabilen Kohlenstoff-Isotope von Steinkohlen und kohlenbürtigem Methan in Nordwestdeutschland', *Z. angew. Geol.*, **14**, 257–265, 17 figs.; Berlin.

Conybeare, C. E. B. (1976): *Geomorphology of Oil and Gas Fields in Sandstone Bodies*, Developments in

Petroleum Science, Vol. 4, 341 pp., 152 figs.; Amsterdam (Elsevier).

Conybeare, C. E. B. (1979): *Lithostratigraphic Analysis of Sedimentary Basins*, 555 pp., 277 figs.; New York (Academic Press).

Cook, H. E. and Enos, P. (eds.) (1977): *Deep-Water Carbonate Environments*, Soc. Econ. Paleontologists Mineralogists, Spec. Publ. 25, 336 pp., 273 figs., 7 tables; Tulsa, Okla.

Cooke, R. U. and Warren, A. (1973): *Geomorphology in Deserts*, 374 pp., 100 figs., 16 tables, 80 pls.; London (B. T. Batsford Ltd.).

Cotillon, P. (1992): *Stratigraphy*, 187 pp., 115 figs.; Berlin (Springer).

Cracknell, A. P. and Hayes, L. W. B. (1991): *Introduction to Remote Sensing*, 293 pp., 133 figs., 18 tables, 3 appendices; London (Taylor & Francis).

Craig, F. F., Wilcox, P. J., Ballard, J. R., and Nation, W. R. (1977): 'Optimized Recovery through Continuing Interdisciplinary Cooperation', *J. Petroleum Technology*, **29**, 755–760, 5 figs.; Dallas, Tex.

Crevello, P. D., Wilson, J. L., Sarg, J. F., and Read, J. F. (eds.) (1989): *Controls on Carbonate Platform and Basin Development*, Soc. Econ. Paleontologists Mineralogists, Spec. Publ. 44, 379 pp., 273 figs.; Tulsa, Okla.

Crimes, T. P. and Harper, J. C. (eds.) (1970): *Trace Fossils (1)*, Spec. Issue, *Geol. J.*, **3**, 547 pp., 114 figs., 9 tables, 87 pls.; Liverpool (Seel House Press).

Crimes, T. P. and Harper, J. C. (eds.) (1977): *Trace Fossils (2)*, Spec. Issue, *Geol. J.*, **9**, 351 pp., 67 figs., 12 tables, 65 pls.; Liverpool (Seel House Press).

Curtis, C. D. (1977): 'Sedimentary Geochemistry: Environments and Processes Dominated by Involvement of an Aqueous Phase', in: J. E. T. Horne and K. Dunham (eds.), *Mineralogy: Towards the Twenty-First Century*, *Phil. Trans. Royal Soc. London (A)*, **286**, 353–372, 3 figs., 3 tables; London.

Curtis, C. D. (1978): 'Possible Links between Sandstone Diagenesis and Depth-Related Geochemical Reactions Occurring in Enclosing Mudstones', *J. Geol. Soc. London*, **135**, 107–117, 4 figs., 3 tables; London.

Curtis, R., Evans, G., Kinsman, D. J. J., and Shearman, D. J. (1963): 'Association of Dolomite and Anhydrite in the Recent Sediments of the Persian Gulf', *Nature*, **197**, 4868, 679–680, 1 fig.; London.

Cushman, J. H. (ed.) (1990): *Dynamics of Fluids in Hierarchical Porous Media*, 505 pp., 158 figs., 4 tables, 1 appendix; London (Academic Press).

Davis, R. A., Jr. (1983): *Depositional Systems. A Genetic Approach to Sedimentary Geology*, 669 pp., 447 figs., 12 tables; Englewood Cliffs, N.Y. (Prentice-Hall).

Davis, R. A., Jr. (ed.) (1985): *Coastal Sedimentary Environments*, 2nd revised, expanded ed., 716 pp., 376 figs., 21 tables; New York, Berlin, Heidelberg, Tokyo (Springer).

Davies, T. A. and Gorsline, D. S. (1976): 'Oceanic Sediments and Sedimentary Processes', Chapter 24: 1–80, 30 figs.; in: J. P. Riley and R. Chester (eds.), *Chemical Oceanography*, 2nd ed., 5, 401 pp., 109 figs., 32 tables; London (Academic Press).

Davies, D. K., Ethridge, F. G., and Berg, R. R. (1971): 'Recognition of Barrier Environment', *Amer. Assoc. Petroleum Geologists, Bull.*, **55**, 550–565, 15 figs., 3 tables; Tulsa, Okla.

Degens, E. T. (1968): *Geochemie der Sedimente*, 282 pp., 75 figs., 20 tables; Stuttgart (Enke).

Demaison, G. J. and Moore, G. T. (1980): 'Anoxic Environments and Oil Source Bed Genesis', *Amer. Assoc. Petroleum Geologists, Bull.*, **64**, 1179–1209, 18 figs.; Tulsa, Okla.

Desbrandes, R. (1985): *Encyclopedia of Well Logging*, 584 pp., 376 figs., 61 tables; Paris (Technip).

Dickey, P. A. (1986): *Petroleum Development Geology*, 3rd ed., 530 pp., 371 figs., 18 tables; Tulsa, Okla. (PennWell Books).

Doeglas, D. J. (1946): 'Interpretation of the Results of Mechanical Analyses', *J. Sediment. Petrol.*, **16**, 19–40, 30 figs., 1 table; Tulsa, Okla.

Dohr, G. (1981): 'Applied Geophysics. Introduction to Geophysical Prospecting', 2nd completely revised ed., 232 pp., 166 figs.; in: H. Beckmann (ed.), *Geology of Petroleum*, Vol. 1; Stuttgart (Enke).

Donaldson, E. C. (ed.) (1991): *Microbial Enhancement of Oil Recovery – Recent Advances*, Developments in Petroleum Science, Vol. 31, 530 pp., 153 figs., 108 tables, appendix with 3 tables and 29 diagrams; Amsterdam (Elsevier).

Donaldson, E. C., Chilingarian, G. V., and Yen, T. F. (eds.) (1985): *Enhanced Oil Recovery, I. Fundamentals and Analyses*, Developments in Petroleum Science, Vol. 17A, 358 pp., 90 figs., 38 tables; Amsterdam (Elsevier).

Donaldson, E. C., Chilingarian, G. V., and Yen, T. F. (eds.) (1989a): *Enhanced Oil Recovery, II. Process and Operations*, Developments in Petroleum Science, Vol. 17B: 604 pp., 241 figs., 51 tables, program listing; Amsterdam (Elsevier).

Donaldson, E. C., Chilingarian, G. V., and Yen, T. F. (eds.) (1989b): *Microbial Enhanced Oil Recovery*, Developments in Petroleum Science, Vol. 22: 227 pp., 62 figs., 31 tables; Amsterdam (Elsevier).

Doveton, J. H. (1989): 'Lithofacies and Geochemical-Facies Profiles from Modern Wireline Logs – New Subsurface Templates for Sedimentary Modeling', 55–56, 1 fig.; in: E. K. Franseen and W. L. Watney (eds.), *Sedimentary Modeling: Computer Simulation of Depositional Sequences*, Kansas Geol. Survey, Subsurface Geol. Ser., 12, 84 pp., 41 figs., 2 tables; Lawrence, Ka.

Dow, W. G. (1974): 'Application of Oil-Correlation and Source-Rock Data to Exploration in Williston Basin', *Amer. Assoc. Petroleum Geologists, Bull.*, **58**, 1253–1262, 16 figs.; Tulsa, Okla.

Dow, W. G. (1977): 'Kerogen Studies and Geological Interpretations', *J. Geochem. Exploration*, **7**, 79–99, 11 figs., 4 tables; Amsterdam.

Dowdeswell, J. A. and Scourse, J. D. (eds.) (1990): *Glacimarine Environments: Processes and Sediments*, Geol. Soc., Spec. Publ. 53, 423 pp., 259 figs., 30 tables, 1 appendix; London.

Drew, L. J. (1990): *Oil and Gas Forecasting. Reflections of a Petroleum Geologist*, Studies in Math. Geol., Vol. 2: 288 pp., 107 figs.; New York/Oxford (Intern. Assoc. Mathematical Geol.).

Drong, H. J. (1979): 'Diagenetische Veränderungen in den Rotliegend Sandsteinen im NW-Deutschen Becken', *Geol. Rundschau*, **68**, 1172–1183, 8 figs.; Stuttgart.

Drong, H. J., Plein, E., Sannemann, D., Schuepback, M. A., and Zimdars, J. (1982): 'Der Schneverdingen-Sandstein des Rotliegenden – eine äolische Sedimentfüllung alter Grabenstrukturen', *Z. dt. geol. Ges.*, **133**, 699–725, 9 figs. 1 table., 5 pls.; Hannover.

Du Dresnay, R. (1977): 'Le milieu récifal fossile du Jurassique inférieur (Lias) dans le domaine des Chaînes atlasiques du Maroc': 2ᵉ Symposium Int. sur.les Coraux et Récifs Coralliens Fossiles, *Mém. Bur. Rech. Géol. Min.*, **89**, 296–312, 8 figs., 4 pls.; Paris.

Dunbar, C. O. and Rodgers, J. (1958): *Principles of Stratigraphy*, 3rd pr., 356 pp., 123 figs., 21 tables; New York (Wiley).

Duncan, D. C. (1967): 'Geologic Setting of Oil Shale Deposits and World Prospects', *7th World Petroleum Congr. Mexico*, Proc., **3**, 659–667, 5 figs.; Barking, Essex (Elsevier).

Dunham, R. J. (1962): 'Classification of Carbonate Rocks According to Depositional Texture', 108–121 pp., 7 pls.; in: W. E. Ham (ed.), *Classification of Carbonate Rocks*, Amer. Assoc. Petroleum Geologists, Mem. 1: 279 pp., 74 figs., 19 tables, 41 pls., 1 chart with appendices; Tulsa, Okla.

Dunoyer de Segonzac, G. (1970): 'The Transformation of Clay Minerals during Diagenesis and Low-Grade Metamorphism: A Review', *Sedimentology*, **15**, 281–346, 15 figs., 1 table; Amsterdam.

Durand, B. and Espitalié, J. (1976): 'Geochemical Studies on the Organic Matter from the Douala Basin (Cameroon) – II. Evolution of Kerogen', *Geochim. Cosmochim. Acta*, **40**, 801–808, 5 figs., 4 tables; Oxford.

Eckhardt, F. J. (1979): 'Der permische Vulkanismus Mitteleuropas', *Geol. Jb.*, **D 35**, 84 pp., 12 figs., 5 tables; Hannover.

Economides, M. J. (1992): *A Practical Companion to Reservoir Stimulation*, Developments in Petroleum Science, Vol. 24, 220 pp., 140 figs., 73 tables; Amsterdam (Elsevier).

Eden, D. N. and Furkert, R. J. (eds.) (1988): 'Loess: Its Distribution, Geology and Soils', *Proc. Int. Symp. on Loess*, New Zealand, 14–21 February 1987, 245 pp., 81 figs., 18 tables, 3 appendices; Rotterdam (Balkema Publ.).

Einsele, G. (1992): *Sedimentary Basins. Evolution, Facies, and Sediment Budget*, 628 pp., 269 figs.; Berlin (Springer).

Eiserbeck, W., Franke, D., Harff, J., Hoffmann, N., Hoth, K., Müller, E. P., and Springer, J. (1990): 'Geologie und Kohlenwasserstoff-Erkundung im Präzechstein der DDR – Nordostdeutsche Senke', *Niedersächsische Akad. Geowiss. Veröff.*, **4**, 4–95, 39 figs., 3 tables; Hannover.

Ekdale, A. A. and Bromley, R. G. (1984): 'Sedimentology and Ichnology of the Cretaceous-Tertiary Boundary in Denmark: Implications for the Causes of the Terminal Cretaceous Extinction', *J. Sediment. Petrol.*, **54**, 681–703, 13 figs., 3 tables; Tulsa, Okla.

Elliott, T. (1986a): 'Deltas', Chapter 6: 113–154, 47 figs.; in: H. G. Reading (ed.), *Sedimentary Environment and Facies*, 2nd ed., 615 pp., 564 figs., 15 tables; Oxford (Blackwell).

Elliott, T. (1986b): 'Siliciclastic Shorelines', Chapter 7: 155–188, 42 figs.; in: H. G. Reading (ed.), *Sedimentary Environment and Facies*, 2nd ed., 615 pp., 564 figs., 15 tables; Oxford (Blackwell).

Ellis, D. V. (1987): *Well Logging for Earth Scientists*, 532 pp., 378 figs., 22 tables; New York, Amsterdam, London (Elsevier).

Ely, J. W. (1985): *Stimulation Treatment Handbook*, 232 pp., 36 figs., 18 tables, 5 appendices; Tulsa, Okla. (PennWell Books).

Emery, K. O. (1968): 'Relict Sediments on Continental Shelves of the World', *Amer. Assoc. Petroleum Geologists, Bull.*, **52**, 445–464, 16 figs.; Tulsa, Okla.

Enos, P. (1977): 'Tamabra Limestone of the Poza Rica Trend. Cretaceous, Mexico', 273–314, 37 figs., 3 tables, in: H. E. Cook and P. Enos (eds.), *Deep-Water Carbonate Environments*, Soc. Econ. Paleontologists Mineralogists, Spec. Publ. 25, 336 pp., 273 figs., 7 tables; Tulsa, Okla.

Enos, P. (1988): 'Evolution of Pore Space in the Poza Rica Trend (Mid-Cretaceous), Mexico', *Sedimentology*, **35**, 287–325, 33 figs., 6 tables; Oxford.

Enos, P. and Moore, C. H. (1983): 'Fore-Reef Slope Environment', Chapter 10. 507-537, 36 figs.; in: P. A. Scholle, D. G. Bebout, and C. H. Moore (eds.), *Carbonate Depositional Environments*, Amer. Assoc. Petroleum Geologists, Mem. 33, 708 pp., 1065 figs., 17 tables; Tulsa, Okla.

Eyles, N. and Miall, A. D. (1984): 'Glacial Facies', in: R. G. Walker (ed.), *Facies Models*, Geoscience Canada Reprint Series, Vol 1, 317 pp., 357 figs., 17 tables; Toronto, Ontario.

Faber, E. and Stahl, W. (1984): 'Geochemical Surface Exploration for Hydrocarbons in North Sea', *Amer. Assoc. Petroleum Geologists, Bull.*, **68**, 363–386, 11 figs., 5 tables; Tulsa, Okla.

Färber, A. (1984): 'Die Temperaturverteilung im Bereich des Salzstockes Dethlingen', *Erdöl Erdgas Z.*, **100**, 276–279, 8 figs., 2 tables; Hamburg, Wien.

Farouq Ali, S. M. (1970): *Oil Recovery by Steam Injection*, 122 pp., 149 figs., 20 tables, Bradford, Pa. (Producers Publ. Co.).

Feazel, C. T. and Farrell, H. E. (1988): 'Chalk from the Ekofisk Area, North Sea, Nannofossils + Micropores = Giant Fields', 155–178, 14 figs.; in: A. J. Lomando and P. M. Harris, *Giant Oil and Gas Fields, A Core Workshop*, SEPM Core Workshop No. 12, Houston, March 19–20, 1988, 853 pp., 403 figs., 16 tables; Tulsa, Okla.

Feigl, F. and Anger, V. (1972): *Spot Tests in Inorganic Analysis*, 6th Engl. ed., completely revised and enlarged, 669 pp., 42 figs., 5 tables; Amsterdam (Elsevier).

Feniak, M. W. (1944): 'Grain Sizes and Shapes of Various Minerals in Igneous Rocks', *Amer. Mineralogist*, **29**, 415–421, 1 fig., 3 tables; Menasha, Wisc.

Ferrar, H. T. (1934): 'The Geology of Dargaville – Rodney Subdivision', *New Zealand Geol. Survey, Bull.*, **34**, 78 pp.; Wellington.

Fertl, W. H. and Timko, D. J. (1970): 'Occurrence and Significance of Abnormal-Pressure Formations', *Oil Gas J.*, Jan. 5, 97–108, 4 figs., 2 tables; Tulsa, Okla.

Fisk, H. N., McFarlan, E., Jr., Kolb, C. R., and Wilbert, L. J. (1954): 'Sedimentary Framework of the Modern Mississippi Delta', *J. Sediment. Petrol.*, **24**, 76–99, 15 figs., 3 tables; Tulsa, Okla.

Fisk, S. (1985): 'A Staining Technique for Barium Silicates in Thin Section', *Miner. Magazine*, **49**, 614–615; London.

FISONS Instruments (1991): *J and W Kapillaren und Zubehör. Katalog '92*, 144 pp., unnumbered figs. and tables; Mainz-Kastel.

Fleet, A. J., Kelts, K., and Talbot, M. R. (eds.) (1988): *Lacustrine Petroleum Source Rocks*, Geol. Soc., London, Spec. Publ. 40, 391 pp., 213 figs., 40 tables, 4 pls.; Oxford (Blackwell).

Flügel, E. (1982): *Microfacies Analysis of Limestones*, 633 pp., 78 figs., 58 tables, 53 pls.; Berlin, Heidelberg, New York (Springer).

Folk, R. L. (1962): 'Spectral Subdivision of Limestone Types', 62–84, 7 figs., 3 tables, 1 pl.; in: W. E. Ham (ed.), *Classification of Carbonate Rocks*, Amer. Assoc. Petroleum Geologists, Mem. 1, 279 pp., 74 figs., 19 tables, 41 pls., 1 chart, with appendices; Tulsa, Okla.

Folk, R. (1968): *Petrology of Sedimentary Rocks*, 170 pp., unnumbered figs. and tables; Austin, Tex. (Hemphill).

Folk, R. (1974): *Petrology of Sedimentary Rocks*, 182 pp.; Austin, Tex. (Hemphill).

Folk, R. and Ward, W. C. (1957): 'Brazos River Bar: A Study in the Significance of Grain Size Parameters', *J. Sediment. Petrol.*, **27**, 3–26, 19 figs., 2 tables; Tulsa, Okla.

Fothergill, C. A. (1955): 'The Cementation of Oil Reservoir Sands and Its Origin', *Proc. 4th World Petroleum Congr.*, Sect. I/B, Geology/Geophysics, 301–314, 5 figs.; Rome (C. Colombo).

Fraser, G. S. (1989): *Clastic Depositional Sequences. Processes of Evolution and Principles of Interpretation*, 459 pp., 261 figs.; Englewood Cliffs, N.J. (Prentice Hall).

Frey, R. W. (ed.) (1975): *The Study of Trace Fossils*, 562 pp., 251 figs., 18 tables; New York (Springer).

Frey, R. W., Pemberton, S. G., and Saunders, T. D. (1990): 'Ichnofacies and Bathymetry: A Passive Relationship', *J. Paleont.*, **64**, 155–158, 1 fig.; Lawrence, Kan.

Friedman, G. M. (1978): 'Staining Techniques', 764–765, 1 figs; in: R. W. Fairbridge and J. Bourgeois (eds.), *The Encyclopedia of Sedimentology*, 911 pp.; Stroudsburg, Pa. (Dowden, Hutchinson & Ross).

Fripiat, J. J. (ed.) (1982): *Advanced Techniques for Clay Mineral Analysis*, Developments in Sedimentology, Vol. 34: 236 pp., 62 figs., 19 tables, 1 appendix; Amsterdam (Elsevier).

Fruit, D. J. and Elmore, R. D. (1988): 'Tide and Storm-Dominated Sand Ridges on a Muddy Shelf: Cottage Grove Sandstone (Upper Pennsylvanian), Northwestern Oklahoma', *Amer. Assoc. Petroleum Geologists, Bull.*, **72**, 1200–1211, 13 figs; Tulsa, Okla.

Füchtbauer, H. (1964a): 'Fazies, Porosität und Gasinhalt der Karbonatgesteine des norddeutschen Zechsteins', *Z. dt. geol. Ges. (für 1962)*, **114**, 484–531, 10 figs., 3 tables, 3 pls.; Hannover.

Füchtbauer, H. (1964b): 'Sedimentpetrographische Untersuchungen in der älteren Molasse nördlich der Alpen', *Eclogae geol. helv.*, **57**, 157–298, 20 figs., 12 tables; Basel.

Füchtbauer, H. (1968): 'Carbonate Sedimentation and Subsidence in the Zechstein Basin (Northern Germany)', 196–204, 4 figs.; in: G. Müller and G. M. Friedman, *Recent Developments in Carbonate Sedimentology in Central Europe*, 255 pp., 168 figs., 19 tables; Berlin, Heidelberg, New York (Springer).

Füchtbauer, H. (1974): *Sedimentary Petrology. 2. Sediments and Sedimentary Rocks*, 2nd revised and enlarged ed., P. 2, 1.3, 464 pp., 199 figs., 38 tables; Stuttgart (Schweizerbart).

Füchtbauer, H. (1988): *Sediment-Petrologie. 2. Sedimente und Sedimentgesteine*, 4., gänzlich neubearb. Aufl., Teil 2, 1141 pp., 660 figs., 113 tables; Stuttgart (Schweizerbart).

Füchtbauer, H. and Reineck, H.E. (1963): 'Porosität und Verdichtung rezenter, mariner Sedimente', *Sedimentology*, **2**, 294–306, 7 figs., 2 tables; Amsterdam (Elsevier).

Fülöp, A. (1991): 'Schichtfläche und Schichtung aus petrophysikalischer Sicht', *Zbl. Geol. Paläont.*, Teil 1, 1990: 895–909, 12 figs.; Stuttgart.

Fuex, A. N. (1977): 'The Use of Stable Carbon Isotopes in Hydrocarbon Exploration', *J. Geochem. Exploration*, **7**, 155–188, 9 figs., 11 tables; Amsterdam.

Gadow, S. and Reineck, H. E. (1969): 'Ablandiger Sandtransport bei Sturmfluten', *Senckenbergiana maritima [1]*, **50**, 63–78, 5 figs., 1 table, 1 pl.; Frankfurt a. M.

Gaida, K. H., Kessel, D. G., Volz, H., and Zimmerle, W. (1987): 'Geologic Parameters of Reservoir Sandstones as Applied to Enhanced Oil Recovery', *SPE Formation Evaluation*, March 1987, 89–96, 21 figs., 2 tables; Dallas, Tex.

Gaida, K.H., Rühl, W., and Zimmerle, W. (1973): 'Rasterelektronenmikroskopische Untersuchungen des Porenraumes von Sandsteinen. Zur Geometrie und Porenstruktur erdöl- und erdgasführender Sandsteine', *Erdöl Erdgas Z.*, **89**, 336–343, 11 figs.; Hamburg, Wien.

Galloway, W. E. (1968): 'Depositional Systems of the Lower Wilcox Group, North-Central Gulf Coast Basin', *Trans. Gulf Coast Assoc. Geol. Societies*, **18**, 275–289, 8 figs.; Houston, Texas.

Galloway, W. E. (1975): 'Process Framework for Describing the Morphologic and Stratigraphic Evolution of Deltaic Depositional Systems', 87–98, 5 figs., 2 tables; in: M. L. Broussard (ed.), *Deltas, Models for Exploration*, 555 pp., 333 figs., 51 tables, Appendix: Ancient deltas, comparison maps; Houston (Geol. Soc. Houston).

Garland, A. N. (1989): 'Microscopical Analysis of Fossil Bone', *Applied Geochemistry*, **4**, 215–229, 10 figs.; Oxford.

Garon, A. M. and Jennings, J. W. (1985): *Thermal Recovery Processes*, SPE Reprint Series, No. 7, Thermal Recovery Processes: 1985 edition, 478 pp., figs., tables; Richardson, Texas (Soc. Petroleum Engineers).

Garrels, R. M. (1960): *Mineral Equilibria at Low Temperature and Pressure*, 254 pp., 65 figs., 11 tables, appendix; New York (Harper & Broth.).

Garrels, R. M. and Christ, C. L. (1965): *Solutions, Minerals and Equilibria*, 450 pp., 128 figs., 26 tables, 2 appendices; New York (Harper & Row).

Gast, R. E. (1988): 'Rifting im Rotliegenden Niedersachsens', *Die Geowissenschaften*, **6**, 115–122, 14 figs.; Weinheim.

Gavish, E. (1974): 'Geochemistry and Mineralogy of a Recent Sabkha along the Coast of Sinai, Gulf of Suez', *Sedimentology*, **21**, 397–414, 14 figs., 2 tables; Oxford.

Gdula, J. E. (1983): 'Reservoir Geology, Structural Framework and Petrophysical Aspects of the De Wijk Gas Field', 191–202, 13 figs.; in: J. P. H. Kaasschieter and T. J. A. Reijers (eds.), *Petroleum Geology of the Southeastern North Sea and the Adjacent Onshore Areas* (The Hague, 1982), *Geol. en Mijnbouw*, **62**, 239 pp., 241 figs., 19 tables, 3 pls., 1 appendix; Den Haag.

Geyer, O. F. (1973): *Grundzüge der Stratigraphie und Fazieskunde. Bd. 1 – Paläontologische Grundlagen I. Das geologische Profil. Stratigraphie und Geochronologie*, 279 pp., 166 figs., 7 tables; Stuttgart (Schweizerbart).

Geyer, O. F. (1977): *Bd. 2 – Paläontologische Grundlagen II. Paläogeographie, Fazieskunde*, 341 pp., 190 figs., 18 tables; Stuttgart (Schweizerbart).

Geyh, M. A. and Schleicher, H. (1990): *Absolute Age Determination. Physical and Chemical Dating Methods and Their Application*, 503 pp., 146 figs., 19 tables, appendices; Berlin (Springer).

Gibbs, R. J. (1967): 'Quantitative X-Ray Diffraction Analysis Using Clay Mineral Standards Extracted from the Samples to Be Analysed', *Clay Minerals*, **7**, 79–90, 4 figs., 1 table; Oxford.

Glennie, K. W. (1970): *Desert Sedimentary Environments*, Developments in Sedimentology, Vol. 14: 222 pp., 146 figs.; Amsterdam (Elsevier).

Glennie, K. W. (1972): 'Permian Rotliegendes of Northwest Europe Interpreted in Light of Modern Desert Sedimentation Studies', *Amer. Assoc. Petroleum Geologists, Bull.*, **56**, 1048–1071, 18 figs.; Tulsa, Okla.

Glennie, K. W. (1982): 'Early Permian (Rotliegendes) Palaeowinds of the North Sea', *Sediment. Geol.*, **34**, 245–265, 9 figs.; Amsterdam.

Glennie, K. W. (1983): 'Lower Permian Rotliegend Desert Sedimentation in the North Sea Area', 521–541, 7 figs.; in: M. E. Brookfield and T. S. Ahlbrandt (eds.), *Eolian Sediments and Processes*, Developments in Sedimentology, Vol. 38: 660 pp., 371 figs., 27 tables; Amsterdam (Elsevier).

Glennie; K. W. (1987): 'Desert Sedimentary Environments, Present and Past – A Summary', *Sediment. Geol.*, **50**, 135–165, 12 figs.; Amsterdam.

Glennie, K. W. (1990): 'Lower Permian-Rotliegend', 120–152, 35 figs., 1 table; in: K. W. Glennie (ed.), *Introduction to the Petroleum Geology of the North Sea*, 3rd ed., 402 pp., 291 figs., 19 tables; London (Blackwell).

Glennie, K. W., Brooks, J., and Brooks, J. R. V. (1987): 'Hydrocarbon Exploration and Geological History of North West Europe', 1–10, 4 figs., 2 tables; in J. Brooks and K. W. Glennie (eds.), *Petroleum Geology of North West Europe*; London (Graham & Trotman).

Glennie, K. W. and Evans, G. (1976): 'A Reconnaissance of the Recent Sediments of the Ranns of Kutch, India', *Sedimentology*, **23**, 625–647, 11 figs., 1 table; Oxford.

Glennie, K. W., Mudd, G. C., and Nagtegaal, P. J. C. (1978): 'Depositional Environment and Diagenesis of Permian Rotliegendes Sandstones in Leman Bank and Sole Pit Areas of the UK Southern North Sea', *J. Geol. Soc. London*, **135**, 25–34, 4 figs., 2 pls.; London.

Goldstein, J. I., Newbury, D. E., Echlin, P., Joy, D. C., Fiori, C., and Lifshin, E. (1981): *Scanning Electron Microscopy and X-Ray Microanalysis. A Text for Biologists, Materials Scientists, and Geologists*, 673 pp., 345 figs., 65 tables; New York (Plenum).

Grabowska-Olszewska, B., Osipov, V., and Sokolov, V. (1984): *Atlas of the Microstructure of Clay Soils*, 414 pp., 86 sample descriptions with 602 figs., 10 pls.; Warszawa (Universytet Warszwawski, Panstwowe Wydawnictwo Naukowe).

Gralla, P. (1988): 'Das Oberrotliegende in NW-Deutschland – Lithostratigraphie und Faziesanalyse', *Geol. Jb.*, **A 106**, 3–59, 34 figs., 3 pls.; Hannover.

Greensmith, J. T. (1989): *Petrology of the Sedimentary Rocks*, 7th ed., 262 pp., 134 figs., 12 tables; London (Unwin Hyman).

Grim, R. E. (1968): *Clay Mineralogy*, 2nd ed., 596 pp., 138 figs., 71 tables; New York (McGraw-Hill).

Grotewold, G., Fuhrberg, H.D. and Philipp, W. (1979): 'Production and Processing of Nitrogen-Rich Natural Gases from Reservoirs in the NE Part of the Federal Republic of Germany', *Proc. 10th World Oil Congr.*, Bucharest 1979, Vol. 4, 47–54, 8 figs.; London–Rheine (Heyden).

Gümbel, C. W. von (1868): *Geognostische Beschreibung des ostbayerischen Grenzgebirges oder des bayerischen und oberpfälzer Waldgebirges*, Bayerisches Oberbergamt Geognostische Abteilung: Geognostische Beschreibung des Koenigreichs Bayern, Vol. 2, 968 pp. Gotha (Justus Perthes).

Häntzschel, W. (1975): 'Trace Fossils and Problematica'; in: C. Teichert (ed.), *Treatise on Invertebrate Paleontology*, 2nd ed. (revised and enlarged); Pt. W., Misc., Suppl. 1, Boulder, Co. and Lawrence, Ks., XXI, 269 pp., 110 figs., 2 tables; (Geol. Soc. America and Univ. Kansas).

Hailwood, E. A. and Kidd, R. B. (eds.) (1993): *High Resolution Stratigraphy*, Geol. Soc. London, Spec. Publ. 70, 360 pp., 171 figs., 17 tables; London.

Halbouty, M. T. (1980): 'Geologic Significance of Landsat Data from 15 Giant Oil and Gas Fields', 8–36, 31 figs.; in: M. T. Halbouty (ed.), *Giant Oil and Gas Fields of the Decade 1968–1978*, Amer. Assoc. Petroleum Geologists, Mem. 30, 596 pp., 513 figs., 38 tables; Tulsa, Okla.

Halbouty, M. T. (1981): 'Methods Used, and Experience Gained, in Exploration for New Oil and Gas Fields in Highly Explored (Mature) Areas', in: J. F. Mason (ed.), *Petroleum Geology in China*. Principal lectures presented to the United Nations International Meeting on Petroleum Geology, Beijing, China 18–25 March 1980, 263 pp., 246 figs., 40 tables; Tulsa, Okla. (PennWell Books).

Halbouty, M. T. (ed.) (1986): *Future Petroleum Provinces of the World*, Amer. Assoc. Petroleum Geologists, Mem. 40, 708 pp., 550 figs., 44 tables; Tulsa, Okla.

Hallam, A. (1981): *Facies Interpretation and the Stratigraphic Record*, 291 pp., 116 figs., 3 tables, appendix; Oxford, San Francisco (Freeman).

Ham, W. E. (ed.) (1962): *Classification of Carbonate Rocks*, Amer. Assoc. Petroleum Geologists, Mem. 1, 279 pp., 74 figs., 19 tables, 41 pls., 1 chart, with appendices; Tulsa, Okla.

Hamblin, W. K. (1962): 'X-Ray Radiography in the Study of Structures in Homogeneous Sediments', *J. Sediment. Petrol.*, **32**, 201–210, 6 figs.; Tulsa, Okla.

Hamblin, W. K. (1971): 'X-Ray Photography', 251–284, 18 figs., 3 tables; in: R. E. Carver (ed.), *Procedures in Sedimentary Petrology*, 653 pp., 188 figs., 72 tables; New York (Wiley-Interscience).

Hammett, D. S. (1980): 'Drilling in Hostile Environments: Offshore', *Proc. 10th World Petroleum Congress, Bucharest 1979, Vol. 2, Exploration, Supply and Demand*, 229–241, 16 figs., 1 table; London, Philadelphia, Rheine (Heyden).

Hancock, N. J. (1978): 'Possible Causes of Rotliegend Sandstone Diagenesis in Northern West Germany', *J. Geol. Soc. London*, **135**, 35–40, 3 figs., 2 pls.; London.

Hardage, B. A. (ed.) (1987): *Seismic Stratigraphy*, Handbook Geophys. Explor., Sect. I, 9, 432 pp., 196 figs., 6 tables; London (Geophysical Press).

Harding, T. P. (1983): 'Graben Hydrocarbon Plays and Structural Styles', 3–23, 26 figs.; in: J. P. H. Kaasschieter and T. J. A. Reijers (eds.) *Petroleum Geology of the Southeastern North Sea and the Adjacent Onshore Areas* (The Hague, 1982), *Geol. en Mijnbouw*, **62**, 239 pp., 241 figs., 19 tables, 3 pls., 1 appendix; Den Haag.

Harding, T. P. (1984): 'Graben Hydrocarbon Occurrences and Structural Style', *Amer. Assoc. Petroleum Geologists, Bull.*, **68**, 333–362, 26 figs.; Tulsa, Okla.

Harris, D. G. and Hewitt, C. H. (1977): 'Synergism in Reservoir Management – The Geologic Perspective', *J. Petroleum Technology*, **29**, 761–770, 14 figs., 1 table; Dallas, Texas.

Harris, P. M. (1979): *Facies Anatomy and Diagenesis of a Bahamian Ooid Shoal*, University of Miami, Comparative Sedimentology Laboratory, Sedimenta 7, 163 pp., 57 figs., 7 tables, 3 appendices with representative core illustrations; Miami, Florida.

Harwood, G. (1988): 'Microscopic Techniques: II. Principles of Sedimentary Petrography', 108–173, 45 figs., 7 tables; in: M. E. Tucker (ed.), *Techniques in Sedimentology*, 394 pp., 223 figs., 38 tables; Oxford (Blackwell).

Harwood, G. M., Smith, D. B., Pattison, J., and Pettigrew, T. (1982): 'Field Excursion Guide EZ82', *Symposium on the English Zechstein*, 53 pp., 22 figs., 1 table, unnumbered profiles; Nottingham.

Hedberg, H. D. (1964): 'Geologic Aspects of Origin of Petroleum', *Amer. Assoc. Petroleum Geologists, Bull.*, **48**, 1755–1803; Tulsa, Okla.

Hedberg, H. D. (1968): 'Significance of High Wax Oils with Respect to Genesis of Petroleum', *Amer. Assoc. Petroleum Geologists, Bull.*, **52**, 736–750, 1 table, 1 appendix; Tulsa, Okla.

Hedemann, H. A. and Teichmüller, R. (1971): 'Die paläogeographische Entwicklung des Oberkarbons', *Fortschr. Geol. Rheinld. Westf.*, **19**, 129–142, 6 figs., 2 tables; Krefeld.

Heim, D. (1990): *Tone und Tonminerale. Grundlagen der Sedimentologie und Mineralogie*, 157 pp., 30 figs., 8 tables, appendix; Stuttgart (Enke).

Heinrich, E. W. (1956): *Microscopic Petrography*, 296 pp., 70 figs., 9 tables; New York (McGraw-Hill).

Hickock IV, W. O. and Moyer, F. T. (1940): 'Geology and Mineral Resources of Fayette County, Pennsylvania', Pennsylvania Geol. Survey: County Report 26, Repr. 1973, 530 pp., 146 figs., unnumbered tables, 2 pls.; Harrisburg, Pa.

Higgs, R. (1979): 'Quartz Grain Surface Features of Mesozoic-Cenozoic Sands from the Labrador and Western Greenland Continental Margins', *J. Sediment. Petrol.*, **49**, 599–610, 9 figs., 1 table; Tulsa, Okla.

Hill, D. (1974): 'An Introduction to the Great Barrier Reef', in: A. M. Cameron *et al.* (eds.), *Proc. Second. Int. Coral Reef Symp.*, **2**, 723–731, 1 fig.; Brisbane, Australia (Great Barrier Reef Committee).

Hill, P. J. and Wood, G. V. (1980): 'Geology of the Forties Field, U.K. Continental Shelf, North Sea', 81–93, 16 figs.; in: M. T. Halbouty (ed.), *Giant Oil and Gas Fields in the Decade 1968–1978*, 596 pp., 514 figs., 38 tables; Tulsa, Okla.

Hobson, G. D. and Tiratsoo, E. N. (1981): *Introduction to Petroleum Geology*, 2nd ed., 352 pp., 94 figs., 58 tables; Beaconsfield, England (Scientific Press Ltd.).

Hoefs, J. (1987): *Stable Isotope Geochemistry*, Minerals and Rocks, Vol. 9, 3rd completely revised and enlarged ed., 241 pp., 62 figs., 23 tables; Berlin, Heidelberg (Springer).

Hollerbach, A. (1985): *Grundlagen der organischen Geochemie*, 190 pp., 35 figs., 25 tables; Berlin (Springer).

Holser, W. T. (1966): 'Bromide Geochemistry of Salt Rocks', in: J. L. Rau (ed.), *2nd Symposium on Salt*, Vol. 1, 248–275, 20 figs., 2 tables; Cleveland, Ohio (Northern Ohio Geol. Soc.).

Holser, W. T. and Kaplan, I. R. (1966): 'Isotope Geochemistry of Sedimentary Sulfates', *Chem. Geol.*, **1**, 93–135, 8 figs., 7 tables; Amsterdam.

Hood, A., Gutjahr, C. C. M., and Heacock, R. L. (1975): 'Organic Metamorphism and the Generation of Petroleum', *Amer. Assoc. Petroleum Geologists, Bull.*, **59**, 986–996, 7 figs; Tulsa, Okla.

Horowitz, A. S. and Potter, P. E. (1971): *Introductory Petrography of Fossils*, 302 pp., 28 figs., 100 pls.; Berlin, Heidelberg, New York (Springer).

Houbolt, J. J. H. C. (1957): *Surface Sediments of the Persian Gulf near the Qatar Peninsula*, 113 pp., 31 photographs, 8 pls.; Utrecht, nat. wiss. Diss. (28 January 1957); Den Haag (Mouton).

Houbolt, J. J. H. C. (1968): 'Recent Sediments in the Southern Bight of the North Sea', *Geol. en Mijnbouw*, **47**, 245–273, 35 figs.; The Hague.

Houghton, H. F. (1980): 'Refined Techniques for Staining Plagioclase and Alkali Feldspars in Thin section', *J. Sediment. Petrol.*, **50**, 629–631, 1 table; Tulsa, Okla.

Hounslow, A. W. (1979): 'Modified Gypsum/Anhydrite Stain', *J. Sediment. Petrol.*, **49**, 636–637; Tulsa, Okla.

Hsü, K. J. (1972): 'Origin of Saline Giants: A Critical Review after the Discovery of the Mediterranean Evaporite', *Earth-Sci. Reviews*, **8**, 371–396, 7 figs., 1 table; Amsterdam.

Hsü, K. J. (1989): *Physical Principle of Sedimentology. A Readable Textbook for Beginners and Experts*, 233 pp., 64 figs., 2 tables, 2 appendices; Berlin (Springer).

Hsü, K. J., Cita, M. B., and Ryan, W. B. (1973): 'The Origin of the Mediterranean Evaporites', *Init. Reports Deep Sea Drilling Project*, **13**, Part 2, 1203–1231, 12 figs., 1 table; Washington, D.C.

Hubbard, D. K., Oertel, G., and Nummedal, D. (1979): 'The Role of Waves and Tidal Currents in the Development of Tidal-Inlet Sedimentary Structures and Sand Body Geometry: Examples from North Carolina, South Carolina, and Georgia', *J. Sediment. Petrol.*, **49**, 1073–1091, 10 figs. 3 tables; Tulsa, Okla.

Hubert, J. F. (1962): 'A Zircon-Tourmaline-Rutile Maturity Index and the Interdependence of the Composition of Heavy Mineral Assemblages with the Gross Composition and Texture of Sandstones', *J. Sediment. Petrol.*, **32**, 440–450, 1 fig., 1 table; Tulsa, Okla.

Huc, A. Y. (ed.) (1990): *Deposition of Organic Facies*, Amer. Assoc. Petroleum Geologists, Studies in Geology, Vol. 30, 234 pp., 164 figs., 9 tables; Tulsa, Okla.

Hudson, J. D. (1977): 'Stable Isotopes and Limestone Lithification', *J. Geol. Soc. London*, **133**, 637–660, 6 figs.; London.

Hunt, B. J. (1991): 'Mechanism of Marble Staining', *J. Engineering Geol. Quarterly*, **24**, 49–53; Oxford.

Hunt, J. M. (1979): *Petroleum Geochemistry and Geology*, 617 pp., 230 figs., 77 tables, appendix, glossary; San Francisco (Freeman).

Hurst, A., Griffiths, C. M., and Worthington, P. F. (eds). (1992): *Geological Applications of Wireline Logs II*, Geol. Soc. London, Spec. Publ. 65, 406 pp., 259 figs., 37 tables, 6 appendices; London.

Hurst, A., Lovell, M. A., and Morton, A. C. (eds.) (1990): *Geological Applications of Wireline Logs*, Geol. Soc. Spec. Publ. 48, 357 pp., 279 figs., 44 tables; London.

Illing, L. V. (1954): 'Bahaman Calcareous Sands', *Amer. Assoc. Petroleum Geologists, Bull.*, **38**, 1–95, 13 figs., 7 tables, 9 pls.; Tulsa, Okla.

Ingle, J. C., Jr. (1966): *The Movement of Beach Sand. An Analysis Using Fluorescent Grains*, Developments in Sedimentology, Vol. 5, 221 pp., 117 figs., 5 tables; Amsterdam (Elsevier).

Institut Français du Pétrole (1986): *Corps Sédimentaires. Exemples Sismiques et Diagraphiques*, 349 pp., 282 figs., 6 tables; Paris (Editions Technip).

International Committee for Coal Petrology (1963/1993): *International Handbook of Coal Petrography*, 2nd ed. (1963), Part I: Nomenclature, Part II: Methods for Analysis; Paris (Centre National de la Recherche Scientifique). Third supplement (1993) to the 2nd edition; Newcastle upon Tyne (University of Newcastle upon Tyne).

Ireland, H. A. (1971): 'Insoluble Residues', 479–498, 4 figs.; in: R. E. Carver (ed.), *Procedures in Sedimentary*

Petrology, 653 pp., 188 figs., 72 tables; New York (Wiley-Interscience).

Irwin, H., Coleman, M., and Curtis, C. D. (1977): 'Isotope Evidence for Several Sources of Carbonate and Distinctive Diagenetic Processes in Organic-Rich Kimmeridgian Sediments', *Nature*, **269**, 5625, 209–213, 2 figs., 1 table; London.

James, D. P. and Leckie, D. A. (eds.) (1988): *Sequences, Stratigraphy, Sedimentology: Surface and Subsurface*, Canad. Soc. Petroleum Geologists, Proc. Techn. Meeting, Sept. 14–16, 1988, Mem. 15: 586 pp., 473 figs., 22 tables; Calgary, Alberta.

James, N. P. (1984): 'Reefs', 229–244, 21 figs.; in: R. G. Walker (ed.), *Facies Models*, 2nd ed., 317 pp., 357 figs., 17 tables, Geoscience Canada, Reprint Series 1, Geol. Assoc. Canada; Toronto, Ontario.

James, N. P. and Ginsburg, R. N. (eds.) (1979): *The Seaward Margin of Belize Barrier and Atoll Reefs. Morphology, Sedimentology, Organism Distribution and Late Quaternary History*, Int. Assoc. Sedimentology, Spec. Publ. 3, 191 pp., 103 figs., 8 tables; Oxford (Blackwell).

Jardine, D., Andrews, D. P., Wishart, J. W., and Young, J. W. (1977): 'Distribution and Continuity of Carbonate Reservoirs', *J. Petroleum Technology*, **29**, July 1977; 873–885, 16 figs.; Dallas, Tex.

Jeans, C. V., Merriman, R. J., Mitchell, J. G., and Bland, D. J. (1982): 'Volcanic Clays in the Cretaceous of Southern England and Northern Ireland', *Clay Minerals*, **17**, 105–156, 23 figs., 5 tables; London.

Jenkyns, H. C. (1986): 'Pelagic Environments', Chapter 11: 343–397, 52 figs., 4 tables, in: H. G. Reading (ed.), *Sedimentary Environments and Facies*, 615 pp., 564 figs., 15 tables; Oxford (Blackwell).

Jenyon, M. K. (1990): *Oil and Gas Traps; Aspects of Their Seismostratigraphy, Morphology, and Development*, 398 pp., figs., tables; Chichester (J. Wiley).

Jin Yusun, Liu Dingzeng, and Luo Changyan (1985): 'Development of Daqing Oil Field by Waterflooding', *J. Petrol. Technology*, **37**, 269–274, 6 figs., 2 tables; Richardson, Texas.

St. John, B. (ed.) (1980): *Sedimentary Basins of the World and Giant Hydrocarbon Accumulations*, (a short text to accompany the map: *Sedimentary Basins of the World*), 23 pp., 7 tables, map as colored appendix; Tulsa, Okla (Amer. Assoc. Petroleum Geologists).

Johnson, H. D. (1978): 'Shallow Siliciclastic Seas', 207–258, 55 figs.; in: H. G. Reading (ed.), *Sedimentary Environments and Facies*, 557 pp., 497 figs., 17 tables; Oxford (Blackwell).

Johnson, H. D. and Baldwin, C. T. (1986): 'Shallow Siliciclastic Seas', Chapter 9: 229–282, 63 figs.; in: H. G. Reading (ed.), *Sedimentary Environments and Facies*, 2nd ed., 615 pp., 564 figs., 15 tables; Oxford (Blackwell).

Johnson, R. J. (1975): 'The Base of the Cretaceous: A Discussion', 389–399, 8 figs.; in: A. W. Woodland (ed.), *Petroleum and the Continental Shelf of North-West Europe, 1, Geology*, 501 pp., 327 figs., 6 tables, 9 pls., 1 appendix; London (Applied Science Publ.).

Jones, M. P. (1987): *Applied Mineralogy – A Quantitative Approach*, 259 pp., 171 figs., 41 tables, 6 appendices; London (Graham & Trotman).

Jopling, A. V. and McDonald, B. C. (eds.) (1975): *Glaciofluvial and Glaciolacustrine Sedimentation*, Soc. Econ. Paleontologists Mineralogists, Spec. Publ. 23, 320 pp., 244 figs., 27 tables, 1 appendix, notations; Tulsa, Okla.

Kaelble, E. F. (ed.) (1967): *Handbook of X-Rays. For Diffraction, Emission, Absorption and Microscopy*, 1120 pp., 638 figs., 240 tables; New York (McGraw-Hill).

Kartsev, A. A. (1963): *Hydrogeology of Oil and Gas Fields* (in Russian), 353 pp.; Moscow (Gosudarstvennoe Nauchno-Teckhnicheskoe Izdatel'stvo Neftyanoi i Gorno-Toplivnoi Literatury).

Karweil, J. (1969): 'Aktuelle Probleme der Geochemie der Kohle', 59–84, 9 figs.; in: P. A. Schenck and I. Havenaar (eds.), *Advances in Organic Geochemistry*, 617 pp., 294 figs., 76 tables; Oxford (Pergamon Press).

Katz, B. J. and Pratt, L. M. (eds.) (1993): *Source Rocks in a Sequence Stratigraphic Framework*, Amer. Assoc. Petroleum Geologists, Studies in Geology, Voll. 37, 246 pp.; Tulsa, Okla.

Katzung, G. (1972): 'Stratigraphie und Paläogeographie des Unterperms in Mitteleuropa', *Geologie*, **21**, 570–584, 3 figs.; Berlin.

Katzung, G. (1975): 'Tektonik, Klima und Sedimentation in der Mitteleuropäischen Saxon-Senke und in angrenzenden Gebieten', *Z. geol. Wiss.*, **3**, 1453–1472, 3 figs.; Berlin.

Kemper, E. (1968): 'Einige Bemerkungen über die Sedimentationsverhältnisse und die fossilen Lebensspuren des Bentheimer Sandsteins (Valanginium)', *Geol. Jb.*, **86**, 49–106, 13 figs., 8 pls.; Hannover.

Kendall, A. C. (1984): 'Evaporites', 259–296, 43 figs., 2 tables, in: R. G. Walker (ed.), *Facies Models*, 2nd ed., 317 pp., 357 figs., 17 tables; Geoscience Canada, Reprint Series 1, Geol. Assoc. Canada; Toronto, Ontario.

Kennedy, J. F. (1978): 'Bed Forms in Alluvial Channels', 56–59, 3 figs.; in: R. W. Fairbridge and J. Bourgeois (eds.), *The Encyclopedia of Sedimentology*, 911 pp.; Stroudsburg, Pa. (Dowden, Hutchinson & Ross).

Kerr, P. F. (1977): *Optical Mineralogy*, 4th ed., 492 pp., 429 figs., 10 tables, charts A–G; New York (McGraw-Hill).

Kettel, D., Devay, L., Block, M. Schröder, L., Porth, H., Cordes, R., Dorn, M., and Stancu-Kristoff, G. (1984): *Untersuchung zur Erdgaslagerstättenbildung in der Nordfortsetzung der unterpermischen Emssenke*, BMFT-FB-T 84-259 Techn. Forsch. u. Entw. – Nichtnukleare Energietechnik, 90 pp., 48 figs.; Hannover (Nieders. Landesamt Bodenforschung).

King, R. E. (ed.) (1972): *Stratigraphic Oil and Gas Fields – Classification, Exploration Methods, and Case Histories*, Amer. Assoc. Petroleum Geologists, Mem. 16: 687 pp., 533 figs., 28 tables; Tulsa, Okla.

Kinsman, D. J. J. (1969): 'Modes of Formation, Sedimentary Associations and Diagnostic Features of Shallow-Water and Supratidal Evaporites', *Amer. Assoc. Petroleum Geologists, Bull.*, **53**, 830–840, 3 figs.; Tulsa, Okla.

Kirkland, D. W. and Evans, R. (1981): 'Source-Rock Potential of Evaporitic Environment', *Amer. Assoc. Petroleum Geologists, Bull.*, **65**, 181–190, 4 figs., 1 table; Tulsa, Okla.

Kleber, W. (1974): *Einführung in die Kristallographie*, 12. durchges. Aufl., 408 pp., 361 figs., 49 pls., 1 appendix; Berlin (VEB Verlag Technik).

Kleinspehn, K. L. and Paola, C. (eds.) (1988): *New Perspectives in Basin Analysis*, 453 pp., 225 figs., 20 tables, 1 appendix; New York (Springer).

Klemme, H. D. (1981): 'Types of Petroliferous Basins', in: J. F. Mason (ed.), *Petroleum Geology in China*, Principal lectures presented to the United Nations International Meeting on Petroleum Geology, Beijing, China, 18–25 March 1980, 101–115, 13 figs., 1 table; Tulsa, Okla. (PennWell Books).

Klug, H. P. and Alexander, L. E. (1974): *X-Ray Diffraction Procedures for Polycrystalline and Amorphous Materials*, 2nd ed. 966 pp., figs., tables, 10 appendices; New York (J. Wiley).

Kobranova, V. N. (1989): *Petrophysics*, Moscow (Mir), translated from the Russian, 375 pp., 168 figs., 24 tables; Berlin (Springer).

Koch, R. (1991): 'Faziesanalyse aus Spülproben', *Zbl. Geol. Paläont.*, Teil 1, 1990: 1029–1043, 6 figs., 3 tables; Stuttgart.

Köster, E. (1964): *Granulometrische und morphometrische Messmethoden an Mineralkörnern, Steinen und sonstigen Stoffen*, 336 pp., 109 figs., 68 tables; Stuttgart (Enke).

Komar, P. D. (1976): *Beach Processes and Sedimentation*, 429 pp., 218 figs., 11 tables; Englewood Cliffs N.J. (Prentice-Hall).

Komar, P. D. (ed.) (1983): *CRC Handbook of Coastal Processes and Erosion*, CRC Series in Marine Science, 2nd print., 305 pp., 179 figs., 18 tables; Boca Raton, Fla.

Kortenshteyn, V. N. (1974): 'Hydrogeologic Characteristics of an Genetic Conditions Affecting the Urengoy and Medvezh'ye Giant Gas Deposits', *Doklady Akad. Nauk SSSR*, **217**, 905–908, 1 fig., Engl. translation from *Dokl. Earth Science Sections*, **217**, 227–229, 1 fig.; Fall Church, Virginia (AGI).

Korvin, G. (1992): *Fractal Models in the Earth Sciences*, 396 pp., 224 figs., 15 tables; Amsterdam (Elsevier).

Kramer, H. J. (1992): *Earth Observation Remote Sensing – Survey of Missions and Sensors*, 251 pp., 67 figs., 41 tables, 2 appendices; Berlin (Springer).

Krauskopf, K. B. (1985): *Introduction to Geochemistry*, 2nd ed., 3rd print., 617 pp., 80 figs., 44 tables, 11 appendices; New York (McGraw-Hill).

Krebs, W. (1969): 'Über Schwarzschiefer und bituminöse Kalke im mitteleuropäischen Variscikum', *Erdöl u. Kohle*, **22**, 2–6, 62–67, 9 figs; Hamburg.

Krebs, W. (1975): 'Geologische Aspekte der Tiefenexploration im Paläozoikum Norddeutschlands und der südlichen Nordsee', *Erdöl Erdgas Z.*, **91**, 277–284, 3 figs.; Hamburg.

Krevelen, D. W. van (1981): *Coal. Typology – Chemistry – Physics – Constitution*, Coal Science and Technology, Vol. 3, 2nd impr., 514 pp., 253 figs., 16 tables, appendix; Amsterdam (Elsevier).

Krumbein, W. C. and Garrels, R. M. (1952): 'Origin and Classification of Chemical Sediments in Terms of pH and Oxidation-Reduction Potentials', *J. Geol.*, **60**, 1–33, 8 figs., 6 tables; Chicago, Ill.

Krumbein, W. C. and Pettijohn, F. J. (1938): *Manual of Sedimentary Petrography. I. Sampling, Preparation for Analysis, Mechanical Analysis, and Statistical Analysis. II. Shape Analysis, Mineralogical Analysis, Chemical Analysis, and Mass Properties*, 549 pp., 265 figs., 53 tables; New York (Appleton-Century-Crafts, Inc.).

Krylov, N. and Korzh, M. (1984): 'Upper Jurassic Black Bituminous Shales in Western Siberia', 211–215, 3 figs.; in: L. F. Jansa, P. F. Burolett, and A. C. Grant (eds.), *Basin Analysis: Principles and Applications*,

Sediment. Geol., Vol. 40, 215 pp., 106 figs., 8 tables; Amsterdam.

Krynine, P. D. (1946): 'The Tourmaline Group in Sediments', *J. Geol.*, **54**, 65–87, 17 figs., 3 tables; Chicago.

Kubanek, F., Nöltner, T., Weber, J., and Zimmerle, W. (1988): 'On the Lithogenesis of the Messel Oil Shale', *Cour. Forsch. Inst. Senckenberg*, **107**, 13–28, 5 figs., 4 tables, 1 pl.; Frankfurt a. M.

Kuenen, Ph. H. (1966): 'Geosynclinal Sedimentation', *Geol. Rundschau*, **56**, 1–19, 1 fig., 4 tables; Stuttgart.

Kühn, R. (1955): 'Die Verwachsungen der Salzmineralien in deutschen Salzgesteinen', *Erzmetall (Zs. f. Erzberg-bau und Metallhüttenwesen)*, **8**, Beiheft 1955: B 93–B 107, 21 figs., 7 tables; Stuttgart (Riederer).

Kulke, H. (1979): 'Sédimentation, Diagenèse et Métamorphisme Léger dans un Milieu Sursalé, Example du Trias Maghrébien', *Sci. de la Terre*, **23**(2), 39–74, 8 figs., 1 table, 2 pls.; Nancy.

Kulpecz, A. A. and Geuns, L. C. van (1990): 'Geological Modeling of a Turbidite Reservoir, Forties Field, North Sea', 489–507, 19 figs.; in: J. H. Barwis, J. G. McPherson, and J. R. J. Studlick (eds.), *Sandstone Petroleum Reservoirs*, 583 pp., 443 figs., 20 tables, 48 color pls.; New York (Springer).

Lake, L. W. and Carroll, H. B., Jr. (eds.) (1986): 'Reservoir Characterization I', *Proc. Reservoir Characterization Technical Conference*, April 29–May 1, 1985, 659 pp., 289 figs., 33 tables, 7 appendices; Dallas, Texas.

Lake, L. W., Carroll, H. B., Jr., and Wesson, T., C. (eds.) (1991): *Reservoir Characterization II*, 726 pp., 354 figs., 37 tables, 4 appendices; San Diego (Academic Press Inc.).

Langier-Kuzniarowa, A. (1973): 'Application of Simultaneous DTA, TG and DTG Methods to Petrographic Investigations', *Hungarian Scientific Instruments*, **1973**, 39–43, 10 figs.; Budapest.

Lanitz, W. (1965): 'Bestimmung der Hydratstufen des Kalziumsulfats durch optische Anfärbung im Phasenkon-trastmikroskop', *Silikattechnik*, **16**, 365–366, 1 fig., 2 tables; Berlin (Verl. Technik).

Latil, M. (ed.) (1980): *Enhanced Oil Recovery*, Transl. from French, 236 pp., 142 figs., 6 tables; Paris (Ed. Technip).

Leeder, M. R. (1982): *Sedimentology – Process and Product*, 344 pp., 377 figs., 29 tables, 8 pls.; London (Allen & Unwin).

Leggewie, R., Füchtbauer, H., and El-Najjar, R. (1977): 'Zur Bilanz des Buntsandsteinbeckens (Korn-größenverteilung und Gesteinsbruchstücke)', *Geol. Rundschau*, **66**, 551–577, 10 figs., 2 tables; Stuttgart.

Lemcke, K. (1977): 'Ölschiefer im Meteoritenkrater des Nördlinger Rieses', *Erdöl-Erdgas Z.*, **93**, 393–397, 6 figs.; Hamburg, Wien.

Le Ribault, L. (1977): *L'Exoscopie des Quartz*, Techniques et Méthodes Sédimentologiques, Collection publiée sous la Direction de A. Rivière: 150 pp., 29 figs., 6 tables, 30 photo pls., annexe with summary charts; Paris (Masson).

Lerman, A. (ed.) (1978): *Lakes: Chemistry, Geology, Physics*, 363 pp., 207 figs., 61 tables; New York, Heidelberg (Springer).

Levandowski, D. W., Kaley, M. E., Silverman, S. R., and Smalley, R. G. (1973): 'Cementation in Lyons Sandstone and Its Role in Oil Accumulation, Denver Basin, Colorado', *Amer. Assoc. Petroleum. Geologists, Bull.*, **57**, 2217–2244, 31 figs., 2 tables; Tulsa, Okla.

Leventhal, J. S. and Kepferle, R. C. (1981): 'Geochemistry and Geology of Strategic Metals and Uranium in Devonian Shales of the Eastern Interior United States', in: P. B. Tarman, (ed.), *Synthetic Fuels from Oil Shale II*, 73–96, 9 tables; Chicago, Ill. (Inst. Gas Technol.).

Liedmann, W. and Quader, H. (1991): 'Optical Analysis of Geological Structures by Confocal Laser Scanning Microscopy', *Naturwissenschaften*, **78**, 413–414, 1 fig.; Berlin.

Lindholm, R. C. (1987): *A Practical Approach to Sedimentology*, 276 pp., 192 figs., 45 tables; London (Allen & Unwin).

Lisitzin, A. P. (1972): *Sedimentation in the World Ocean. With Emphasis on the Nature, Distribution and Behaviour of Marine Suspensions*, Soc. Econ. Paleontologists Mineralogists, Spec. Publ. 17, K. S. Rodolfo (ed.): 218 pp., 181 figs., 24 tables; Tulsa, Okla.

Littmann, W. (1988): *Polymer Flooding*, Developments in Petroleum Science, Vol. 24, 212 pp., 101 figs., 26 tables; Amsterdam (Elsevier).

Liu Hefu (1986): 'Geodynamic Scenario and Structural Styles of Mesozoic and Cenozoic Basins in China', *Amer. Assoc. Petroleum Geologists, Bull.*, **70**, 377–395, 22 figs., 1 table; Tulsa, Okla.

Loon, J. C. van and Barefoot, R. R. (1988): *Analytical Methods for Geochemical Exploration*, 344 pp., 46 figs., 123 tables; London (Academic Press).

Loucks, R. G. and Sarg, J. F. (eds.) (1993): *Carbonate Sequence Stratigraphy – Recent Developments and Applications*, Amer. Assoc. Petroleum Geologists, Mem. 57, 546 pp., 338 figs., 13 tables, 11 foldouts; Tulsa, Okla.

Ludwig, G. and Rosenbaum, A. (1966): 'Zur diagnostischen Anfärbung von Kalimineralien in Bergbau und Aufbereitung', *Bergakademie*, **18**, 628–630, 3 figs.; Leipzig (VEB Deutscher Verlag Grundstoffindustrie).

Lutz, M., Kaasschieter, J. P. H., and Wijhe, D. H. van (1975): 'Geological Factors Controlling Rotliegend Gas Accumulations in the Mid-European Basin', *Proc. 9th World Petroleum Congr.* Tokyo, 1975, Proc. 2, 93–103, 7 figs.; London.

Mack, G. H. (1984): 'Exceptions to the Relationship between Plate Tectonics and Sandstone Composition', *J. Sediment. Petrol.*, **54**, 212–220, 5 figs.; Tulsa, Okla.

Mack, G. H., Thomas, W. A., and Horsey, C. A. (1983): 'Composition of Carboniferous Sandstones and Tectonic Framework of Southern Appalachian-Ouachita Orogen', *J. Sediment. Petrol.*, **53**, 931–946, 12 figs.; Tulsa, Okla.

Magara, K. (1986): *Geological Models of Petroleum Entrapment*, 328 pp., 246 figs., 15 tables; London New York (Elsevier Applied Science Publ.).

Majewske, O. P. (1969): *Recognition of Invertebrate Fossil Fragments in Rocks and Thin Sections*, 101 pp., 7 tables, 280 pls., 19 diagr.; Leiden (E. J. Brill).

Ma Li (1985): 'Subtle Oil Pools in Xingshugang Delta, Songliao Basin', *Amer. Assoc. Petroleum Geologists, Bull.*, **69**, 1123–1132, 17 figs.; Tulsa, Okla.

Ma Li, Yang Jiliang, and Ding Zhengyan (1989): 'Songliao Basin – An Intracratonic Continental Sedimentary Basin of Combination Type', 77–87, 10 figs.; in: X. Zhu (ed.), *Chinese Sedimentary Basins*, Sedimentary Basins of the World, Vol. 1, 238 pp., 150 figs., 15 tables; Amsterdam (Elsevier).

Mange, M. A. and Maurer, H. F. W. (1991): *Schwerminerale in Farbe*, 150 pp., 178 figs. in color; Stuttgart (Enke).

Mansfield, G. R. (1922): 'Potash in the Greensands of New Jersey', *U.S. Geol. Survey, Bull.* **727**, 146 pp., 6 figs., unnumbered tables, 10 pls.; Washington.

Marschall, R. (1982): *Verbesserung der seismischen Erkundungsmethoden für die Seismogrammbearbeitung durch Einbeziehung des seismischen Signals. Bundesministerium für Forschung und Technologie*, Forschungsbericht T 83-230, Technologische Forschung und Entwicklung – Nichtnukleare Energietechnik, 361 pp., 123 figs., equations, 2 appendices; Hannover.

Marshall, D. J. (1988): *Cathodoluminescence of Geological Materials*, 146 pp., 85 figs., 12 tables, 13 color pls.; Boston (Unwin & Hyman).

Mason, J. F. and Dickey, P. A. (eds.) (1989): *Oil Field Development Techniques: Proceedings of the Daqing International Meeting, 1982*, Amer. Assoc. Petroleum Geologists, Studies in Geology, Vol. 28, 247 pp., 284 figs., 42 tables; Tulsa, Okla.

Matter, A. and Tucker, M. E. (eds.) (1978): *Modern and Ancient Lake Sediments*, Internat. Assoc. Sedimentologists, Spec. Publ. 2, 290 pp., 171 figs., 10 tables; Oxford (Blackwell).

Maxwell, W. G. H. (1968): *Atlas of the Great Barrier Reef*, 258 pp., 166 figs., 3 tables; Amsterdam (Elsevier).

Maynard, J. B. (1983): *Geochemistry of Sedimentary Ore Deposits*, 302 pp., 149 figs.; Berlin (Springer).

McIver, R. D. (1967): *Composition of Kerogen – Clue to Its Role in the Origin of Petroleum*, 7th World Petroleum Congr. Mexico, Proc., Vol. 2, 25–36, 8 figs., 3 tables; Barking, Essex (Elsevier).

McKee, E. D. (1982): *Sedimentary Structures in Dunes of the Namib Desert, South West Africa*, Geol. Soc. America, Spec. Pap. 188, 64 pp., 47 figs., 2 tables; Boulder, Colo.

McKee, E. D. and Weir, G. W. (1953): 'Terminology for Stratification and Cross-Stratification in Sedimentary Rocks', *Geol. Soc. America, Bull.*, **64**, 381–390, 2 figs., 4 tables; New York.

McManus, J. (1988): 'Grain Size Determination and Interpretation', 63–85, 17 figs., 5 tables, in: M. E. Tucker (ed.), *Techniques in Sedimentology*, 394 pp., 223 figs., 38 tables; Oxford (Blackwell).

Meischner, K. D. (1964): 'Allodapische Kalke, Turbidite in Riff–nahen Sedimentations-Becken', in: A. H. Bouma and A. Brouwer (eds.), *Turbidites, Development in Sedimentology*, **3**, 156–191, 5 figs., 1 table, 3 pls; Amsterdam (Elsevier).

Merin, I. S. and Moore, W. R. (1986): 'Application of Landsat Imagery to Oil Exploration in Niobrara Formation, Denver Basin, Wyoming', *Amer. Assoc. Petroleum Geologists, Bull.*, **70**, 351–359, 15 figs.; Tulsa, Okla.

Merkel, R. H. (1979): *Well Log Formation Evaluation*, Amer. Assoc. Petrolleum Geologists, Education Department, Continuing Education Course Note Series No. 14, 82 pp., 38 figs., 2 tables; Tulsa, Okla.

Miall, A. D. (1977): 'A Review of the Braided-River Depositional Environment', *Earth Sci. Rev.*, **13**, 1–62, 16 figs., 6 tables; Amsterdam.

Miall, A. D. (1990): *Principles of Sedimentary Basin Analysis*, 2nd ed., 668 pp., 466 figs., 26 tables; New York (Springer).

Michelsen, O. (ed.) (1982): *Geology of the Danish Central Graben*, Geol. Surv. Denmark. Ser. B., No. 8, 133 pp., 59 figs., 2 sets of tables; Copenhagen (Reitzel).

Middleton, G. V. and Hampton, M. A. (1976): 'Subaqueous Sediment Transport and Deposition by Sediment Gravity Flows', 197–218, 10 figs.; in: D. J. Stanley and D. J. P. Swift (eds.), *Marine Sediment Transport and Environmental Management*, 602 pp., 382 figs., 32 tables; New York (J. Wiley).

Miller, J. (1988): 'Microscopical Techniques: I. Slices, Slides, Stains and Peels', 86–107, 11 figs.; in: M. E. Tucker (ed.), *Techniques in Sedimentology*, 394 pp., 223 figs., 38 tables; Oxford (Blackwell).

Millot, G. (1970): *Geology of Clays. Weathering. Sedimentology. Geochemistry*, 429 pp., 85 figs., 15 tables, 2 pls.; New York, Heidelberg, Berlin (Springer), Paris (Masson et Cie.), London (Chapman & Hall).

Milner, H. B. (1962): *Sedimentary Petrography. Vol. I: Methods in Sedimentary Petrography. Vol. II: Principles and Applications*, 643 pp., 82 figs., 20 tables, 71 pls.; London (Allen & Unwin Ltd.).

Minnis, M. M. (1984): 'An Automatic Point-Counting Method for Mineralogical Assessment', *Amer. Assoc. Petroleum Geologists, Bull.*, **68**, 744–752, 7 figs., 8 tables; Tulsa, Okla.

Mitchell, J. G., Lehman, P. J., Cantrell, D. L., Al-Jallal, I. A., and Al-Thagafy, M. A. R. (1988): 'Lithofacies, Diagenesis, and Depositional Sequence; Arab-D Member, Ghawar Field, Saudi Arabia', 459–514, 20 figs.; in: A. J. Lomando and P. M. Harris (eds.), *Giant Oil and Gas Fields. A Core Workshop*, SEPM Core Workshop No. 12, Houston, March 19–20, 1988, 853 pp., 403 figs., 16 tables; Tulsa, Okla.

Mitchum, R. M. (1977): 'Seismic Stratigraphy and Global Changes of Sea Level, Part 1: Glossary of Terms Used in Seismic Stratigraphy', 205–212; in C. E. Payton (ed.), *Seismic Stratigraphy – Applications to Hydrocarbon Exploration*, Amer. Assoc. Petroleum Geologists, Mem. 26: 516 pp., 420 figs., 12 tables; Tulsa, Okla.

Mitchum, R. M. (1985): 'Seismic Stratigraphic Expression of Submarine Fans', 117–138, 13 figs., 2 tables; in: O. R. Berg and D. G. Woolverton (eds.), *Seismic Stratigraphy II: An Integrated Approach to Hydrocarbon Exploration*, Amer. Assoc. Petroleum Geologists, Mem. 39, 276 pp., 255 figs., 8 tables; Tulsa, Okla.

Mitchum, R. M., Jr., Vail, P. R., and Thompson, S., III (1977): 'Seismic Stratigraphy and Global Changes of Sea Level, Part 2: The Depositional Sequence as a Basic Unit for Stratigraphic Analysis', 53–62, 4 figs.; in: C. E. Payton (ed.), *Seismic Stratigraphy – Applications to Hydrocarbon Exploration*, Amer. Assoc. Petroleum Geologists, Mem. 26, 516 pp., 420 figs., 12 tables; Tulsa, Okla.

Möller, P. (1986): *Anorganische Geochemie – Eine Einführung*, Heidelberger Taschenbücher, Vol. 240, 326 pp., 141 figs., 79 tables; Berlin Heidelberg (Springer).

Moenke, H. (1962): *Mineralspektren I (aufgenommen mit dem Jenaer Spektralphotometer UR 10)*, Deutsche Akademie der Wissenschaften zu Berlin, Kommission für Spektroskopie, 40 pp., 5 figs., 2 tables, 355 infrared mineral spectra on cards; Berlin (Akademie-Verlag).

Moenke, H. (1966): *Mineralspektren II (aufgenommen mit dem Jenaer Spektralphotometer UR 10)*, Deutsche Akademie der Wissenschaften zu Berlin, Kommission für Spektroskopie. 22 pp., 2 tables, 150 infrared mineral spectra on cards; Berlin (Akademie-Verlag).

Moenke, H. and Moenke-Blankenburg, L. (1965): *Optische Bestimmungsverfahren und Geräte für Mineralogen und Chemiker. Goniometrie, Refraktometrie, Mikroskopie, Kolometrie, Photometrie, Spektrometrie und Polarometrie*, 568 pp., 264 figs., 14 tables; Leipzig (Akademische Verlagsgesellschaft Geest and Portig K.G.).

Momper, J. (1963): 'Nomenclature, Lithofacies and Genesis of Permo-Pennsylvanian Rocks – Northern Denver Basin', 41–67, 16 figs.; in: D. W. Bolyard and P. J. Katich (eds.), *Geology of the northern Denver Basin*, Rocky Mtn. Assoc. Geologists, 14th Ann. Field Conf. Guidebook, 295 pp.; Denver, Colo.

Moore, C. H. (1989): *Carbonate Diagenesis and Porosity*, Developments in Sedimentology, Vol. 46, 338 pp., 176 figs., 17 tables; Amsterdam (Elsevier).

Moore, D. G. and Scruton, P. C. (1957): 'Minor Internal Structures of Some Recent Unconsolidated Sediments', *Amer. Assoc. Petroleum Geologists, Bull.*, **41**, 2723–2751, 16 figs., 1 table; Tulsa, Okla.

Moore, D. M. and Reynolds, R. C. (1989): *X-Ray Diffraction and the Identification and Analysis of Clay Minerals*, 332 pp., 105 figs., 42 tables; Oxford (University Press).

Moorkens, T. L. (1991): 'Depositional Environments of Organic-Rich Sediments and the Recognition of the Different Types of Source Rocks', *Zbl. Geol. Paläont.*, Teil 1, 1990, 1073–1089, 4 figs.; Stuttgart.

Morgan, J. P. and Shaver, R. H. (eds.) (1970): *Deltaic Sedimentation Modern and Ancient*, Soc. Econ. Paleontologists Mineralogists, Spec. Publ. 15, 312 pp., 232 figs., 24 tables, 6 pls.; Tulsa, Okla.

Morgan, K. M. and Koger, D. G. (1986): 'Latest Advances in the Application of Remote Sensing to Petroleum Exploration', 305–311, 4 figs., 2 tables; in: M. J. Davidson (ed.), *Unconventional Methods in Exploration for Petroleum and Natural Gas*, 4th Symposium held on May 1 and 2, 1985 in Dallas, 350 pp., 175 figs., 14 tables; Dallas (Southern Methodist University Press).

Morris, K. (1979): 'A Classification of Jurassic Marine Shale Sequences; An Example from the Toarcian (Lower Jurassic) of Great Britain', *Palaeogr. Palaeoclim. Palaeoecol.*, **26**, 117–126, 4 figs., 2 tables; Amsterdam.

Morton, A. C., Todd, S. P., and Haughton, P. D. W. (eds.) (1991): *Developments in Sedimentary Provenance Studies*, Geol. Soc., Spec. Publ. 57, 370 pp., 206 figs., 52 tables; London (Geol. Soc.).

Müller, G. (1967): *Methods in Sedimentary Petrology*, Part I of Sedimentary Petrology by W. von Engelhardt, H. Füchtbauer, and G. Müller, 283 pp., 91 figs., 31 tables, 2 diagrams, appendix with 2 tables; Stuttgart (Schweizerbart).

Multer, H. G. (1977): *Field Guide to Some Carbonate Rock Environments, Florida Keys and Western Bahamas*, 415 pp., 294 figs., 5 tables, 7 pls., 4 appendices, index maps 1–10; Dubuque, Iowa (Kendall/Hunt Publ. Co.).

Murphy, C. P. (1986): *Thin Section Preparation of Soils and Sediments*, 149 pp., 62 figs., 3 tables, 2 appendices with informative data on polyester resin systems and thin-section preparation; Berkhamsted, Herts. (AB Academic Publ.).

Murris, R. J. (1980): 'Middle East: Stratigraphic Evolution and Oil Habitat', *Amer. Assoc. Petroleum Geologists, Bull.*, **64**, 597–618, 25 figs.; Tulsa, Okla.

Murris, R. J. (1981): 'Seals for Major Middle East Fields (Abstract)', *Amer. Assoc. Petroleum Geologists, Bull.*, **65**, 964; Tulsa, Okla.

Mutti, E. and Ricci Lucchi, F. (1978): 'Turbidites of the Northern Appennines: Introduction to Facies Analysis', *Intern. Geol. Review*, **20**, 125–166, 30 figs., 1 table; Falls Church.

Naeser, N. D. and McCulloh, T. H. (1989): *Thermal History of Sedimentary Basins. Methods and Case Histories*, 319 pp., 197 figs., 28 tables, 3 appendices; New York, Berlin, Heidelberg (Springer).

Nagtegaal, P. J. C. (1979): 'Relationships of Facies and Reservoir Quality in Rotliegendes Desert Sandstones, Southern North Sea Region', *J. Petroleum Geol.*, **2**, 145–158, 9 figs., 3 tables; Beaconsfield.

Nagtegaal, P. J. C. (1980): 'Diagenetic Models for Predicting Clastic Reservoir Quality', *Revista Inst. Invest. geol. Disputacion Provincial, Universidad de Barcelona*, **34**, 5–19, 11 figs.; Barcelona.

Nanz, R. H., Jr. (1954): 'Genesis of Oligocene Sandstone Reservoir, Seeligson Field, Jim Wells and Kleberg Counties, Texas', *Amer. Assoc. Petroleum Geologists, Bull.*, **38**, 96–117, 21 figs., 4 tables; Tulsa, Okla.

Negendank, J. F. W., Irion, G., and Linden, J. (1982): 'Ein eozänes Maar bei Eckfeld nordöstlich Manderscheid (SW-Eifel)', *Mainzer geowiss. Mitt.*, **11**, 157–172, 12 figs., 2 tables; Mainz.

Nesterov, I. I. (1979): 'New Type of Oil and Gas Reservoir', *Petroleum Geol.*, **17**, 476–479, 4 figs.; (translation from *Geologiya Neft i Gaza*, **10**, 26–29, 1979); Virginia (McLean).

Nesterov, I. I. *et al.* (1991): 'Formation of West Siberian Basins Related to the Lithosphere Evolution', 157–172, 4 figs.; in A. H. Bouma and R. M. Carter (eds.), *Facies Models in Exploration and Development of Hydrocarbon and Ore Deposits*, 254 pp., 101 figs., 4 tables; Utrecht (VSP).

Newell, N. D., Rigby, J. K., Fischer, A. G., Whiteman, A. J., Hickox, J. E., and Bradley, J. S. (1953): *The Permian Reef Complex of the Guadalupe Mountains Region, Texas and New Mexico*, 236 pp., 32 pls.; San Francisco (W. H. Freeman).

Nio, S. D. (1976): 'Marine Transgressions as a Factor in the Formation of Sandwave Complexes', *Geol. en Mijnbouw*, **55**, 18–40, 15 figs.; The Hague.

Nöltner, T. (1988): *Submikroskopische Komponenten und Mikrotextur klastischer Sedimente*, 170 pp., 20 figs., 10 tables, 35 pls.; Stuttgart (Enke).

Normark, W. R. (1978): 'Fan Valleys, Channels and Depositional Lobes on Modern Submarine Fans: Characters for Recognition of Sandy Turbidite Environments', *Amer. Assoc. Petroleum Geologists, Bull.*, **62**, 912–931, 12 figs.; Tulsa, Okla.

Nygaard, E., Lieberkind, K., and Frykman, P. (1983): 'Sedimentology and Reservoir Parameters of the Chalk Group in the Danish Central Graben', in: J. P. H. Kaasschieter and T. J. A. Reijers (eds.), *Petroleum Geology of the Southeastern North Sea and the Adjacent Onshore Areas* (The Hague, 1982), *Geol. en Mijnbouw*, **62**, 177–190, 31 figs.; Den Haag.

O'Brien, N. R. and Slatt, R. M. (1990): *Argillaceous Rock Atlas*, 141 pp., 242 figs. (46 in full color), 2 tables; New York, Berlin, Heidelberg (Springer).

Odin, G. S. (ed.) (1988): *Green Marine Clays – Oolithic Ironstone Facies, Verdine Facies, Glaucony Facies, and Celadonite-Bearing Rock Facies. A Comparative Study*, Developments in Sedimentology, Vol. 45, 445 pp., 168 figs., 59 tables; Amsterdam (Elsevier).

Oehmig, R. (1988): 'Petrographie und Log-Daten einer klastischen Rotliegend/Buntsandstein- Folge', *Heidelberger geowiss. Abh.*, **14**, 219 pp., 126 figs., 17 tables, 4 pls.; Heidelberg.

Oosthuyzen, E. J. (1980): *An Elementary Introduction to Image Analysis – A New Field of Interest at the National Institute for Metallurgy*, National Institute for Metallurgy, Mineralogy Division, Report No. 2058, 26 pp., 12 figs., 3 tables, 1 pl.; Randburg, S.A. (Nat. Inst. Metallurgy).

Ortoleva, P. J. (1990): 'Self-Organization in Geological Systems', *Earth-Sci. Reviews*, **29**, 417 pp., 251 figs., 10 tables, 4 pls.; Amsterdam.

Ostwald, W. (1921): *Der Farbenatlas*, about 2 500 colors in more than 100 plates, with instructions for use and scientific description; Leipzig (Unesma).

Otto, G. H. (1938): 'The Sedimentation Unit and Its Use in Field Sampling', *J. Geol.*, **46**, 569–582, 1 table; Chicago, Ill.

Ovanesov, G. P., Alexin, O. K., Glotov, S. G., Sarkisyan, R. O., Khachatryan, R. O., and Pashkov, Yu. V. (1980): 'Developments in Exploration of Non-Tectonic Traps', *10th World Petr. Congr. Proc.*, Bucharest, 1979, Vol. 2, *Exploration, Supply and Demand*, 391–400, 7 figs.; London (Heyden).

Ovchinnikov, S. I. (1976): 'Effects of Gas on the Post-Sedimentational Transformation of Rock Traps', *Lithology and Mineral Resources* (a translation of Litologiya i Poleznye Iskopaemye) **10**(5), 658–662, 1 fig., 2 tables; New York (Plenum).

Pagel, M., Walgenwitz, F., and Dubessy, J. (1986): 'Fluid Inclusions in Oil and Gas-Bearing Sedimentary Formations', 565–583, 9 figs.; in: J. Burrus (ed.), *Thermal Modeling in Sedimentary Basins*, 1st IFP Exploration Research Conference, Carcans, France, June 3–7, 1985; Paris (Editions Technip).

Palaz, I. and Sengupta, S. (eds.) (1992): *Automated Pattern Analysis in Petroleum Exploration*, 295 pp., 199 figs., 20 tables, 14 pls.; New York (Springer).

Parasnis, D. S. (1986): *Principles of Applied Geophysics*, 4th ed., 402 pp., 195 figs., 18 tables, 13 appendices; London, New York (Chapman and Hall).

Parfenoff, A., Pomerol, Ch., and Tourenq, J. (1970): *Les Minéraux en Grains – Méthodes d'Étude et Détermination*, 578 pp., 123 figs., 34 tables, 2 pls.; Paris (Masson).

Parker, J. R. (1977): 'Deep-Sea Sands', Chapter 7: 225–242, 7 figs., 1 table; in: G. D. Hobson (ed.), *Development in Petroleum Geology – 1*, 335 pp., 18 figs., 5 tables; London (Applied Science).

Passega, R. (1957): 'Texture as a Characteristic of Clastic Deposition', *Amer. Assoc. Petroleum Geologists, Bull.*, **41**, 1952–1984, 17 figs.; Tulsa, Okla.

Patijn, R. J. H. (1964a): 'Die Entstehung von Erdgas infolge der Nachinkohlung im Nordosten der Niederlande', *Erdöl u. Kohle*, **17**, 2–9, 5 figs., 1 table; Hamburg.

Patijn, R. J. H. (1964b): 'La Formation de Gaz Due à des Rehouillifications dans le Nord-Est des Pays-Bas' (with discussion), *5th Congr. Int. Stratigr. Géol. Carbonifère*, C. R. T. 2, 631–645, 8 figs., 1 table; Paris.

Pavlov, N. D., Salov, Yu. A., Gogonenkov, G. N., Akopov, Yu. I., Tolstykh, A. A., and Denisyuk, R. S. (1988): 'Geological-Geophysical Model of the Tengiz Paleo-Atoll Based on Seismostratigraphy' *Intern. Geol. Review*, **30**, 1057–1069, 6 figs.; Silver Spring, Maryland (Winston & Sons).

Payton, C. E. (ed.) (1977): *Seismic Stratigraphy – Applications to Hydrocarbon Exploration*, Amer. Assoc. Petroleum Geologists, Mem. 26, 516 pp., 420 figs., 12 tables; Tulsa, Okla.

Pettijohn, F. J. (1963): 'Chemical Composition of Sandstones – Excluding Carbonate and Volcanic Sands'; in: M. Fleischer (ed.), *Data of Geochemistry*, 6th ed., Chapter S, U.S. Geol. Survey, Prof. Paper, 440–S, 21 pp., 4 figs., 14 tables; Washington, D.C.

Pettijohn, F. J. (1975): *Sedimentary Rocks*, 3rd ed., 628 pp., 354 figs., 88 tables; New York (Evanston), San Francisco, London (Harper & Row).

Pettijohn, F. J., Potter, P. E., and Siever, R. (1987): *Sand and Sandstone*, 2nd ed., 553 pp., 355 figs., 80 tables; New York, Berlin, Heidelberg (Springer).

Philipp, W. (1961): 'Struktur- und Lagerstättengeschichte des Erdölfeldes Eldingen', *Z. dt. geol. Ges. (für 1960)*, **112**, 414–482, 52 figs., 3 tables; Hannover.

Philipp, W., Drong, H. J., Füchtbauer, H., Haddenhorst, H.G., and Jankowsky, W. (1963): 'The History of Migration in the Gifhorn Trough (NW-Germany)', *6th World Oil Congr.*, Sect. I, Pap. 19, PD 2: 457–481, 15 figs., 2 tables; Frankfurt a. M.

Philpotts, A. R. (1989): *Petrography of Igneous and Metamorphic Rocks*, 178 pp., 242 figs., 6 tables, 62 mineral descriptions; Englewood Cliffs, N. J. (Prentice Hall).

Picard, M. D. and Lee, R. H., Jr. (1972): 'Criteria for Recognizing Lacustrine Rocks', 108–145, 24 figs., 12 tables; in: K. J. Rigby and W. K. Hamblin (eds.), *Recognition of Ancient Sedimentary Environments*, Soc.

Econ. Paleontologists Mineralogists, Spec. Publ. 16, 340 pp., 217 figs., 30 tables; Tulsa, Okla.

Pickering, K. T., Hiscott, R. N., and Hein, F. J. (1989): *Deep-Marine Environments. Clastic Sedimentation and Tectonics*, 416 pp., 282 figs., 24 tables; London (Unwin Hyman).

Pilkey, O. H. and Noble, D. (1966): 'Carbonate and Clay Mineralogy of the Persian Gulf', *Deep-Sea Research*, **13**, 1–16, 14 figs., 1 table; Oxford.

Pirson, S. J. (1983): *Geologic Well Log Analysis*, 3rd ed., 475 pp., 156 figs., 7 tables, Appendices 1–4; Houston, Texas (Gulf Publ. Comp.).

Plas, L. van der (1966): 'The Identification of Detrital Feldspars', *Developments in Sedimentology*, **6**, 305 pp., 66 figs., 40 tables; Amsterdam (Elsevier).

Plein, E. (1978): 'Rotliegend-Ablagerungen im Norddeutschen Becken', *Z. dt. geol. Ges.*, **129**, 71–97, 10 figs., 6 pls.; Hannover.

Plint, A. G., Walker, R. G., and Bergman, K. M. (1986): 'Cardium Formation – 6. Stratigraphic Framework of the Cardium in Subsurface', *Canad. Petroleum Geol., Bull.*, **34**, 213–225, 10 figs.; Calgary.

Pluman, I. I. (1975): 'Distribution of Uranium, Thorium and Potassium in the Sedimentary Rocks of the West Siberian Platform', *Geochemistry International*, **12**(3), 97–107, 6 figs., 4 tables; Washington.

Poh-Hsi Pan (1981): 'Direct Detection of Hydrocarbons by the Seismic Reflection Method', 193–201, 18 figs.; in: J. F. Mason (ed.), *Petroleum Geology in China*, Principal lectures presented to the United Nations International Meeting on Petroleum Geology, Beijing, China 18–25 March 1980, 263 pp., 246 figs., 40 tables; Tulsa, Okla. (PennWell Books).

Polesny, H. (1983): 'Verteilung der Öl- und Gasvorkommen in der oberösterreichischen Molasse', *Erdöl Erdgas*, **99**, 90–102, 16 figs.; Hamburg, Wien.

Posamentier, H. W., Jervey, M. T., and Vail, P. R. (1988): 'Eustatic Controls on Clastic Deposition, I – Conceptual Framework', 109–124, 19 figs., 1 table; in: C. K. Wilgus, B. S. Hastings, H. Posamentier, J. van Wagoner, C. A. Ross, and C. G. St. G. Kendall (eds.), *Sea-Level Changes – An Integrated Approach*, Soc. Econ. Paleontologists Mineralogists, Spec. Publ. 42, 407 pp., 319 figs., 9 tables; Tulsa, Okla.

Posamentier, H. W., Summerhayes, C. P., Haq, B. U., and Allen, G. P. (eds.) (1993): *Sequence Stratigraphy and Facies Associations*, Intern. Assoc. Sedimentologists, Spec. Publ. 18, 644 pp., 405 figs., 12 tables, 2 appendices; Oxford (Blackwell).

Potter, P. E. and Pettijohn, F. J. (1977): *Paleocurrents and Basin Analysis*, 2nd corrected and updated ed., 425 pp., 167 figs., 30 pls.; New York (Springer).

Potter, P. E., Maynard, J. B., and Pryor, W. A. (1980): *Sedimentology of Shale – Study Guide and Reference Source*, 310 pp., 154 figs., 25 tables; New York, Heidelberg, Berlin (Springer).

Potter, P. E., Maynard, J. B., and Pryor, W. A. (1982): 'Appalachian Gas Bearing Devonian Shales: Statements and Discussions', *Oil Gas J.*, Jan. 25, 1982, 290–318, 23 figs.; Tulsa, Okla.

Powers, R. W. (1962): 'Arabian Upper Jurassic Carbonate Reservoir Rocks', 122–192, 23 figs., 1 table, 7 pls.; in: W. E. Ham (ed.), *Classification of Carbonate Rocks*, Amer. Assoc. Petroleum Geologists, Mem. 1, 279 pp., 74 figs., 19 tables, 41 pls., 1 chart, with appendices; Tulsa, Okla.

Powers, R. W. (1968): 'Saudi Arabia', in: L. Dubertret (ed.), *Lexique Stratigraphique International*, Vol. III, Asie, Arabie Saoudite, fasc 10b1, 177 pp., 1 table, 4 pls.; Paris (CNRS).

Pratsch, J. C. (1983): 'Gasfields, NW German Basin: Secondary Gas Migration as a Major Geologic Parameter', *J. Petroleum Geol.*, **5**, 229–244, 22 figs.; Beaconsfield, Bucks, England.

Protsvetalova, T. N. (1983): 'Lower Cretaceous Stratigraphic Oil- and Gas-Bearing Complexes of the West Siberian Platform', *Intern. Geol. Rev.*, **25**, 447–454, 1 fig.; Falls Church, Va. (Amer. Geol. Inst.).

Pryor, W. A. (1971): 'Petrology of the Weißliegendes Sandstones in the Harz and Werra-Fulda areas, Germany', *Geol. Rundschau.*, **60**, 524–552, 13 figs., 3 tables; Stuttgart.

Purdy, E. G. (1963a): 'Recent Calcium Carbonate Facies of the Great Bahama Bank. I. Petrography and Reaction Groups', *J. Geol.*, **71**, 334–355, 5 figs., 2 tables, 5 pls.; Chicago.

Purdy, E. G. (1963b): 'Recent Calcium Carbonate Facies of the Great Bahama Bank. 2. Sedimentary Facies', *J. Geol.*, **71**, 472–497, 4 figs., 7 tables, 1 pl.; Chicago.

Purdy, E. G. (1974): 'Reef Configurations: Cause and Effect', in: L. F. Laporte (ed.), *Reefs in Time and Space*, Soc. Econ. Paleontologists Mineralogists, Spec. Publ. 18, 9–76, 43 figs.; Tulsa, Okla.

Purser, B. H. (1973): *The Persian Gulf. Holocene Carbonate Sedimentation and Diagenesis in a Shallow Epicontinental Sea*, 471 pp., 250 figs., 7 pls., 3 maps; Berlin (Springer).

Pye, K. (1987): *Aeolian Dust and Dust Deposits*, 334 pp., 161 figs., 22 tables; London (Academic Press).

Quester, H. (1964): 'Petrographie des erdgashöffigen Hauptdolomits im Zechstein 2 zwischen Weser und Ems',

Z. dt. geol. Ges. (für 1962), **114**, 461–483, 4 figs., 5 pls.; Hannover.

Raaf, J. F. M. de, Boersma, J. R., and Gelder, A. van (1977): 'Wave-Generated Structures and Sequences from a Shallow Marine Succession, Lower Carboniferous, County Cork, Ireland', *Sedimentology*, **24**, 451–483, 23 figs.; Oxford.

Rachocki, A. H. and Church, M. (eds.) (1990): *Alluvial Fans – A Field Approach*, 391 pp., 224 figs., 22 tables; Chichester (J. Wiley).

Rahmani, R. A. and Flores, R. M. (eds.) (1985): *Sedimentology of Coal and Coal-Bearing Sequences*, Int. Assoc. Sedimentologists, Spec. Publ. 7, 418 pp., 304 figs., 30 tables; Oxford (Blackwell).

Ramsay, A. T. S. (1974): 'The Distribution of Calcium Carbonate in Deep Sea Sediments', 58–76, 10 figs., 1 table; in W. W. Hay (ed.), *Studies in Paleo-Oceanography*, Soc. Econ. Paleontologists, Mineralogists, 218 pp., figs., tables; Tulsa, Okla.

Reading, H. G. (ed.) (1978): *Sedimentary Environments and Facies*, 557 pp., 497 figs., 17 tables; Oxford (Blackwell).

Reading, H. G. (ed.) (1986): *Sedimentary Environments and Facies*, 2nd ed., 615 pp., 564 figs., 15 tables; Oxford (Blackwell).

Reeckmann, A. and Friedman, G. M. (1982): *Exploration for Carbonate Petroleum Reservoirs*, Elf-Aquitaine Centres de Recherches de Boussens et de Pau, 213 pp., 125 figs., 31 tables; New York (J. Wiley).

Reineck, H. E. (1984): *Aktuogeologie klastischer Sedimente*, Senckenberg-Buch, Vol. 61, 348 pp., 250 figs., 12 tables; Frankfurt a. M. (W. Kramer).

Reineck, H.E. and Singh, I. B. (1980): *Depositional Sedimentary Environments with Reference to Terrigenous Clastics*, 2nd revised and updated ed., 549 pp., 683 figs., Berlin, Heidelberg (Springer).

Reinson, G. E. (1984): 'Barrier-Island and Associated Strand-Plain Systems', 119–140, 28 figs.; in: R. G. Walker (ed.), *Facies Models*, Geoscience Canada, Reprint Ser. 1, 2nd ed., 317 pp., 357 figs., 17 tables; Toronto, Ontario (Geol. Assoc. Canada).

Riba, O., Villena, J., and Quirantes, J. (1967): 'Nota preliminar sobre la sedimentación en paleocanales terciarios de la zona de Caspe-Chiprana (Província de Zaragoza)', *Anales Edafología Agrobiología*, **26**, 617–634, 12 figs., 2 tables; Madrid.

Rice, R. B. *et al.* (1981): 'Developments in Exploration Geophysics, 1975–1980', *Geophysics*, **46**, 1088–1099; Tulsa, Okla.

Rieke, H. H. and Chilingarian, G. V. (1974): *Compaction of Argillaceous Sediments*, Developments in Sedimentology, Vol. 16, 424 pp., 217 figs., 49 tables; Amsterdam, London, New York (Elsevier).

Rieken, R. (1988): 'Lösungs-Zusammensetzung und Migrationsprozesse von Paläo-Fluidsystemen in Sedimentgesteinen des Norddeutschen Beckens (Mikrothermometrie, Laser-Raman-Spektroskopie und Isotopen-Geochemie)', *Göttinger Arb. Geol. Paläont.*, **37**, 116 pp., 37 figs., 22 tables, 5 pls.; Göttingen.

Rieken, R. and Gaupp, R. (1991): 'Fluideinschluss-Untersuchungen in Sandsteinen des Gasfeldes Thönse', 68–99, 15 figs., 5 tables, 2 pls., Niedersächsische Akad. Geowiss., Veröff., Vol. 6, *Das Gasfeld Thönse in Niedersachsen – Ein Unikat – Thermische Geschichte, Speicher- und Kohlenwasserstoff-Entwicklung*, 139 pp., 85 figs., 24 tables, 5 pls.; Hannover.

Risk, M. J. and Szczuczko, R. B. (1977): 'A Method for Staining Trace Fossils', *J. Sediment. Petrol.*, **47**, 855–859, 2 figs., 1 table; Tulsa, Okla.

Robertson, A. H. F. and Hudson, J. D. (1974): 'Pelagic Sediments in the Cretaceous and Tertiary History of the Troodos Massif, Cyprus', 403–436, 11 figs., 1 table; in: K. J. Hsü and H. C. Jenkyns (eds.), *Pelagic Sediments: On Land and under the Sea*, Intern. Assoc. Sedimentologists, Spec. Publ. 1, 447 pp., 204 figs., 17 tables; Oxford.

Robinson, R. B. (1966): 'Classification of Reservoir Rocks by Surface Texture', *Amer. Assoc. Petroleum Geologists, Bull.*, **50**, 547–559, 14 figs., 2 tables; Tulsa, Okla.

Rock-color chart (1963): Prepared by the Rock-Color Chart Committee, Reprint 1975, 8 pp.; New York (Geol. Soc. America).

Rock-color chart (1984): Rock-Color Chart Committee, Geol. Soc. America, Repr., 8 unnumbered pages, figs. in 1 appendix; Boulder, Colo.

Rodriguez, J. and Gutschick, R. C. (1970): 'Late Devonian-Early Mississippian Ichnofossils from Western Montana and Northern Utah', 407–438, 6 figs., 10 pls.; in: T. P. Crimes and J. C. Harper (eds.), *Trace fossils (1)*, Spec. Issue, *Geol. J.*, **3**, 547 pp., 114 figs., 9 tables, 87 pls.; Liverpool (Seel House Press).

Roehl, P. O. and Choquette, P. W. (eds.) (1985): *Carbonate Petroleum Reservoirs*, 622 pp., 396 figs., 27 tables, 35 summary charts, glossary, appendix; New York (Springer).

374 *References*

Rose, A. W., Hawkes, H. E., and Webb, J. S. (1979): *Geochemistry in Mineral Exploration*, 2nd ed., 657 pp., 259 figs., 70 tables, appendix; London (Academic Press).

Rose, W. and Pfannkuch, H. O. (1982): 'Unconventional Ideas about Unconventional Gas', 417–426,. 2 figs., 1 table; SPE/DOE 10836 presented at the 1982 SPE/DOE, *Proc. Unconventional Gas Recovery Symposium*, Pittsburgh, Pa., May 16–18; Dallas, Texas.

Rupke, N. A. (1978): 'Deep Clastic Seas', Chapter 12: 372–415, 59 figs., 3 tables; in: H. G. Reading (ed.), *Sedimentary Environments and Facies*, 557 pp., 497 figs., 16 tables; Oxford (Blackwell).

Russell, P. L. (1990): *Oil Shales of the World, Their Origin, Occurrence and Exploitation*, 753 pp., 314 figs., 54 tables; Oxford (Pergamon Press).

Rust, B. R. and Koster, E. H. (1984): 'Coarse Alluvial Deposits', 53–69, 18 figs., 2 tables; in: R. G. Walker (ed.), *Facies Models*, 2nd ed., 317 pp., 357 figs., 17 tables, Geoscience Canada, Reprint Series 1, Toronto, Ontario (Geol. Assoc. Canada).

Ruzyla, K. (1986): 'Characterization of Pore Space by Quantitative Image Analysis', *Soc. Petroleum Engineers Formation Evaluation*, **1**(4), 389–398, 14 figs., 3 tables; Dallas.

Sabins, F. F., Jr. (1962): 'How Do Bisti and Dead Horse Creek Strat Traps Compare?', *Oil Gas J.*, **60**, No. 33: 192–198, 10 figs., 1 table; Tulsa, Okla.

Sabins, F. F., Jr. (1980): 'Oil Occurrence and Plate Tectonics as Viewed on Landsat Images', *10th Petroleum World Congress*, Bucharest 1979, Proc. Vol. 2, *Exploration, Supply and Demand*, 105–109, 7 figs.; London, Philadelphia, Rheine (Heyden).

Sahu, Basanta K. (1964): 'Depositional Mechanisms from the Size Analysis of Clastic Sediments', *J. Sediment. Petrol.*, **34**, 73–83, 1 fig., 2 tables; Tulsa, Okla.

Sander, B., Athen, K., Billmann, S,. and Strubelt, A. (1993): 'Zur Präparation und Färbung von Blattkutikulen der Fossillagerstätte Grube Messel', *Der Präparator*, **39**, 17–18, 4 figs.; Bochum.

Sangree, J. B. and Widmier, J. M. (1977): 'Seismic Stratigraphy and Global Changes of Sea Level. Part 9: Seismic Interpretation of Clastic Depositional Facies', 165–184, 14 figs., 1 table; in: C. E. Payton (ed.), *Seismic Stratigraphy – Applications to Hydrocarbon Exploration*, Amer. Assoc. Petroleum Geologists, Mem. 26, 516 pp., 420 figs., 12 tables; Tulsa, Okla.

Sarg, J. F. (1988): 'Carbonate Sequence Stratigraphy', 155–181, 23 figs., 1 table; in: C. K. Wilgus et al. (eds.), *Sea-Level Change – An Integrated Approach*, Soc. Econ. Paleontologists Mineralogists, Spec. Publ. 42, 407 pp., 318 figs., 7 tables, 2 tables in appendix; Tulsa, Okla.

Schäfer, A. (1982): 'The Kontron Videoplan, A New Device for Determination of Grain Size Distributions from Thin Sections', *N. Jb. Geol. Paläont. Mh.*, **2**, 115–128, 7 figs.; Stuttgart.

Schettler, D., Jr. (1979): 'Gas Content and Transport in Shale and Their Implications for Well Productivity', in: H. Barow (ed.), *Proc. Third Eastern Gas Shales Symposium*, Oct. 1–3, 1979, U.S. Department of Energy, Morgantown Energy Technology Center, Morgantown, W.V., METC/SP–79/6, 27-30.

Schlager, W. (1992): *Sedimentology and Sequence Stratigraphy of Reefs and Carbonate Platforms. A Short Course*, Amer. Assoc. Petroleum Geologists, Continuing Education Course Note Series, Vol. 34; Tulsa, Okla.

Schlumberger Ltd. (1981): *Dipmeter Interpretation, Vol. 1, Fundamentals*, 61 pp., 118 figs., New York (Schlumberger Ltd.).

Schlumberger (1983): *Well Evaluation Conference, India*, 263 pp., 356 figs., unnumbered photographs and tables; Singapore (Vincent Printing Press).

Schmidt, V. (1965): 'Facies, Diagenesis, and Related Reservoir Properties in the Gigas Beds (Upper Jurassic), Northwestern Germany', 124–168, 32 figs., 10 tables; in: L. C. Pray and R. C. Murray (eds.), *Dolomitization and Limestone Diagenesis*, Soc. Econ. Paleontologists Mineralogists, Spec. Publ. 13, 180 pp., 98 figs., 22 tables; Tulsa, Okla.

Schmidt, V., McDonald, D. A., and Platt, R. L. (1977): 'Pore Geometry and Reservoir Aspects of Secondary Porosity in Sandstones', *Canad. Petroleum Geol., Bull.*, **25**, 271–290, 11 figs., 15 photomicrographs; Calgary, Alberta.

Schneider, W. (1977): 'Diagenese devonischer Karbonatkomplexe Mitteleuropas', *Geol. Jb.*, **D21**, 3–107, 11 figs., 7 tables, 8 pls.; Hannover.

Schneiderhöhn, P. (1954): 'Eine vergleichende Studie über Methoden zur quantitativen Bestimmung von Abrundung und Form an Sandkörnern (im Hinblick auf die Verwendung an Dünnschliffen)', *Heidelb. Beitr. Miner. Petrogr.* **4**, 172–191, 10 figs., 2 tables; Berlin, Göttingen, Heidelberg.

Schoch, R. M. (1989): *Stratigraphy – Principles and Methods*, 375 pp., 24 figs., 1 table, 4 appendices; New

York (van Nostrand).

Schönfeld, M. (1979): 'Stratigraphische, fazielle, paläogeographische und tektonische Untersuchungen im Oberen Malm des Deisters, Osterwaldes und Süntels (NW-Deutschland),' *Clausthaler geol. Abh.*, **35**, 270 pp., 15 figs., 6 tables, 7 pls.; Clausthal-Zellerfeld.

Schreiber, B. C., Friedman, G. M., Decima, A., and Schreiber, E. (1976): 'Depositional Environments of Upper Miocene (Messinian) Evaporite Deposits of the Sicilian Basin', *Sedimentology*, **23**, 729–760, 22 figs.; Oxford.

Schreiber, B. C., Roth, M. S., and Helman, M. L. (1982): 'Recognition of Primary Facies Characteristics of Evaporite and the Differentiation of These Forms from Diagenetic Overprints', 1–32; in: C. R. Handford, R. G. Loucks, and G. R. Davies (eds.), *Depositional and Diagenetic Spectra of Evaporites. A Core Workshop*, Soc. Econ. Paleontologists, Mineralogists, Core Workshop 3, 395 pp., 190 figs., 5 tables, 1 appendix with core photographs; Calgary.

Schroeder, J. H. and Purser, B. H. (eds.) (1986): *Reef Diagenesis*, 455 pp., 187 figs., 22 tables; Berlin (Springer).

Schröder, L. and Schöneich, H. (1986): *International Map of Natural Gas Fields in Europe*, 1 : 2 500 000, explanatory notes, 2nd ed., 3–175, 17 figs., 5 text tables, 8 tables as appendix; Hannover.

Scotchman, I. C. and Johnes, L. H. (1990): 'Wave-Dominated Deltaic Reservoirs of the Brent Group, Northwest Hutton Field, North Sea', 227–261, 28 figs., 1 table; in: J. H. Barwis, J. G. McPherson and J. R. J. Studlick (eds.), *Sandstone Petroleum Reservoirs*, 583 pp., 443 figs., 20 tables, 48 color pls.; New York (Springer).

Seemann, U. (1979): 'Diagenetically Formed Interstitial Clay Minerals as a Factor in Rotliegend Sandstone Reservoir Quality in the Dutch Sector of the North Sea', *J. Petroleum Geol.*, **1**(3), 55–62, 6 figs.; Beaconsfield, Bucks.

Seemann, U. (1982): 'Depositional Facies, Diagenetic Clay Minerals and Reservoir Quality of Rotliegend Sediments in the Southern Permian Basin (North Sea): A Review', *Clay Minerals*, **17**, 55–67, 10 figs.; London.

Seemann, U. and Scherer, M. (1984): 'Volcaniclastics as Potential Hydrocarbon Reservoirs', *Clay Minerals*, **19**, 457–470, 13 figs., 1 table; London.

Seemann, W. (1980): 'Das Kohlenwasserstoffpotential karbonatischer Gesteine', *Erdöl u. Kohle*, **33**, 481; Hamburg.

Seilacher, A. (1962): 'Paleontological Studies on Turbidite Sedimentation and Erosion', *J. Geol.*, **70**, 227–234, 1 fig., 1 table, 1 pl.; Chicago.

Seilacher, A. (1964): 'Biogenic Sedimentary Structures', in: J. Imbrie and N. Newell (eds.), *Approaches to Paleoecology*, 296–316, 8 figs., 1 table, 1 pl.; New York (J. Wiley & Sons).

Seilacher, A. (1967): 'Bathymetry of Trace Fossils', *Marine Geol.*, **5**, 413–428, 4 figs., 2 pls.; Amsterdam.

Seim, R. and Tischendorf, G. (eds.) (1990): *Grundlagen der Geochemie*, 632 pp., 274 figs., 130 tables, 13 appendices; Leipzig (VEB Deutscher Verlag für Grundstoffindustrie).

Selley, R. C. (1970): *Ancient Sedimentary Environments. A Brief Survey*, 237 pp., 71 figs., 8 tables; London (Chapman and Hall Ltd.).

Selley, R. C. (1977): 'Deltaic Facies and Petroleum Geology', Chapter 6: 197–224, 9 figs., 1 table; in: G. D. Hobson (ed.), *Development in Petroleum Geology – 1*, 333 pp., 18 figs., 5 tables; London (Applied Science).

Selley, R. C. (1978): 'Porosity Gradients in North Sea Oil-Bearing Sandstones', *J. Geol. Soc. London*, **135**, 119–132, 16 figs., 2 tables; London.

Selley, R. C. (1982): *An Introduction to Sedimentology*, 2nd ed., 417 pp., 197 figs., 38 tables; London (Academic Press).

Selley, R. C. (1985): *Elements of Petroleum Geology*, 449 pp., 286 figs., 28 tables, 7 pls., 3 appendices; New York (Freeman).

Selley, R. C. (1988): *Applied Sedimentology*, 3rd ed., 446 pp., 284 figs., 37 tables, 8 color pls.; London (Academic Press).

Sellwood, B. W. (1978): 'Biogenic Sedimentary Structures', 61–64, 2 figs.; in: R. W. Fairbridge and J. Bourgeois (eds.), *The Encyclopedia of Sedimentology*, 901 pp.; Stroudsburg Pa. (Dowden, Hutchinson & Ross, Inc.).

Sellwood, B. W. (1986): 'Shallow-Marine Carbonate Environments', Chapter 10: 283–342, 69 figs., 3 tables; in: H. G. Reading (ed.), *Sedimentary Environments and Facies*, 615 pp., 564 figs., 15 tables; Oxford (Blackwell).

SEPM Spec. Publ. No. 35 (1985): *Biogenic Structures: Their Use in Interpreting Depositional Environments*, Soc. Economic Paleontologists Mineralogists, Spec. Publ. 35, 347 pp.; Tulsa, Okla.

Serra, J. (1982): *Image Analysis and Mathematical Morphology*, 628 pp.; London (Academic Press).

Serra, O. (1984): *Fundamentals of Well-Log Interpretation – 1. The Acquisition of Logging Data*, Developments in Petroleum Science, Vol. 15A, 423 pp., 497 figs., 45 tables, 7 appendices with 32 figs. and 25 tables, index and glossary; Amsterdam (Elsevier).

Serra, O. (1985): *Diagraphies Différées – Bases de l'Interprétation, Tome 2, Interprétation des Données Diagraphiques*, Bull. Centres Recherches Exploration-Production Elf Aquitaine, Mém. 7, 631 pp., 888 figs., 71 tables, appendix with 10 tables; Paris (Technip).

Serra, O. (1986): *Fundamentals of Well-Log Interpretation – 2. The Interpretation of Logging Data*, Developments in Petroleum Science, Vol. 15B, 684 pp., 1025 figs., 76 tables, 1 appendix with 14 tables; Amsterdam, Oxford etc. (Elsevier).

Serra, O., and Abbott, H. T. (1982): 'The Contribution of Logging Data to Sedimentology and Stratigraphy', *J. Soc. Petroleum Engineers*, **22**, 117–131, 21 figs., 1 table; Dallas, Texas.

Serra, O. and Sulpice, L. (1975): *Sedimentological Analysis of Shale-Sand Series from Well Logs*, SPWLA, 16th Ann. Log Symp. Trans., paper W, 23 pp., 13 figs.; Houston, Texas.

Shannon, P. M. and Naylor, D. (1989): *Petroleum Basin Studies*, 206 pp., 108 figs., 1 table; London (Graham & Trotman).

Sharma, P. V. (1986): *Geophysical Methods in Geology*, 2nd ed., 442 pp., 265 figs., 21 tables; Amsterdam (Elsevier).

Shearman, D. J. (1970): 'Recent Halite Rock, Baja California, Mexico', *Inst. Mining Metallurgy Trans. Sect. B*, **79**, B155–B162, 6 figs.; London.

Shepard, F. P. (ed.) (1963): *Submarine Geology*, 2nd ed., 557 pp., 222 figs., 15 tables, 1 chart of the world; New York (Harper and Row).

Sheriff, R. E. (1976): 'Inferring Stratigraphy from Seismic Data', *Amer. Assoc. Petroleum Geologists Bull.* **60**, 528–542, 10 figs.; Tulsa, Okla.

Sheriff, R. E. (1980a): *Seismic Stratigraphy*, 227 pp., 115 figs., 2 tables, 8 color pls.; Boston (Intern. Human Resources Develop. Corp.).

Sheriff. R. E. (1980b): 'Seismic Stratigraphy', Chapter 5, 189–206, 9 figs., 1 table; in: G. D. Hobson (ed.), *Developments in Petroleum Geology – 2*, 345 pp., 152 figs.; London (Applied Science).

Shumaker, R. C. (1980): 'The Importance of Regional and Local Structure to Devonian Shale Gas Production from the Appalachian Basin', *Energies fossiles – Les Hydrocarbures, 26e Congr. Géol. Intern.*, Sect. 14, 147–174, 19 figs.; Paris (Technip).

Shumaker, R. C. (1982): 'A Detailed Geologic Study of Three Fractured Devonian Shale Gas Fields', *Unconventional Gas Recovery Symposium Proc.*, May 16–18, 1982, 9–28; Pittsburgh, Pa.

Siemers, R. W., Tillman, R. W., and Williamson, C. R. (eds.) (1981): *Deep Water Clastic Sediments, A Core Workshop*, Soc. Econ. Paleontologists Mineralogists, Core Workshop No. 2, 416 pp., 156 figs., 12 tables, 1 appendix; Tulsa, Okla.

Siever, R. (1962): 'Silica Solubility, 0°–200°C, and the Diagenesis of Siliceous Sediments', *J. Geol.*, **70**, 127–150, 6 figs., 1 table; Chicago, Ill.

Sindowski, K. H. (1957): 'Die synoptische Methode des Kornkurven-Vergleiches zur Ausdeutung fossiler Sedimentationsräume', *Geol. Jb.*, **73**, 235–275, 68 figs., 1 table; Hannover.

Slingerland, R., Harbaugh, J. W., and Furlong, K. (1994): *Simulting Clastic Sedimentary Basins. Physical Fundamentals and Computer Programs for Creating Dynamic Systems*, Englewood Cliffs, N.J. (Prentice Hall).

Sloss, L. L. (1963): 'Sequences in the Cratonic Interior of North America', *Geol. Soc. America, Bull.*, **74**, 93–113, 6 figs.; New York.

Smart, P. and Tovey, N. K. (1981): *Electron Microscopy of Soils and Sediments: Examples*, 178 pp., 167 figs.; Oxford (Clarendon Press).

Smart, P. and Tovey, N. K. (1982): *Electron Microscopy of Soils and Sediments: Techniques*, 264 pp., 145 figs., 18 tables; Oxford (Clarendon Press).

Smith, D. B. (1980): (a) 'The Shelf-Edge Reef of the Middle Magnesian Limestone (English Zechstein Cycle 1) of Northeastern England – A Summary', 3–5, 1 fig.; (b) 'The Evolution of the English Zechstein Basin', 7–34, 2 figs., 1 table; in: H. Füchtbauer and T. Peryt (eds.), *The Zechstein Basin with Emphasis on Carbonate Sequences*, Contrib. to *Sedimentology* 9, 328 pp., 168 figs., 5 tables, 3 folders; Stuttgart (Schweizerbart).

Smith, D. B. (1982): 'The Early Permian Desert in Britain (Abstract)', *11th Intern. Congress on Sedimentology*, McMaster University, Hamilton, Ontario, Canada, August 22–27, 1982, p. 68; Hamilton, Canada.

Sneider, R. M., King, H. R., Hawkes, H. E., and Davis, T. B. (1984): 'Reservoir Rock Studies Improve Discovery

Success', *World Oil*, **198**(6), 87–102, 11 figs., 1 table; Houston, Texas.

Sneider, R. M., Tinker, C. N., and Meckel, L. D. (1978): 'Deltaic Environment Reservoir Types and Their Characteristics', *J. Petroleum Technol.*, **30**, 1538–1546, 16 figs.; Dallas, Tex.

Soeparjadi, R. A., Nayoan, G. A. S., Beddoes, L. R., Jr., and James, W. V. (1975): 'Exploration Play Concepts in Indonesia', 51–64, 8 figs., 2 tables; *Proc. 9th World Petroleum Congress* Tokyo 1975, Vol. 3, *Exploration and Transportation*, 382 pp., 165 figs., 38 tables; London (Applied Science Publ.).

Sonnenberg, S. A. and Weimer, R. J. (1981): 'Tectonics, Sedimentation, and Petroleum Potential, Northern Denver Basin, Colorado, Wyoming, and Nebraska', *Colorado School Mines, Quart.*, **76**, 45 pp., 39 figs., 1 table, 2 pls.; Golden, Colo.

Spencer, A. M. (ed.) (1991): *Generation, Accumulation, and Production of Europe's Hydrocarbons*, Spec. Publ. European Assoc. Petroleum Geoscientists, No. 1, 459 pp., 449 figs., 23 tables, 2 appendices; Oxford (University Press).

Stach, E., Mackowsky, M.T., Teichmüller, M., Taylor, G. H., Chandra, D., and Teichmüller, R. (eds.) (1982): *Stach's Textbook of Coal Petrology*, 3. revised and enlarged ed., transl. and English rev. by D. G. Murchison, G. H. Taylor, and F. Zierke, 535 pp., 204 figs., 49 tables, 6 pls.; Berlin, Stuttgart (Borntraeger).

Stahl, W. (1968): *Kohlenstoff-Isotopenanalysen zur Klärung der Herkunft nordwestdeutscher Erdgase*, TH Clausthal naturw. Diss., 9 February 1968: 98 pp., 19 figs., 40 tables; Clausthal-Zellerfeld.

Stalder, P. J. (1973): 'Influence of Crystallographic Habit and Aggregate Structure of Authigenic Clay Minerals on Sandstone Permeability', *Geol. en Mijnbouw*, **52**, 217–220, 3 figs.; Culemborg.

Stäuble, A. J. and Milius, G. (1970): 'Geology of Groningen Gas Field, Netherlands', 359–369, 8 figs., 1 table; in: M. T. Halbouty (ed.), *Geology of Giant Petroleum Fields*, Amer. Assoc. Petroleum Geologists, Mem. 14, 575 pp., 417 figs., 34 tables; Tulsa, Okla.

Stancu-Kristoff, G. and Stehn, O. (1984): 'Ein grossregionaler Schnitt durch das nordwestdeutsche Oberkarbon-Becken vom Ruhrgebiet bis in die Nordsee', *Fortschr. Geol. Rheinl. u. Westf.*, **32**, 35–38, 1 pl., Krefeld.

Stanley, D. J. and Swift, D. J. P. (eds.) (1976): *Marine Sediment Transport and Environmental Management*, 602 pp., 382 figs., 38 tables; New York (J. Wiley).

Steele, R. P. (1982): 'Conclusions from a Study of British Aeolian Sandstones (Abstract)', *11th Intern. Congress on Sedimentology*, McMaster University, Hamilton, Ontario, Canada, August 22–27, 1982, p. 69; Ontario, Canada.

Stehli, F. G., Creath, W. B., Upshaw, C. F., and Forgotson, J. M. (1972): 'Depositional History of Gulfian Cretaceous of East Texas Embayment', *Amer. Assoc. Petroleum Geologists, Bull.*, **56**, 38–67, 29 figs.; Tulsa, Okla.

Steineke, M., Bramkamp, R. A., and Sander, N. J. (1958): 'Stratigraphic Relations of Arabian Jurassic Oil', 1294–1329, 6 figs., 1 table; in: L. G. Weeks (ed.), *Habitat of Oil, A Symposium*, Amer. Assoc. Petroleum Geologists, 1384 pp.; Tulsa, Okla.

Stevens, R. E. and Carron, M. K. (1948): 'Simple Field Test for Distinguishing Minerals by Abrasion pH', *Amer. Mineralogist*, **33**, 31–49, 2 figs., 2 tables as appendix; Menasha, Wisc.

Storz, M. (1931): *Die sekundäre authigene Kieselsäure in ihrer petrographisch-geologischen Bedeutung. II. Teil: Die Einwirkung der sekundären authigenen Kieselsäure auf vorhandene Gesteine (Einkieselung und Verkieselung)*, Monogr. Geol. Paläont., Ser. II, H. 5, 139–479, 147 figs., 15 pls.; Berlin (Borntraeger).

Stow, D. A. V. (1986): 'Deep Clastic Seas', 399–444, 45 figs., 3 tables; in: H. G. Reading (ed.), *Sedimentary Environments and Facies*, 2nd ed., 615 pp., 544 figs., 15 tables; Oxford (Blackwell).

Stow, D. A. V. and Piper, D. J. W. (eds.) (1984): *Fine-Grained Sediments: Deep-Water Processes and Facies*, Geol. Soc. Spec. Publ., 664 pp., 280 figs.; Oxford (Blackwell Scientific).

Stride, A. H. (1963): 'Current-Swept Sea Floors near the Southern Half of Great Britain', *J. Geol. Soc. London Quart.*, **119**, 175–199, 13 figs., 7 pls.; London.

Stride, A. H., Belderson, R. H., Kenyon, N. H., and Johnson, M. A. (1982): 'Offshore Tidal Deposits: Sand Sheet and Sand Bank Facies', 95–125, 22 figs., 2 tables; in: A. H. Stride (ed.), *Offshore Tidal Sands, Processes and Deposits*, 222 pp., 91 figs., 15 tables, 35 pls.; London (Chapman and Hall).

Sundborg, A. (1956): 'The River Klarälven – A Study of Fluvial Processes', *Geografiska Annaler*, **38**, 125–316, 60 figs., 8 tables, 1 pl., 1 map; Stockholm.

Swennen, R., Poot, B., and Marchal, G. (1991): 'Computerized Tomography as a Tool in Reservoir Characterization', *Zbl. Geol. Paläont.*, Teil 1, 1990, 1105–1124, 1 table, 6 pls.; Stuttgart.

Swift, D. J. P., Duane, D. B., and Pilkey, O. H. (eds.) (1972): *Shelf Sediment Transport: Process and Pattern*, 656 pp., 262 figs., 38 tables; Stroudsburg, Pa. (Dowden, Hutchinson & Ross).

Swift, D. J. P., Oertel, G. F., Tillman, R. W., and Thorne, J. A. (eds.) (1991): *Shelf Sand and Sandstone Bodies – Geometry, Facies and Sequence Stratigraphy*, Int. Assoc. Sedimentologists, Spec. Publ. 14, 532 pp., 292 figs., 26 tables, 6 appendices; Oxford (Blackwell).

Szymansky, A. (1989): *Technical Mineralogy and Petrography. An Introduction to Materials Technology*, Part A: 716 pp., 611 figs., 123 tables, Part B: 223 pp., 1 fig., 152 tables; Warszawa (PWN – Polish Scientific Publ.); Amsterdam (Elsevier).

Taylor, J. C. M. (1977): 'Sandstones as Reservoir Rocks', Chapter 5: 147–196, 2 figs., in: G. D. Hobson (ed.), *Development in Petroleum Geology – 1*, 335 pp., 18 figs., 5 tables; London (Applied Science).

Taylor, J. C. M. (1983): 'Bit-Metamorphism, Illustrated by Lithological Data from German North Sea Wells', 211–219, 11 figs., 2 tables; in: J. P. H. Kaasschieter and T. J. A. Reijers (eds.), *Petroleum Geology of the Southeastern North Sea and the Adjacent Onshore Areas*, (The Hague 1982), *Geol. en Mijnbouw*, **62**, 239 pp., 241 figs., 19 tables, 3 pls., 1 appendix; Den Haag.

Teichmüller, M. and Teichmüller, R. (1968): 'Geological Aspects of Coal Metamorphism', 233–267, 21 figs., 1 table, 3 pls.; in: D. Murchison and T. S. Westoll (eds.), *Coal and Coal-Bearing Strata*, 418 pp., 123 figs., 23 tables, 41 pls.; Edinburg (Oliver & Boyd).

Teyssen, T. A. L. (1984): 'Sedimentology of the Minette Oolitic Ironstones of Luxembourg and Lorraine: A Jurassic Subtidal Sandwave Complex', *Sedimentology*, **31**, 195–211, 13 figs.; Oxford.

Tietze, K. W. (1979): *Dynamik und Sedimenthaushalt einer Gezeitenlagune an der Nordküste der Bretagne*, Habilitationsschrift Universität Marburg, 170 pp., 69 figs., 7 tables, 6 pls.; Marburg.

Tillman, J. E. and Barnes, H. L. (1983): 'Deciphering Fracturing and Fluid Migration Histories in the Northern Appalachian Basin', *Amer. Assoc. Petroleum Geologists, Bull.*, **67**, 692–705, 11 figs., 1 table; Tulsa, Okla.

Tillman, R. W., Swift, D. J. P., and Walker, R. G. (eds.) (1985): *Shelf Sand and Sandstone Reservoirs*, Soc. Econ. Paleontologists Mineralogists, Short course 13, 708 pp., 467 figs., 12 tables, with outcrop description; Tulsa, Okla.

Tillman, R. W. and Weber, K. J. (1987): *Reservoir Sedimentology*, Soc. Econ. Paleontologists Mineralogists, Spec. Publ. 40, 357 pp., 326 figs., 31 tables; Tulsa Okla.

Tissot, B. P. and Welte, H. D. (1984): *Petroleum Formation and Occurrence*, 2nd revised and enlarged ed., 699 pp., 327 figs., 68 tables; Berlin, Heidelberg, New York (Springer).

Trewin, N. (1988): 'Use of the Scanning Electron Microscope in Sedimentology', 229–273, 25 figs., 2 tables; in: M. E. Tucker (ed.), *Techniques in Sedimentology*, 394 pp., 223 figs., 38 tables, Oxford etc. (Blackwell).

Tröger, W. E. and Braitsch, O. (1967): *Optische Bestimmung der gesteinsbildenden Minerale. Teil 2. Textband*, 822 pp., 259 figs., 16 tables; Stuttgart (Schweizerbart).

Tröger, W. E. (1979): *Optical Determination of Rock-Forming Minerals, Part 1. Determinative Tables*, English ed. of the 4th German ed. by H. U. Bambauer, F. Taborszky, and H. D. Trochim, 1979, 188 pp., 264 figs., 1 chart, 2 stereogr., 112 diagr.; Stuttgart (Schweizerbart).

Tucker, M. E. (1973): 'Sedimentology and Diagenesis of Devonian Pelagic Limestones (Cephalopodenkalk) and Associated Sediments of the Rhenohercynian Geosyncline, West Germany', *N. Jb. Geol. Paläont., Abh.*, **142**, 320–350, 24 figs., 1 table; Stuttgart.

Tucker, M. E. (1974): 'Sedimentology of Paleozoic Pelagic Limestones: The Devonian Griotte (Southern France) and Cephalopodenkalk (Germany)', in: K. J. Hsü and H. C. Jenkyns (eds.), *Pelagic Sediments: On Land and under the Sea*, Intern. Assoc. Sedimentologists, Spec. Publ. 1, 71–92, 18 figs.; Oxford (Blackwell).

Tucker, M. E. (ed.) (1988): *Techniques in Sedimentology*, 394 pp., 223 figs., 38 tables; Oxford, London etc. (Blackwell).

Tucker, M. E. and Wright, V. P. (1990): *Carbonate Sedimentology*, 482 pp., 392 figs., 6 tables; Oxford, London, etc. (Blackwell).

Ushatinskiy, I. N. (1982): 'Lithology and Petroleum Prospects of Jurassic-Neocomian Bituminous Deposits of West Siberia', *Intern. Geol. Review*, **24**, 1211–1221, 6 figs., 5 tables; translated from 'Litologiya perspektivy neftenosnosti yursko-neokomskikh bituminoznyk otlozheny Zapadnoy Sibiri', *Sovetskaya Geologiya*, 1981, **2**, 11–22; Falls Church, Virginia (AGI).

Vahl, F. (1956): 'Der Anfärbetest nach *J. Leonhard* und *F. Vahl* als eine mikroskopisches Untersuchungsverfahren für Tonminerale', *Photographie und Wissenschaft*, **5**, 17–18, 2 figs.; Leverkusen (Agfa).

Vail, P. R., Mitchum, R. M., Jr., and Thompson, S., III (1977): 'Seismic Stratigraphy and Global Changes of Sea Level, Part 3: Relative Changes of Sea Level from Coastal Onlap', 63–81, 15 figs.; in C. E. Payton (ed.), *Seismic Stratigraphy – Applications to Hydrocarbon Exploration*, Amer. Assoc. Petroleum Geologists, Mem. 26, 516 pp., 420 figs. 12 tables; Tulsa, Okla.

Vail, P. R., Todd, R. G., and Sangree, J. B. (1977): 'Chronostratigraphic Significance of Seismic Reflections', 99–116, 17 figs., in: C. E. Payton (ed.), *Seismic Stratigraphy – Applications to Hydrocarbon Exploration*, Amer. Assoc. Petroleum Geologists, Mem. 26, 516 pp., 420 figs., 12 tables; Tulsa, Okla.

Valloni, R. and Mezzadri, G. (1984): 'Compositional Suites of Terrigenous Deep-Sea Sands of the Present Continental Margins', *Sedimentology*, **3**, 353–364, 10 figs., 2 tables; Oxford.

Vinopal, R. J., Nuhfer, E. B., and Hohn, M. E. (1982): 'Classification of Fabric Types and Fracture Types in Upper Devonian Mudrocks from West Virginia and Virginia, USA (Abstract)', *11th Intern. Congress on Sedimentology*, McMaster University, August 22–27, 1982, p. 8; Hamilton, Ontario.

Visher, G. S. (1965): 'Use of Vertical Profile in Environmental Reconstruction', *Amer. Assoc. Petroleum Geologists, Bull.*, **49**, 41–61, 16 figs.; Tulsa, Okla.

Visher, G. (1990): *Exploration Stratigraphy*, 2nd ed., 433 pp., 788 figs., 53 tables, 2 appendices; Tulsa, Okla. (PennWell Books).

Vyšemirskij, V. S. (ed.) (1986): *Baženovskij gorizont Zapa dnoj Sibiri stratigrafija, paleogeografija, ekosistema, neftenosost'* (The Bashenov Horizon of Western Siberia; Stratigraphy, Paleogeography, Ecosystem, Oil Potential), 216 pp., 24 figs., 4 tables, 56 pls.; Novosibirsk.

Wagoner, J. C. van, Mitchum, R. M., Jr., Posamentier, H. W., and Vail, P. R. (1987): 'Key Definitions of Sequence Stratigraphy (Part 2)', 11–14, 4 figs.; in: A. W. Bally (ed.), *Atlas of Seismic Stratigraphy, Vol. 1*. Amer. Assoc. Petroleum Geologists, Studies in Geology, Vol. 27, 125 pp., 179 figs., 1 table; Tulsa, Okla.

Wagoner, J. C. van, Mitchum, R. M., Campion, K. M., and Rahmanian, V. D. (1990): *Siliciclastic Sequence Stratigraphy in Well Logs, Cores, and Outcrops: Concepts for High-Resolution Correlation of Time and Facies*, Amer. Assoc. Petroleum Geologists, Methods in Exploration Series, No. 7, 55 pp., 40 figs., 2 tables; Tulsa, Okla.

Walker, R. G. (1978): 'Deep-Water Sandstone Facies and Ancient Submarine Fans: Models for Exploration for Stratigraphic Traps', *Amer. Assoc. Petroleum Geologists, Bull.*, **62**, 932–966, 23 figs., 2 tables; Tulsa, Okla.

Walker, R. G. (ed.) (1984): *Facies Models*, 2nd ed., 317 pp., 357 figs., 17 tables, Geoscience Canada, Reprint Series 1; Toronto, Ontario (Geol. Assoc. Canada).

Walker, T. R. (1967): 'Formation of Red Beds in Modern and Ancient Deserts', *Geol. Soc. America, Bull.*, **78**, 353–368, 5 figs., 1 table, 2 pls.; New York.

Walmsley, P. J. (1975): 'The Forties Field', 477–484, 8 figs.; in: A. W. Woodland (ed.), *Petroleum and the Continental Shelf of North-West Europe, 1, Geology*, 501 pp., 327 figs., 6 tables, 9 pls., 1 appendix; London (Applied Science Publ.).

Walther, J. (1894): 'Einleitung in die Geologie als historische Wissenschaft. Beobachtungen über die Bildung der Gesteine und ihrer organischen Einschlüsse', Th. 1.–3. 3. *Lithogenesis der Gegenwart. Beobachtungen über die Bildung der Gesteine an der heutigen Erdoberfläche*, 535–1055, 2 figs., unnumbered tables; Jena (Fischer).

Waples, D. W. (1985): *Geochemistry in Petroleum Exploration*, Geol. Sci. Ser., 232 pp., 194 figs., 27 tables, glossary; Dordrecht (D. Reidel).

Warne, S. St. J. (1962): 'A Quick Field or Laboratory Staining Scheme for the Differentiation of the Major Carbonate Minerals', *J. Sediment. Petrol.*, **32**, 29–38, 3 figs., 3 tables; Tulsa, Okla.

Warren, G. (1974): 'Simplified Form of the Folk–Ward Skewness Parameter', *J. Sediment. Petrol.*, **44**, 259; Tulsa, Okla.

Warren, J. K. (1989): *Evaporite Sedimentology – Importance in Hydrocarbon Accumulation*, 285 pp., 109 figs., 8 tables; Englewood Cliffs, N.J. (Prentice Hall).

Weaver, C. E. (1989): *Clays, Muds, and Shales*, Developments in Sedimentology, Vol. 44, 820 pp., 325 figs., 75 tables; Amsterdam (Elsevier).

Weber, K. J. (1971): 'Sedimentological Aspects of Oil Fields in the Niger Delta', *Geol. en Mijnbouw*, **50**, 559–576, 17 figs.; Rotterdam.

Weber, K. J. and Daukoru, E. (1975): 'Petroleum Geology of the Niger Delta', *Proc. 9th World Petroleum Conf.*, Tokyo 1975, Vol. 2, *Geology*, 209–221, 7 figs.; London (Applied Science Publ.).

Wedepohl, K. H. and Correns, C. W. (eds.) (1969–1978): *Handbook of Geochemistry*, Vol. 1, 442 pp., unnumbered figs. and tables; Vol. 2 Part 1–5, Elements, loose page editions; Berlin, Heidelberg (Springer).

Weimer, R. J., Porter, K. W., and Land, C. B. (1985): *Depositional Modeling of Detrital Rocks – With Emphasis on Cored Sequences of Petroleum Reservoirs*, Soc. Econ. Paleontologists Mineralogists, Core Workshop No. 8, Golden, Colorado, August 10–11, 1985: 252 pp., 131 figs., 1 table; Tulsa, Okla.

Weismann, T. J. (1980): 'Developments in Geochemistry and Their Contribution to Hydrocarbon Exploration',

10th World Petroleum Congr. Bucharest, 1979, Vol. 2, Exploration Supply and Demand, 369–386, 7 figs., 2 tables; London, Philadelphia, Rheine (Heyden).

Welton, J. E. (1984): *SEM Petrology Atlas,* Amer. Assoc. Petroleum Geologists, Methods in Exploration Series 4, 237 pp., 4 figs., numerous scanning electron micrographs and EDX spectra of 119 sedimentary mineral occurrences in plates, 1 appendix, glossary; Tulsa, Okla.

Werner, F. and Wetzel, A. (1982): 'Interpretation of Biogenic Structures in Oceanic Sediments', *Actes Colloque Internat. CNRS,* Bordeaux, September 1981, no. 31, 275–288; 15 figs.; Paris.

Wetzel, A. (1981): 'Ökologische und stratigraphische Bedeutung biogener Gefüge in quartären Sedimenten am NW-afrikanischen Kontinentalrand. Reihe C.: Geologie und Geophysik'; in: M. Sarnthein and H. J. Dürbaum (eds.), *Meteor Forschungsergebnisse,* No. 34, IV: 64 pp., 44 figs., 10 pls.; Berlin (Borntraeger).

Wetzel, A. (1991): 'Ecological Interpretation of Deep-Sea Trace Fossil Communities', *Palaeogeography, Palaeoclimatology, Palaeoecology,* **85,** 47–69, 10 figs., 6 tables; Amsterdam.

Wetzel, A. and Aigner, T. (1986): 'Stratigraphic Completeness: Tiered Trace Fossils Provide a Measuring Stick', *Geology,* **14,** 234–237, 5 figs.; Boulder, Colo.

Wetzel, K. (1983): 'Über die Stellung der Isotopengeochemie im Rahmen der geologischen Wissenschaften', *Z. geol. Wiss.,* **11,** 531–540, 1 fig.; Berlin.

Whalley, W. B. (ed.) (1978): *Scanning Electron Microscopy in the Study of Sediments,* Symp. held at Swansea in Sept. 1977: 414 pp., 252 figs., 11 tables; Norwich (Geo Abstracts).

Whateley, M. K. G. and Pickering, K. T. (eds.) (1989): *Deltas; Sites and Traps for Fossil Fuels,* April 21–22, 1987, London, United Kingdom, Geol. Soc. London, Spec. Publ. 41, 368 pp.; Oxford.

White, P. D. and Moss, J. T. (1983): *Thermal Recovery Methods,* 361 pp., figs., tables; Tulsa, Okla. (PennWell Books).

Wijhe, D. H. van and Bless, M. J. M. (1974): 'The Westphalian of the Netherlands with Special Reference to Miospore Assemblages', *Geol. en Mijnbouw,* **53,** 295–328, 9 figs., 2 tables, 6 pls.; Leiden.

Wijhe, D. H. van, Lutz, M., and Kaasschieter, J. P. H. (1980): 'The Rotliegend in the Netherlands and Its Gas Accumulation', *Geol. en Mijnbouw,* **59,** 3–24, 18 figs.; Den Haag.

Wilgus, C. K., Hastings, B. S., Posamentier, H., Wagoner, J. van, Ross, C. A., and Kendall, C. S. C. (eds.) (1988): *Sea-Level Changes: An Integrated Approach,* Soc. Econ. Paleontologists Mineralogists, Spec. Publ. 42, 407 pp., 319 figs., 9 tables, 2 tables in appendix, 1 pl.; Tulsa, Okla.

Williams, G. D. and Dobb, A. (eds.) (1993): *Tectonics and Seismic Sequence Stratigraphy,* Geol. Soc. London, Spec. Publ. 71, 225 pp., 115 figs., 3 tables; London.

Williams, P. J. and Smith, M. W. (1989): *The Frozen Earth. Fundamentals of Geocryology,* 300 pp.; Cambridge.

Wilson, H. H. (1977): '"Frozen-In"-Hydrocarbon Accumulations or Diagenetic Traps – Exploration Targets', *Amer. Assoc. Petroleum Geologists, Bull.,* **61,** 483–491, 1 fig.; Tulsa, Okla.

Wilson, H. H. (1987): 'Structural Evolution of the Golden Lane, Tampico Embayment, Mexico', *J. Petroleum Geology,* **10**(1), 5–40, 21 figs.; Beaconsfield, Bucks, England.

Wilson, J. L. (1970): 'Depositional Facies across Carbonate Shelf Margins', *Trans. Gulf Coast Assoc. Geol. Soc.,* **20,** 229–233, 1 fig.; Shreveport, Louisiana.

Wilson, J. L. (1974): 'Characteristics of Carbonate-Platform Margins', *Amer. Assoc. Petroleum Geologists, Bull.,* **58,** 810–824, 6 figs.; Tulsa, Okla.

Wilson, J. L. (1975): *Carbonate Facies in Geologic History,* 471 pp., 183 figs., 30 pls.; Berlin (Springer).

Wilson, J. L. (1980): 'A Review of Carbonate Reservoirs', 95–117, 14 figs.; in: A. D. Miall (ed.), *Facts and Principles of World Petroleum Occurrence,* Canad. Soc. Petroleum Geologists, Mem. 6, 1003 pp., 661 figs., 48 tables; Calgary, Alberta.

Wilson, M. D. (1982): *Reservoir Quality and Formation Damage Susceptibility in Sandstones,* Typescript, June 1 and 2, 1982, Hilton Intern. London, 115 pp., 124 figs.; London.

Wilson, M. D. and Pittman, E. D. (1977): 'Authigenic Clays in Sandstones: Recognition and Influence on Reservoir Properties and Paleoenvironmental Analysis', *J. Sediment. Petrol.,* **47,** 3–31, 21 figs., 4 tables; Tulsa, Okla.

Wittenhagen, S. (1980): *Sedimentologisch-petrographische Untersuchungen des Bentheimer Sandsteins im Erdölfeld Bramberge/Emsland,* Clausthaler geowiss. Diss. 5, 132 pp., 75 figs., 30 tables, 2 appendices; Clausthal-Zellerfeld.

Woick, R. (1986): *Sedimentologie und Diagenese des Lias alpha am Westrand des Gifhorner Troges;* Diss. Clausthal-Zellerfeld, 1986, 303 pp., 88 figs., 16 tables, 4 pls.; Clausthal-Zellerfeld.

Wolf, K. H. (1973): 'Conceptual Models, 1. Examples in Sedimentary Petrology, Environmental and Stratigraphic Reconstruction, and Soil, Reef, Chemical and Placer Sedimentary Ore Deposits', *Sedimentary Geol.*, **9**, 153–193, 14 figs.; Amsterdam.

Wolff, M. and Pelissier-Combescure, J. (1982): 'FACIOLOG – Automatic Electrofacies Determination', *SPWLA 23rd Annual Logging Symposium*, July 6–9, 1982 Corpus Christi, Trans., FF: 1–23, 13 figs., glossary; Houston, Texas.

Wu Chungyu (1982): *Mesozoic and Cenozoic lake deltas and oil-gas distribution in eastern China*, Scientific Research Inst. Petroleum Explor. and Development, 22 pp., 15 figs., 1 table; Beijing, China.

Wurster, P. (1964): 'Geologie des Schilfsandsteins', *Mitt. geol. Staatsinst. Hamburg*, **33**, 140 pp., 57 figs., 4 pls., 15 maps; Hamburg.

Xu Shice and Wang Hengjian (1981): 'Deltaic Deposits of a Large Lake Basin', 202–213, 10 figs., 4 tables; in: J. F. Mason, (ed.), *Petroleum Geology in China*, 263 pp., 246 figs., 40 tables; Tulsa, Okla. (PennWell Books).

Yang Puhua (1980): 'Effect of Pore Structure on Mechanism of Water Drive of Oil', *Acta Petrolei Sinica*, Special Issue in Commemoration of the 20th Anniversary of Daqing Oilfield, 103–112; Beijing (in Chinese).

Yang Wanli (1983): 'A Formation Model of a Giant Nonmarine Oil Field' (Abstract), Annual Convention in Dallas, *Amer. Assoc. Petroleum Geologists, Bull.*, **67**, 573–574; Tulsa, Okla.

Yang Wanli (1985): Daqing Oil Field, People's Republic of China: A Giant Field with Oil of Nonmarine Origin', *Amer. Assoc. Petroleum Geologists, Bull.*, **69**, 1101–1111, 10 figs., 9 tables; Tulsa, Okla.

Yariv, S. and Cross, H. (1979): *Geochemistry of Colloid Systems – For Earth Scientists*, 450 pp., 86 figs., 32 tables; Berlin, Heidelberg, New York (Springer).

Yastrebova, T. A. (1976): 'Complexes Minéralogiques Terrigènes des Formations Productives du Néocomien du Gisement de l'Urengoj' (in Russian), *Bull. Moskov. Obshchest. Ispytatelej Prirody, Otd. Geol. (S.S.S.R.)*, **51**, 125–128.

Yurewicz, D. A. (1977): 'Sedimentology of Mississippian Basin-Facies Carbonates, New Mexico and West Texas – The Rancheria Formations', 203–219, 10 figs.; in: H. E. Cook and P. Enos (eds.), *Deep-Water Carbonate Environments*, Soc. Econ. Paleontologists Mineralogists, Spec. Publ. 25, 336 pp., 273 figs., 7 tables; Tulsa, Okla.

Yurkova, R. M. (1971): 'Application of Epigenetic Intralayer Solution of Some Accessory Minerals in Determining the Formation Time of Oil Deposits' (in Russian); in: *Epigenesis and Its Mineral Indicators*, Akad. Nauk SSSR, Geol. Inst. Tr., No. 221: 154–167; Moscow (Nauka) .

Zajic, J. E., Cooper, D. G., Jack, T. R., and Kosaric, N. (eds.) (1983): *Microbial Enhanced Oil Recovery*, 174 pp., 84 figs., 75 tables; Tulsa, Okla. (PennWell Books).

Zankl, H. (1971): 'Upper Triassic Carbonate Facies in the Northern Limestone Alps', 147–185, 20 figs., 1 table; in: G. Müller (ed.), *Sedimentology of Parts of Central Europe*, Guidebook to excursions held during the VIII International Sedimentological Congress 1971 in Heidelberg, Germany, 344 pp., figs., tables; Frankfurt a. M. (Verlag Waldemar Kramer).

Zhai Guangming, Zhang Wenzhao, and Hu Chaoyuan (1982): 'Oil, Gas Accumulations in China's Continental Basins', *Oil & Gas J.*, December 13, 1982, 129–136, 8 figs., 1 table; Tulsa, Okla.

Zhu Guohua (1982): 'Effects of Diagenesis on Pore Texture and Petrophysical Properties of Sandstone Reservoir and Its Geological Significance', in: D. G. Russel (ed.), *Proc. Int. Meeting on Petroleum Engineering*, Soc. Petroleum Engineers, 181–220, 9 tables; Dallas.

Ziegler, P. A. (1990): *Geological Atlas of Western and Central Europe*, 2nd and completely revised ed., 239 pp., 100 figs., 56 encl.; The Hague (Shell Intern. Petroleum Maatschappij, B.V.).

Zimdars, J. H. (1982): 'Desert Sedimentation in the Permian Rotliegendes in NW-Germany (Abstract)', *11th Intern. Congress on Sedimentology*, McMaster University, Hamilton, Ontario, Canada, August 22–27, 1982, p. 69, Hamilton.

Zimmerle, W. (1985): 'New Aspects on the Formation of Hydrocarbon Source Rocks', *Geol. Rundschau*, **74**, 385–416, 8 figs., 3 pls.; Stuttgart.

Zimmerle, W. (1991): 'Thin-Section Petrography of Pelites, A Promising Approach in Sedimentology', *Geol. en Mijnbouw*, **70**, 163–174, 2 tables, 4 pls.; Dordrecht.

Zimmerle, W. (1994): 'Detrital Quartz as Guide to Provenance', *Zbl. Geol. Paläont.* 1993; Stuttgart (in print).

Zingg, Th. (1935): 'Beitrag zur Schotteranalyse. Die Schotteranalyse und ihre Anwendung auf die Glattalschotter', *Schweiz. mineral. petrogr. Mitt.*, **15**, 39–140, 20 figs., 16 tables; Zürich.

Zinkernagel, U. (1978): 'Cathodoluminescence of Quartz and Its Application to Sandstone Petrology', *Contr. Sedimentology*, **8**, 69 pp., 14 figs., 5 tables, appendix; Stuttgart (Schweizerbart).

Zuffa, G. G. (ed.) (1984): *Provenance of Arenites*, 408 pp., 122 figs., 27 tables, 4 pls.; Dordrecht (D. Reidel Publ.)

SUBJECT INDEX

383

GEOGRAPHICAL INDEX

403

AUTHOR INDEX